Elementary Number Theory
Third Edition

David M. Burton
University of New Hampshire

WCB **Wm. C. Brown Publishers**
Dubuque, Iowa•Melbourne, Australia•Oxford, England

Book Team

Editor *Earl McPeek*
Developmental Editor *Theresa Grutz*
Production Editor *Kay Driscoll*

Wm. C. Brown Publishers
A Division of Wm. C. Brown Communications, Inc.

Vice President and General Manager *Beverly Kolz*
National Sales Manager *Vincent R. Di Blasi*
Director of Marketing *John W. Calhoun*
Marketing Manager *Elizabeth Robbins*
Advertising Manager *Amy Schmitz*
Director of Production *Colleen A. Yonda*
Manager of Visuals and Design *Faye M. Schilling*
Design Manager *Jac Tilton*
Art Manager *Janice Roerig*
Publishing Services Manager *Karen J. Slaght*
Permissions/Records Manager *Connie Allendorf*

Wm. C. Brown Communications, Inc.

President and Chief Executive Officer *G. Franklin Lewis*
Corporate Vice President, President of WCB Manufacturing *Roger Meyer*
Vice President and Chief Financial Officer *Robert Chesterman*

Copyedited by Robert J. Minteer

To Martha

CONTENTS

PREFACE

Plato said, "God is a geometer." Jacobi changed this to, "God is an arithmetician." Then came Kronecker and fashioned the memorable expression, "God created the natural numbers, and all the rest is the work of man."

FELIX KLEIN

The purpose of the present volume is to give a simple account of classical number theory, as well as to impart some of the historical background in which the subject evolved. While primarily intended for use as a textbook in a one-semester course at the undergraduate level, it is designed to be used in teachers' institutes or as supplementary reading in mathematics survey courses. The work is well suited for prospective secondary school teachers for whom a little familiarity with number theory may be particularly helpful.

The theory of numbers has always occupied a unique position in the world of mathematics. This is due to the unquestioned historical importance of the subject: it is one of the few disciplines having demonstrable results which predate the very idea of a university or an academy. Nearly every century since classical antiquity has witnessed new and fascinating discoveries relating to the properties of numbers; and, at some point in their careers, most of the great masters of the mathematical sciences have contributed to this body of knowledge. Why has number theory held such an irresistible appeal for the leading mathematicians and for thousands of amateurs? One answer lies in the basic nature of its problems. While many questions in the field are extremely hard to decide, they can be formulated in terms simple enough to arouse the interest and curiosity of those with little mathematical training. Some of the simplest sounding questions have withstood the intellectual assaults for ages and remain among the most elusive unsolved problems in the whole of mathematics.

It therefore comes as something of a surprise to find that many students look upon number theory with good-humored indulgence, regarding it as a frippery on the edge of mathematics. This no doubt stems from the widely held view that it is the purest branch of pure mathematics and from the attendant suspicion that it can have few substantive applications to real-world problems. Some of the worst offenders, when it comes to celebrating the uselessness of their subject, have been number theorists themselves. G. H. Hardy, the best known figure of twentieth century British mathematics, once wrote, "Both Gauss and lesser mathematicians may be justified in rejoicing that there is one science at any rate, and that their own, whose very remoteness from ordinary human activities should keep it clean and gentle." The prominent role that this "clean and gentle" science played in the newly invented public-key cryptosystems (Section 7.5) may serve as something of a reply to Hardy. Leaving practical applications aside, the importance of number theory derives from its central position in mathematics; its concepts and problems have been instrumental in the creation of large parts of mathematics. Few branches of the discipline have absolutely no connection with the theory of numbers.

The last decade has seen a dramatic shift in focus in the undergraduate curriculum away from the more abstract areas of mathematics and toward applied and computational mathematics. With the increasing latitude in course choices, one commonly encounters the mathematics major who knows little or no number theory. This is especially unfortunate, since the elementary theory of numbers should be one of the very best subjects for early mathematical instruction. It requires no long preliminary training, the content is tangible and familiar, and—more than in any other part of mathematics—the methods of inquiry adhere to the scientific approach. The student working in the field must rely to a large extent upon trial and error, in combination with his own curiosity, intuition, and ingenuity; nowhere else in the mathematical disciplines is rigorous proof so often preceded by patient, plodding experiment. If the going occasionally becomes slow and difficult, one can take comfort in the fact that nearly every noted mathematician of the past has traveled the same arduous road.

There is a dictum which says that anyone who desires to get at the root of a subject should first study its history. Endorsing this, we have taken pains to fit the material into the larger historical frame. In addition to enlivening the theoretical side of the text, the historical remarks woven into the presentation bring out the point that number theory is not a dead art, but a living one fed by the efforts of many practitioners. They reveal that the discipline developed bit by bit, with the work of each individual contributor built upon the research of many others; often centuries of endeavor were required before significant steps were made. A student who is aware of how people of genius stumbled and groped their way through the creative process to arrive piecemeal at their results is less likely to be discouraged by his own fumblings with the homework problems.

A word about the problems. Most sections close with a substantial number of them ranging in difficulty from the purely mechanical to challenging theoretical questions. These are an integral part of the book and require the reader's active participation, for nobody can learn number theory without solving problems. The computational exercises develop basic techniques and test understanding of concepts, while those of a theoretical nature give practice in constructing proofs. Besides conveying additional information about the material covered earlier, the problems introduce a variety of ideas not treated in the body of the text. We have on the whole resisted the temptation to use the problems to introduce results that will be needed thereafter. As a consequence, the reader need not work all the exercises in order to digest the rest of the book. Problems whose solutions do not appear straightforward are frequently accompanied by hints.

Although the text was written with the mathematics major in mind, very little is demanded in the way of formal prerequisites; it can be profitably read by anyone having a sound background in high school mathematics. In particular, a knowledge of the concepts of abstract algebra is not assumed. When used for students who have had such a course (say, at the level represented by the author's own *Abstract Algebra*), much of the first four chapters can be omitted.

From the table of contents, it is apparent that our treatment includes more material than can be covered satisfactorily during a one-semester course. This should provide the flexibility desirable for a diverse audience; it permits the instructor to choose topics in accordance with personal tastes and it gives students the opportunity for further reading in the subject. Experience indicates that a standard course can be built up from Chapters 1 through 9; if the occasion demands, Sections 6.2, 6.3, 7.4, 7.5, 8.4, and 9.4 may be deleted from the program without destroying the continuity. Since the last six chapters are entirely independent of each other, they may be taken up at pleasure.

NEW TO THIS EDITION

Readers familiar with the previous edition will find that this one has the same general organization and content. Nevertheless, the preparation of this third edition has provided the opportunity for making a number of small improvements, as well as several more significant ones. Among the larger changes is the splitting of the former Chapter 13 into two distinct chapters. The new chapters are entitled Fibonacci Numbers (Chapter 13) and Continued Fractions (Chapter 14). The Prime Number Theorem has been moved from the Appendix and now appears in Some Twentieth-Century Developments (Chapter 15). The chapter restructuring also involved the addition of two entirely new sections on the lives and work of Srinivasa Ramanujan (Section 14.1) and Hardy, Dickson, and Erdös (Section 15.1).

Another notable difference is the consideration in Section 7.5 of a further type of public-key cryptosystem, one based on what is commonly described as the "knapsack problem." The present edition also pays increased attention to Fibonacci and Lucas numbers, both through textual development and additional problems. Also, various pages have been rewritten and their content has been amplified to improve clarity and, in some cases, their correctness.

Beyond these specific modifications, there are a number of relatively minor changes: portraits of several prominent number theorists have been included, the references and suggested readings brought up to date, and certain numerical information kept current in light of recent findings. An attempt has been made to correct any minor errors which crept into the earlier edition.

I would like to take the opportunity to express my deep appreciation to those mathematicians who read the manuscript in its various editions and offered valuable suggestions leading to its improvement. The advice of the following reviewers was particularly helpful:

L. A. Best, The Open University
Jack Ceder, University of California at Santa Barbara
Robert A. Chaffer, Central Michigan University
Howard Eves, University of Maine
Frederick Hoffman, Florida Atlantic University
Neal McCoy, Smith College
David Outcalt, University of California at Santa Barbara
Michael Rich, Temple University
David Roeder, Colorado College
William W. Smith, University of North Carolina
Virginia Taylor, Lowell Technical Institute
Paul Vicknair, California State University at San Bernardino

This third edition has also been influenced by mathematicians who commented on my proposed changes:

Ruth Berger, Memphis State University
A. B. Billings, Waynesburg College
Gregory S. Call, Amherst College
William Eppright, Northwestern College
Monty Kester, Liberty University
Kevin McDougal, University of Wisconsin–Oshkosh
James E. Nymann, The University of Texas at El Paso
William Ramaley, Fort Lewis College
David Sanker, Holy Names College
Paul Schaefer
Stephen Tanny, University of Toronto
Ray Theis, St. Mary of the Plains College
Al Tirman, East Tennessee State
Eddie Warren, University of Texas at Arlington
Anne L. Young, Loyola College
Norman E. Young, Concordia University

A special debt of gratitude must go to my wife, Martha, whose generous assistance with the book at all stages of development was indispensable.

It remains to acknowledge the fine cooperation of the staff of Wm. C. Brown Publishers and the usual high quality of their work. The author must, of course, accept the responsibility for any errors or shortcomings that remain.

Durham, New Hampshire   DAVID M. BURTON

1

Some Preliminary Considerations

"Number was born in superstition and reared in mystery, . . . numbers were once made the foundation of religion and philosophy, and the tricks of figures have had a marvellous effect on a credulous people."

F. W. PARKER

1.1 MATHEMATICAL INDUCTION

The theory of numbers is concerned, at least in its elementary aspects, with properties of the integers and more particularly with the positive integers 1, 2, 3, . . . (also known as the natural numbers). The origin of this misnomer harks back to the early Greeks for whom the word "number" meant positive integer, and nothing else. The natural numbers have been known to us for so long that the mathematician Leopold Kronecker once remarked, "God created the natural numbers, and all the rest is the work of man." Far from being a gift from Heaven, number theory has had a long and sometimes painful evolution, a story which is told in the ensuing pages.

We shall make no attempt to construct the integers axiomatically, assuming instead that they are already given and that any reader of the book is familiar with many elementary facts about them. Among these is the Well-Ordering Principle. To refresh the memory, it states:

WELL-ORDERING PRINCIPLE.
Every nonempty set S of nonnegative integers contains a least element; that is, there is some integer a in S such that a ≤ b for all b belonging to S.

Since this principle will play a critical role in the proofs here and in subsequent chapters, let us use it to show that the set of positive integers has what is known as the Archimedean Property.

THEOREM 1-1 (Archimedean Property).
> *If a and b are any positive integers, then there exists a positive integer n such that na ≥ b.*

> *Proof:* Assume that the statement of the theorem is not true, so that for some a and b, $na < b$ for every positive integer n. Then the set
>
> $$S = \{b - na \mid n \text{ a positive integer}\}$$
>
> consists entirely of positive integers. By the Well-Ordering Principle S will possess a least element, say $b - ma$. Notice that $b - (m + 1)a$ also lies in S, since S contains all integers of this form. Furthermore, we have
>
> $$b - (m + 1)a = (b - ma) - a < b - ma,$$
>
> contrary to the choice of $b - ma$ as the smallest integer in S. This contradiction arose out of our original assumption that the Archimedean property did not hold, hence this property is proven true.

> With the Well-Ordering Principle available, it is an easy matter to derive the Principle of Finite Induction. The latter principle provides a basis for a method of proof called "mathematical induction." Loosely speaking, the Principle of Finite Induction asserts that if a set of positive integers has two specific properties, then it is the set of all positive integers. To be less cryptic:

THEOREM 1-2 (Principle of Finite Induction).
> *Let S be a set of positive integers with the properties*

> (i)   *1 belongs to S, and*
> (ii) *whenever the integer k is in S, then the next integer k + 1 must also be in S.*

> *Then S is the set of all positive integers.*

> *Proof:* Let T be the set of all positive integers not in S, and assume that T is nonempty. The Well-Ordering Principle tells us that T possesses a least element, which we denote by a. Since 1 is in S, certainly $a > 1$ and so $0 < a - 1 < a$. The choice of a as the smallest positive integer in T implies that $a - 1$ is not a member of T, or equivalently, that $a - 1$ belongs to S. By hypothesis, S must also contain $(a - 1) + 1 = a$, which contradicts the fact that a lies in T. We conclude that the set T is empty, and in consequence that S contains all the positive integers.

> Here is a typical formula that can be established by mathematical induction:

$$(1) \qquad 1^2 + 2^2 + 3^2 + \cdots + n^2 = \frac{n(2n + 1)(n + 1)}{6}$$

for $n = 1, 2, 3, \ldots$. In anticipation of using Theorem 1-2, let S denote the set of all positive integers n for which formula (1) is true. We observe that when $n = 1$, the formula becomes

$$1^2 = \frac{1(2 + 1)(1 + 1)}{6} = 1;$$

this means that 1 is in S. Next, assume that k belongs to S (where k is a fixed but unspecified integer) so that

(2) $$1^2 + 2^2 + 3^2 + \cdots + k^2 = \frac{k(2k + 1)(k + 1)}{6}.$$

To obtain the sum of the first $k + 1$ squares, we merely add the next one, $(k + 1)^2$, to both sides of equation (2). This gives

$$1^2 + 2^2 + \cdots + k^2 + (k + 1)^2 = \frac{k(2k + 1)(k + 1)}{6} + (k + 1)^2.$$

After some algebraic manipulation, the right-hand side becomes

$$(k + 1)\left[\frac{k(2k + 1) + 6(k + 1)}{6}\right] = (k + 1)\left[\frac{2k^2 + 7k + 6}{6}\right]$$

$$= \frac{(k + 1)(2k + 3)(k + 2)}{6},$$

which is precisely the right-hand member of formula (1) when $n = k + 1$. Our reasoning shows that the set S contains the integer $k + 1$ whenever it contains the integer k. By Theorem 1-2, S must be all the positive integers; that is, the given formula is true for $n = 1, 2, 3, \ldots$.

   While mathematical induction provides a standard technique for attempting to prove a statement about the positive integers, one disadvantage is that it gives no aid in formulating such statements. Of course, if we can make an "educated guess" at a property which we believe might hold in general, then its validity can often be tested by the induction principle. Consider, for instance, the list of equalities

$$1 = 1,$$
$$1 + 2 = 3,$$
$$1 + 2 + 2^2 = 7,$$
$$1 + 2 + 2^2 + 2^3 = 15,$$
$$1 + 2 + 2^2 + 2^3 + 2^4 = 31,$$
$$1 + 2 + 2^2 + 2^3 + 2^4 + 2^5 = 63.$$

What is sought is a rule which gives the integers on the right-hand side. After a little reflection, the reader might notice that

$$1 = 2 - 1, \quad 3 = 2^2 - 1, \quad 7 = 2^3 - 1,$$
$$15 = 2^4 - 1, \quad 31 = 2^5 - 1, \quad 63 = 2^6 - 1$$

(how one arrives at this observation is hard to say, but experience helps). The pattern emerging from these few cases suggests a formula for obtaining the value of the expression $1 + 2 + 2^2 + 2^3 + \cdots + 2^{n-1}$; namely,

(3) $$1 + 2 + 2^2 + 2^3 + \cdots + 2^{n-1} = 2^n - 1$$

for every positive integer n.

To confirm that our guess is correct, let S be the set of positive integers n for which formula (3) holds. For $n = 1$, (3) is certainly true, whence 1 belongs to the set S. We assume that (3) is true for a fixed integer k, so that for this k

$$1 + 2 + 2^2 + \cdots + 2^{k-1} = 2^k - 1$$

and we attempt to prove the validity of the formula for $k + 1$. Addition of the term 2^k to both sides of the last-written equation leads to

$$1 + 2 + 2^2 + \cdots + 2^{k-1} + 2^k = 2^k - 1 + 2^k$$
$$= 2 \cdot 2^k - 1 = 2^{k+1} - 1.$$

But this says that formula (3) holds when $n = k + 1$, putting the integer $k + 1$ in S; so that $k + 1$ is in S whenever k is in S. According to the induction principle, S must be the set of all positive integers.

> REMARK: When giving induction proofs, we shall usually shorten the argument by eliminating all reference to the set S, and proceed to show simply that the result in question is true for the integer 1, and if true for the integer k is then also true for $k + 1$.

We should inject a word of caution at this point, to wit, that one must be careful to establish both conditions of Theorem 1-2 before drawing any conclusions; neither is sufficient alone. The proof of condition (i) is usually called the *basis for the induction*, while the proof of (ii) is called the *induction step*. The assumptions made in carrying out the induction step are known as the *induction hypotheses*. The induction situation has been likened to an infinite row of dominoes all standing on edge and arranged in such a way that when one falls it knocks down the next in line. If either no domino is pushed over (that is, there is no basis for the induction) or if the spacing is too large (that is, the induction step fails), then the complete line will not fall.

The validity of the induction step does not necessarily depend on the truth of the statement which one is endeavoring to prove. Let us look at the false formula

(4) $$1 + 3 + 5 + \cdots + (2n - 1) = n^2 + 3.$$

Assume that this holds for $n = k$; in other words,

$$1 + 3 + 5 + \cdots + (2k - 1) = k^2 + 3.$$

Knowing this, we then obtain

$$1 + 3 + 5 + \cdots + (2k - 1) + (2k + 1) = k^2 + 3 + 2k + 1$$
$$= (k + 1)^2 + 3,$$

which is precisely the form that formula (4) should take when $n = k + 1$. Thus, if (4) holds for a given integer, then it also holds for the succeeding integer. It is not possible, however, to find a value of n for which the formula is true.

There is a variant of the induction principle that is often used when Theorem 1-2 by itself seems ineffective. As with the first version, this Second Principle of Finite Induction gives two conditions which guarantee that a certain set of positive integers actually consists of all positive integers. What happens is this: we retain requirement (i), but (ii) is replaced by

(ii′)   *If k is a positive integer such that $1, 2, \ldots, k$ belong to S, then $k + 1$ must also be in S.*

The proof that S consists of all positive integers has the same flavor as that of Theorem 1-2. Again, let T represent the set of positive integers not in S. Assuming that T is nonempty, we pick n to be the smallest integer in T. Then $n > 1$, by supposition (i). The minimal nature of n allows us to conclude that none of the integers $1, 2, \ldots, n - 1$ lies in T, or, if one prefers a positive assertion, $1, 2, \ldots, n - 1$ all belong to S. Property (ii') then puts $n = (n - 1) + 1$ in S, which is an obvious contradiction. The result of all this is to make T empty.

The First Principle of Finite Induction is used more often than the Second, but there are occasions when the Second is favored and the reader should be familiar with both versions. It sometimes happens that in attempting to show that $k + 1$ is a member of S, one requires proof of the fact that not only k, but all positive integers which precede k, lie in S. Our formulation of these induction principles has been for the case in which the induction begins with 1. Each form can be generalized to start with any positive integer n_0. In this circumstance, the conclusion reads, "Then S is the set of all positive integers $n \geq n_0$."

Mathematical induction is often used as a method of definition as well as a method of proof. For example, a common way of introducing the symbol $n!$ (pronounced "n factorial") is by means of the inductive definition

(a) $1! = 1$,
(b) $n! = n \cdot (n - 1)!$ for $n > 1$.

This pair of conditions provides a rule whereby the meaning of $n!$ is specified for each positive integer n. Thus, by (a), $1! = 1$; (a) and (b) yield

$$2! = 2 \cdot 1! = 2 \cdot 1;$$

while by (b) again,

$$3! = 3 \cdot 2! = 3 \cdot 2 \cdot 1.$$

Continuing in this manner, using condition (b) repeatedly, the numbers $1!$, $2!$, $3!$, $\ldots, n!$ are defined in succession up to any chosen n. In fact,

$$n! = n \cdot (n - 1) \cdots 3 \cdot 2 \cdot 1.$$

Induction enters in showing that $n!$, as a function on the positive integers, exists and is unique; we shall make no attempt however to give the argument.

It will be convenient to extend the definition of $n!$ to the case in which $n = 0$ by stipulating that $0! = 1$.

Example 1-1

To illustrate a proof which requires the Second Principle of Finite Induction, consider the so-called *Lucas sequence*

$$1, 3, 4, 7, 11, 18, 29, 47, 76, \ldots .$$

Except for the first two terms, each term of this sequence is the sum of the preceding two, so that the sequence may be defined inductively by

$$a_1 = 1,$$

$$a_2 = 3,$$

$$a_n = a_{n-1} + a_{n-2}, \qquad\qquad \text{for all } n \geq 3.$$

We contend that the inequality

$$a_n < (7/4)^n$$

holds for every positive integer n. The argument used is interesting because in the inductive step, it is necessary to know the truth of this inequality for two successive values of n in order to establish its truth for the following value.

First of all, for $n = 1$ and 2, we have

$$a_1 = 1 < (7/4)^1 = 7/4 \text{ and } a_2 = 3 < (7/4)^2 = 49/16,$$

whence the inequality in question holds in these two cases. This provides a basis for the induction. For the induction step, pick an integer $k \geq 3$ and assume that the inequality is valid for $n = 1, 2, \ldots, k - 1$. Then, in particular,

$$a_{k-1} < (7/4)^{k-1} \text{ and } a_{k-2} < (7/4)^{k-2}.$$

By the way in which the Lucas sequence is formed, it follows that

$$a_k = a_{k-1} + a_{k-2} < (7/4)^{k-1} + (7/4)^{k-2}$$

$$= (7/4)^{k-2}(7/4 + 1)$$

$$= (7/4)^{k-2}(11/4)$$

$$< (7/4)^{k-2}(7/4)^2 = (7/4)^k.$$

Since the inequality is true for $n = k$ whenever it is true for the integers 1, 2, $\ldots, k - 1$, we conclude by the second induction principle that $a_n < (7/4)^n$ for all $n \geq 1$.

Among other things, this example suggests that if objects are defined inductively, then mathematical induction is an important tool for establishing the properties of these objects.

PROBLEMS 1.1

1. Establish the formulas below by mathematical induction:

(a) $1 + 2 + 3 + \cdots + n = \dfrac{n(n + 1)}{2}$ for all $n \geq 1$;

(b) $1 + 3 + 5 + \cdots + (2n - 1) = n^2$ for all $n \geq 1$;

(c) $1 \cdot 2 + 2 \cdot 3 + 3 \cdot 4 + \cdots + n(n + 1) = \dfrac{n(n + 1)(n + 2)}{3}$ for all $n \geq 1$;

(d) $1^2 + 3^2 + 5^2 + \cdots + (2n - 1)^2 = \dfrac{n(4n^2 - 1)}{3}$ for all $n \geq 1$;

(e) $1^3 + 2^3 + 3^3 + \cdots + n^3 = \left[\dfrac{n(n + 1)}{2}\right]^2$ for all $n \geq 1$.

2. If $r \neq 1$, show that

$$a + ar + ar^2 + \cdots + ar^n = \frac{a(r^{n+1} - 1)}{r - 1}$$

for any positive integer n.

3. Use the Second Principle of Finite Induction to establish that

$$a^n - 1 = (a - 1)(a^{n-1} + a^{n-2} + a^{n-3} + \cdots + a + 1)$$

for all $n \geq 1$.
[*Hint:* $a^{n+1} - 1 = (a + 1)(a^n - 1) - a(a^{n-1} - 1)$.]

4. Prove that the cube of any integer can be written as the difference of two squares.
[*Hint:* Notice that

$$n^3 = (1^3 + 2^3 + \cdots + n^3) - (1^3 + 2^3 + \cdots + (n - 1)^3).]$$

5. (a)   Find the values of $n \leq 7$ for which $n! + 1$ is a perfect square (it is unknown whether $n! + 1$ is a square for any $n > 7$).
   (b)   True or false? For positive integers m and n, $(mn)! = m!n!$ and $(m + n)! = m! + n!$.

6. Prove that $n! > n^2$ for every integer $n \geq 4$, while $n! > n^3$ for every integer $n \geq 6$.

7. Use mathematical induction to derive the formula

$$1(1!) + 2(2!) + 3(3!) + \cdots + n(n!) = (n + 1)! - 1$$

for all $n \geq 1$.

8. (a)   Verify that

$$2 \cdot 6 \cdot 10 \cdot 14 \cdot \ldots \cdot (4n - 2) = \frac{(2n)!}{n!}$$

for all $n \geq 1$.
   (b)   Use part (a) to obtain the inequality $2^n(n!)^2 \leq (2n)!$ for all $n \geq 1$.

9. Establish the Bernoulli inequality: if $1 + a > 0$, then

$$(1 + a)^n \geq 1 + na$$

for all $n \geq 1$.

10. For all $n \geq 1$, prove by mathematical induction that

(a) $\dfrac{1}{1^2} + \dfrac{1}{2^2} + \dfrac{1}{3^2} + \cdots + \dfrac{1}{n^2} \leq 2 - \dfrac{1}{n}$,

(b) $\dfrac{1}{2} + \dfrac{2}{2^2} + \dfrac{3}{2^3} + \cdots + \dfrac{n}{2^n} \leq 2 - \dfrac{n}{2^n}$.

11. Show that the expression $\dfrac{(2n)!}{2^n n!}$ is an integer for all $n \geq 0$.

12. Consider the function defined by

$$T(n) = \begin{cases} \dfrac{3n + 1}{2} & \text{for } n \text{ odd} \\[2ex] \dfrac{n}{2} & \text{for } n \text{ even} \end{cases}$$

The $3n+1$ conjecture is the claim that starting from any integer $n > 1$ the sequence of iterates $T(n)$, $T(T(n))$, $T(T(T(n)))$, . . . , eventually reaches the integer 1, and subsequently runs through the values 1 and 2. This has been verified for all $n \le 10^{12}$. Confirm the conjecture in the cases $n = 21$ and $n = 23$.

1.2 THE BINOMIAL THEOREM

Closely connected with the factorial notation are the *binomial coefficients* $\begin{pmatrix} n \\ k \end{pmatrix}$. For any positive integer n and any integer k satisfying $0 \le k \le n$, these are defined by

$$\begin{pmatrix} n \\ k \end{pmatrix} = \frac{n!}{k!(n - k)!}.$$

By cancelling out either $k!$ or $(n - k)!$, $\begin{pmatrix} n \\ k \end{pmatrix}$ can be written as

$$\begin{pmatrix} n \\ k \end{pmatrix} = \frac{n(n - 1) \cdots (k + 1)}{(n - k)!} = \frac{n(n - 1) \cdots (n - k + 1)}{k!}.$$

For example, with $n = 8$ and $k = 3$, we have

$$\begin{pmatrix} 8 \\ 3 \end{pmatrix} = \frac{8!}{3!5!} = \frac{8 \cdot 7 \cdot 6 \cdot 5 \cdot 4}{5!} = \frac{8 \cdot 7 \cdot 6}{3!} = 56.$$

Also observe that if $k = 0$ or $k = n$, the quantity $0!$ appears on the right-hand side of the definition of $\begin{pmatrix} n \\ k \end{pmatrix}$; since we have taken $0!$ as 1, these special values of k give

$$\begin{pmatrix} n \\ 0 \end{pmatrix} = \begin{pmatrix} n \\ n \end{pmatrix} = 1.$$

There are numerous useful identities connecting binomial coefficients. One that we require here is *Pascal's rule:*

$$\begin{pmatrix} n \\ k \end{pmatrix} + \begin{pmatrix} n \\ k - 1 \end{pmatrix} = \begin{pmatrix} n + 1 \\ k \end{pmatrix}, \qquad\qquad 1 \le k \le n.$$

Its proof consists of multiplying the identity

$$\frac{1}{k} + \frac{1}{n - k + 1} = \frac{n + 1}{k(n - k + 1)}$$

by $\dfrac{n!}{(k - 1)!(n - k)!}$ in order to obtain

$$\frac{n!}{k(k-1)!(n-k)!} + \frac{n!}{(k-1)!(n-k+1)(n-k)!}$$
$$= \frac{(n+1)n!}{k(k-1)!(n-k+1)(n-k)!}.$$

Falling back on the definition of the factorial function, this says that

$$\frac{n!}{k!(n-k)!} + \frac{n!}{(k-1)!(n-k+1)!} = \frac{(n+1)!}{k!(n+1-k)!},$$

from which Pascal's rule follows.

This relation gives rise to a configuration, known as *Pascal's triangle,* in which the binomial coefficient $\binom{n}{k}$ appears as the $(k+1)$th number in the nth row:

$$\begin{array}{ccccccccccccc}
 & & & & & 1 & & 1 & & & & & \\
 & & & & 1 & & 2 & & 1 & & & & \\
 & & & 1 & & 3 & & 3 & & 1 & & & \\
 & & 1 & & 4 & & 6 & & 4 & & 1 & & \\
 & 1 & & 5 & & 10 & & 10 & & 5 & & 1 & \\
1 & & 6 & & 15 & & 20 & & 15 & & 6 & & 1
\end{array}$$
$$\cdot \ \cdot \ \cdot$$

The rule of formation should be clear. The borders of the triangle are composed of 1's; a number not on the border is the sum of the two numbers nearest it in the row above.

The so-called *Binomial Theorem* is in reality a formula for the complete expansion of $(a+b)^n$, $n \geq 1$, into a sum of powers of a and b. This expression appears with great frequency in all phases of number theory and it is well worth our time to look at it now. By direct multiplication, it is easy to verify that

$$(a+b)^1 = a + b,$$
$$(a+b)^2 = a^2 + 2ab + b^2,$$
$$(a+b)^3 = a^3 + 3a^2b + 3ab^2 + b^3,$$
$$(a+b)^4 = a^4 + 4a^3b + 6a^2b^2 + 4ab^3 + b^4, \text{ etc.}$$

The question is how to predict the coefficients. A clue lies in the observation that the coefficients of these first few expansions form the successive rows of Pascal's triangle. This would lead one to suspect that the general binomial expansion will take the form

$$(a+b)^n = \binom{n}{0}a^n + \binom{n}{1}a^{n-1}b + \binom{n}{2}a^{n-2}b^2$$
$$+ \cdots + \binom{n}{n-1}ab^{n-1} + \binom{n}{n}b^n$$

or, written more compactly,

$$(a+b)^n = \sum_{k=0}^{n} \binom{n}{k}a^{n-k}b^k.$$

Mathematical induction provides the best means for confirming this guess. When $n = 1$, the conjectured formula reduces to

$$(a + b)^1 = \sum_{k=0}^{1} \binom{1}{k} a^{1-k} b^k = \binom{1}{0} a^1 b^0 + \binom{1}{1} a^0 b^1 = a + b,$$

which is certainly correct. Assuming that the formula holds for some fixed integer m, we go on to show that it also must hold for $m + 1$. The starting point is to notice that

$$(a + b)^{m+1} = a(a + b)^m + b(a + b)^m.$$

Under the induction hypothesis,

$$a(a + b)^m = \sum_{k=0}^{m} \binom{m}{k} a^{m-k+1} b^k = a^{m+1} + \sum_{k=1}^{m} \binom{m}{k} a^{m+1-k} b^k$$

and

$$b(a + b)^m = \sum_{j=0}^{m} \binom{m}{j} a^{m-j} b^{j+1}$$

$$= \sum_{k=1}^{m} \binom{m}{k-1} a^{m+1-k} b^k + b^{m+1}.$$

Upon adding these expressions, we obtain

$$(a + b)^{m+1} = a^{m+1} + \sum_{k=1}^{m} \left[\binom{m}{k} + \binom{m}{k-1} \right] a^{m+1-k} b^k + b^{m+1}$$

$$= \sum_{k=0}^{m+1} \binom{m+1}{k} a^{m+1-k} b^k,$$

which is the formula in the case $n = m + 1$. This establishes the binomial theorem by induction.

Before abandoning these ideas, we might remark that the first acceptable formulation of the method of mathematical induction appears in the treatise *Traité du Triangle Arithmetiqué*, by the 17th century French mathematician and philosopher Blaise Pascal. This short work was written in 1653, but not printed until 1665 because Pascal had withdrawn from mathematics (at the age of 25) to dedicate his talents to religion. His careful analysis of the properties of the binomial coefficients helped lay the foundations of probability theory.

Problems 1.2

1. (a) Derive Newton's Identity

$$\binom{n}{k} \binom{k}{r} = \binom{n}{r} \binom{n-r}{k-r}, \quad n \geq k \geq r \geq 0.$$

(b) Use part (a) to express $\binom{n}{k}$ in terms of its predecessor:

$$\binom{n}{k} = \frac{n - k + 1}{k} \binom{n}{k - 1}, \quad n \geq k \geq 1.$$

2. If $2 \leq k \leq n - 2$, show that

$$\binom{n}{k} = \binom{n - 2}{k - 2} + 2\binom{n - 2}{k - 1} + \binom{n - 2}{k}, \qquad n \geq 4.$$

3. For $n \geq 1$, derive each of the identities below:

(a) $\binom{n}{0} + \binom{n}{1} + \binom{n}{2} + \cdots + \binom{n}{n} = 2^n;$

[*Hint:* Let $a = b = 1$ in the Binomial Theorem.]

(b) $\binom{n}{0} - \binom{n}{1} + \binom{n}{2} + \cdots + (-1)^n \binom{n}{n} = 0;$

(c) $\binom{n}{1} + 2\binom{n}{2} + 3\binom{n}{3} + \cdots + n\binom{n}{n} = n2^{n-1};$

[*Hint:* After expanding $n(1 + b)^{n-1}$ by the Binomial Theorem, let $b = 1$; note also that

$$n\binom{n - 1}{k} = (k + 1)\binom{n}{k + 1}.]$$

(d) $\binom{n}{0} + 2\binom{n}{1} + 2^2\binom{n}{2} + \cdots + 2^n\binom{n}{n} = 3^n;$

(e) $\binom{n}{0} + \binom{n}{2} + \binom{n}{4} + \binom{n}{6} + \cdots$

$$= \binom{n}{1} + \binom{n}{3} + \binom{n}{5} + \cdots = 2^{n-1};$$

[*Hint:* Use parts (a) and (b).]

(f) $\binom{n}{0} - \frac{1}{2}\binom{n}{1} + \frac{1}{3}\binom{n}{2} - \cdots + \frac{(-1)^n}{n + 1}\binom{n}{n} = \frac{1}{n + 1}.$

[*Hint:* The left-hand side equals

$$\frac{1}{n + 1}\left[\binom{n + 1}{1} - \binom{n + 1}{2} + \binom{n + 1}{3} - \cdots + (-1)^n\binom{n + 1}{n + 1}\right].]$$

4. Prove that for $n \geq 1$:

(a) $\binom{n}{r} < \binom{n}{r + 1}$ if and only if $0 \leq r < \frac{1}{2}(n - 1).$

(b) $\binom{n}{r} > \binom{n}{r + 1}$ if and only if $n - 1 \geq r > \frac{1}{2}(n - 1).$

(c) $\binom{n}{r} = \binom{n}{r + 1}$ if and only if n is an odd integer, and $r = \frac{1}{2}(n - 1).$

5. (a) For $n \geq 2$, prove that

$$\binom{2}{2} + \binom{3}{2} + \binom{4}{2} + \cdots + \binom{n}{2} = \binom{n+1}{3}.$$

[*Hint:* Use induction, and Pascal's rule.]

(b) From part (a), and the relation $m^2 = 2\binom{m}{2} + m$ for $m \geq 2$, deduce the formula

$$1^2 + 2^2 + 3^2 + \cdots + n^2 = \frac{n(n+1)(2n+1)}{6}.$$

(c) Apply the formula in part (a) to obtain a proof that

$$1 \cdot 2 + 2 \cdot 3 + \cdots + n(n+1) = \frac{n(n+1)(n+2)}{3}.$$

[*Hint:* Observe that $(m-1)m = 2\binom{m}{2}$.]

6. Derive the binomial identity

$$\binom{2}{2} + \binom{4}{2} + \binom{6}{2} + \cdots + \binom{2n}{2} = \frac{n(n+1)(4n-1)}{6}, \, n \geq 2.$$

[*Hint:* For $m \geq 2$, $\binom{2m}{2} = 2\binom{m}{2} + m^2$.]

7. For $n > 1$, verify that

$$1^2 + 3^2 + 5^2 + \cdots + (2n-1)^2 = \binom{2n+1}{3}.$$

8. Show that for $n \geq 1$,

$$\binom{2n}{n} = \frac{1 \cdot 3 \cdot 5 \cdots (2n-1)}{2 \cdot 4 \cdot 6 \cdots 2n} 2^{2n}.$$

9. Establish the inequality $2^n < \binom{2n}{n} < 2^{2n}$ for $n > 1$.

[*Hint:* Put $x = 2 \cdot 4 \cdot 6 \cdots (2n)$, $y = 1 \cdot 3 \cdot 5 \cdots (2n-1)$ and $z = 1 \cdot 2 \cdot 3 \cdots n$; show that $x > y > z$, hence $x^2 > xy > xz$.]

10. The *Catalan numbers,* defined by

$$C_n = \frac{1}{n+1}\binom{2n}{n} = \frac{(2n)!}{n! \quad (n+1)!} \qquad n = 0,1,2, \cdots.$$

form the sequence $1,1,2,5,14,42,132,429,1430,4862, \cdots$. They first appeared in 1838 when Eugène Catalan (1814–1894) showed that there are C_n ways of parenthesizing a nonassociative product of $n + 1$ factors. [For instance, when $n = 3$ there are five ways: $((ab)c)d$, $(a(bc)d)$, $a((bc)d)$, $a(b(cd))$, $(ab)(ac)$.] For $n \geq 1$, prove that C_n can be given inductively by

$$C_n = \frac{2(2n-1)}{n+1} C_{n-1}.$$

1.3 EARLY NUMBER THEORY

Before becoming weighted down with detail, we should say a few words about the origin of number theory. The theory of numbers is one of the oldest branches of mathematics; an enthusiast, by stretching a point here and there, could extend its roots back to a surprisingly remote date. While it seems probable that the Greeks were largely indebted to the Babylonians and ancient Egyptians for a core of information about the properties of the natural numbers, the first rudiments of an actual theory are generally credited to Pythagoras and his disciples.

Our knowledge of the life of Pythagoras is scanty and little can be said with any certainty. According to the best estimates, he was born between 580 and 562 B.C. on the Aegean island of Samos. It seems that he studied not only in Egypt, but even may have extended his journeys as far east as Babylonia. When Pythagoras reappeared after years of wandering, he sought out a favorable place for a school, and finally settled upon Croton, a prosperous Greek settlement on the heel of the Italian boot. The school concentrated on four *mathemata,* or subjects of study: *arithmetica* (arithmetic, in the sense of number theory, rather than the art of calculating), *harmonia* (music), *geometria* (geometry), and *astrologia* (astronomy). This fourfold division of knowledge became known in the Middle Ages as the *quadrivium,* to which was added the *trivium* of logic, grammar, and rhetoric. These seven liberal arts came to be looked upon as the necessary course of study for an educated person.

Pythagoras divided those who attended his lectures into two groups: the Probationers (or listeners) and the Pythagoreans. After three years in the first class, a listener could be initiated into the second class, to whom were confided the main discoveries of the school. The Pythagoreans were a closely knit brotherhood, holding all worldly goods in common and bound by an oath not to reveal the founder's secrets. Legend has it that a talkative Pythagorean was drowned in a shipwreck as the gods' punishment for publicly boasting that he had added the dodecahedron to the number of regular solids enumerated by Pythagoras. For a time the autocratic Pythagoreans succeeded in dominating the local government in Croton, but a popular revolt in 501 B.C. led to the murder of many of its prominent members, and Pythagoras himself was killed shortly thereafter. Although the political influence of the Pythagoreans was thus destroyed, they continued to exist for at least two centuries more as a philosophical and mathematical society. To the end, they remained a secret order, publishing nothing and, with a noble self-denial, ascribing all their discoveries to the Master.

The Pythagoreans believed that the key to an explanation of the universe lay in number and form, their general thesis being that "Everything is Number." (By number, they meant of course a positive integer.) For a rational understanding of nature, they considered it sufficient to analyze the properties of certain numbers. Pythagoras himself, we are told, "seems to have attached supreme importance to the study of arithmetic, which he advanced and took out of the realm of commercial utility."

The Pythagorean doctrine is a curious mixture of cosmic philosophy and number-mysticism, a sort of supernumerology which assigned to everything material or spiritual a definite integer. Among their writings, we find that 1 represented reason, for reason could produce only one consistent body of truth; 2 stood for man and 3 for woman; 4 was the Pythagorean symbol for justice, being the first number which is the product of equals; 5 was identified with marriage, since it is formed by the union of 2 and 3; and so forth. All the even numbers, after the first one, were capable of separation into other numbers; hence, they were prolific and were considered as feminine and earthy—and somewhat less highly regarded in general. Being a predominantly male society, the Pythagoreans classified the odd numbers, after the first two, as masculine and divine.

Although these speculations about numbers as models of "things" appear frivolous today, it must be borne in mind that the intellectuals of the classical Greek period were largely absorbed in philosophy and that these same men, because they had such intellectual interests, were the very ones who were engaged in laying the foundations for mathematics as a system of thought. To Pythagoras and his followers, mathematics was largely a means to an end, the end being philosophy. Only with the founding of the School of Alexandria do we enter a new phase in which the cultivation of mathematics is pursued for its own sake.

We might digress here to point out that mystical speculation about the properties of numbers was not unique to the Pythagoreans. One of the most absurd yet widely spread forms which numerology took during the Middle Ages was a pseudo-science known as *gematria* or *arithmology*. By assigning numerical values to the letters of the alphabet in some order, each name or word was given its own individual number. From the standpoint of gematria, two words were considered equivalent if the numbers represented by their letters when added together gave the same sum. All this probably originated with the early Greeks for whom the natural ordering of the alphabet provided a perfect way of recording numbers; α standing for 1, β for 2, and so forth. For example, the word "amen" is $\alpha\mu\eta\nu$ in Greek; these letters have the values 1, 40, 8, and 50, respectively, which total 99. In many old editions of the Bible, the number 99 appears at the end of a prayer as a substitute for amen. The most famous number was 666, the "number of the beast," mentioned in the Book of Revelations. A favorite pastime among certain Catholic theologians during the Reformation was devising alphabet schemes in which 666 was shown to stand for the name of Martin Luther, thereby supporting their contention that he was the Antichrist. Luther replied in kind: he connected a system in which 666 became the number assigned to the reigning Pope, Leo X.

It was at Alexandria, not Athens, that a science of numbers divorced from mystic-philosophy first began to develop. For nearly a thousand years, until its destruction by the Arabs in 641 A.D., Alexandria stood at the cultural and commercial center of the Hellenistic world. (After the fall of Alexandria, most of its scholars migrated to Constantinople. During the next 800 years, while formal learning in the West all but disappeared, this enclave at Constantinople preserved for us the mathematical works of the various Greek Schools.) The so-called Alexandrian Museum, a forerunner of the modern university, brought together the leading poets and scholars of the day; adjacent to it there was established an enormous library, reputed to hold over 700,000 volumes—hand-copied—at its height. Of all the distinguished names connected with the Museum, that of Euclid (circa 350 B.C.), founder of the School of Mathematics, is in a special class. Posterity has come to know him as the author of the *Elements,* the oldest Greek treatise on mathematics to reach us in its entirety. The *Elements* is a compilation of much of the mathematical knowledge available at that time, organized into thirteen parts or Books, as they are called. The name of Euclid is so often associated with geometry that one tends to forget that three of the Books, VII, VIII, and IX, are devoted to number theory.

Euclid's *Elements* constitute one of the great success stories of world literature. Scarcely any other book save the Bible has been more widely circulated or studied. Over a thousand editions of it have appeared since the first printed version in 1482, and before that manuscript copies dominated much of the teaching of mathematics in Western Europe. Unfortunately no copy of the work has been found that actually dates from Euclid's own time; the modern editions are descendants of a revision prepared by Theon of Alexandria, a commentator of the fourth century A.D.

PROBLEMS 1.3

1. Each of the numbers

$$1 = 1, 3 = 1 + 2, 6 = 1 + 2 + 3, 10 = 1 + 2 + 3 + 4, \ldots$$

   represents the number of dots that can be arranged evenly in an equilateral triangle:

   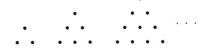

   This led the ancient Greeks to call a number *triangular* if it is the sum of consecutive integers, beginning with 1. Prove the following facts concerning triangular numbers:
   (a) A number is triangular if and only if it is of the form $n(n + 1)/2$ for some $n \geq 1$. (Pythagoras, circa 550 B.C.)
   (b) The integer n is a triangular number if and only if $8n + 1$ is a perfect square. (Plutarch, circa 100 A.D.)
   (c) The sum of any two consecutive triangular numbers is a perfect square. (Nicomachus, circa 100 A.D.)
   (d) If n is a triangular number, then so are $9n + 1$, $25n + 3$, and $49n + 6$. (Euler, 1775)

2. If t_n denotes the nth triangular number, prove that in terms of the binomial coefficients

$$t_n = \binom{n + 1}{2}, \qquad n \geq 1.$$

3. Derive the following formula for the sum of triangular numbers, attributed to the Hindu mathematician Aryabhatta (circa 500 A.D.):

$$t_1 + t_2 + t_3 + \cdots + t_n = \frac{n(n + 1)(n + 2)}{6}, \qquad n \geq 1.$$

   [*Hint:* Group the terms on the left-hand side in pairs, noting the identity $t_{k-1} + t_k = k^2$.]

4. Prove that the square of any odd multiple of 3 is the difference of two triangular numbers; specifically, that

$$9(2n + 1)^2 = t_{9n + 4} - t_{3n + 1}.$$

5. In the sequence of triangular numbers, find
   (a) two triangular numbers whose sum and difference are also triangular numbers;
   (b) three successive triangular numbers whose product is a perfect square;
   (c) three successive triangular numbers whose sum is a perfect square.

6. (a) If the triangular number t_n is a perfect square, prove that $t_{4n(n + 1)}$ is also a square.
   (b) Use part (a) to find three examples of squares which are also triangular numbers.

7. Show that the difference between the squares of two consecutive triangular numbers is always a cube.

8. Prove that the sum of the reciprocals of the first n triangular numbers is less than 2; that is,

$$1/1 + 1/3 + 1/6 + 1/10 + \cdots + 1/t_n < 2.$$

[*Hint:* Observe that $\dfrac{2}{n(n+1)} = 2\left(\dfrac{1}{n} - \dfrac{1}{n+1}\right).$]

9. (a) Establish the identity $t_x = t_y + t_z$, where

$$x = 1/2\, n(n+3) + 1, \, y = n+1, \, z = 1/2\, n(n+3),$$

and $n \geq 1$, thereby proving that there are infinitely many triangular numbers which are the sum of two other such numbers.

   (b) Find three examples of triangular numbers which are sums of two other triangular numbers.

2

Divisibility Theory in the Integers

"Integral numbers are the fountainhead of all mathematics."
H. MINKOWSKI

2.1 THE DIVISION ALGORITHM

We have been exposed to relationships between integers for several pages and as yet not a single divisibility property has been derived. It is time to remedy this situation. One theorem, the Division Algorithm, acts as the foundation stone upon which our whole development rests. The result is familiar to most of us; roughly, it asserts that an integer a can be "divided" by a positive integer b in such a way that the remainder is smaller in size than b. The exact statement of this fact is

THEOREM 2-1 (Division Algorithm).
   Given integers a and b, with $b > 0$, there exist unique integers q and r satisfying

$$a = qb + r, \qquad\qquad 0 \le r < b.$$

The integers q and r are called, respectively, the quotient and remainder in the division of a by b.

   Proof: We begin by proving that the set

$$S = \{a - xb \mid x \text{ an integer}; \ a - xb \ge 0\}$$

is nonempty. For this, it suffices to exhibit a value of x making $a - xb$ nonnegative. Since the integer $b \ge 1$, we have $|a|b \ge |a|$ and so

$$a - (-|a|)b = a + |a|b \ge a + |a| \ge 0.$$

Hence, for the choice $x = -|a|$, $a - xb$ will lie in S. This paves the way for an application of the Well-Ordering Principle, from which we infer that the set S contains a smallest integer; call it r. By the definition of S, there exists an integer q satisfying

$$r = a - qb, \qquad\qquad 0 \leq r.$$

We argue that $r < b$. If this were not the case, then $r \geq b$ and

$$a - (q + 1)b = (a - qb) - b = r - b \geq 0.$$

The implication is that the integer $a - (q + 1)b$ has the proper form to belong to the set S. But $a - (q + 1)b = r - b < r$, leading to a contradiction of the choice of r as the smallest member of S. Hence, $r < b$.

We next turn to the task of showing the uniqueness of q and r. Suppose that a has two representations of the desired form; say

$$a = qb + r = q'b + r',$$

where $0 \leq r < b$, $0 \leq r' < b$. Then $r' - r = b(q - q')$ and, owing to the fact that the absolute value of a product is equal to the product of the absolute values,

$$|r' - r| = b|q - q'|.$$

Upon adding the two inequalities $-b < -r \leq 0$ and $0 \leq r' < b$, we obtain $-b < r' - r < b$ or, in equivalent terms, $|r' - r| < b$. Thus, $b|q - q'| < b$, which yields

$$0 \leq |q - q'| < 1.$$

Since $|q - q'|$ is a nonnegative integer, the only possibility is that $|q - q'| = 0$, whence $q = q'$; this in its turn gives $r = r'$, ending the proof.

A more general version of the Division Algorithm is obtained on replacing the restriction that b be positive by the simple requirement that $b \neq 0$.

COROLLARY.

If a and b are integers, with $b \neq 0$, then there exist unique integers q and r such that

$$a = qb + r, \qquad\qquad 0 \leq r < |b|.$$

Proof: It is enough to consider the case in which b is negative. Then $|b| > 0$ and the theorem produces unique integers q' and r for which

$$a = q'|b| + r, \qquad\qquad 0 \leq r < |b|.$$

Noting that $|b| = -b$, we may take $q = -q'$ to arrive at $a = qb + r$, with $0 \leq r < |b|$.

To illustrate the Division Algorithm when $b < 0$, let us take $b = -7$. Then, for the choices of $a = 1, -2, 61$, and -59, one gets the expressions

$$1 = 0(-7) + 1,$$
$$-2 = 1(-7) + 5,$$
$$61 = (-8)(-7) + 5,$$
$$-59 = 9(-7) + 4.$$

We wish to focus attention, not so much on the Division Algorithm, as on its applications. As a first illustration, note that with $b = 2$ the possible remainders are $r = 0$ and $r = 1$. When $r = 0$, the integer a has the form $a = 2q$ and is called *even;* when $r = 1$, the integer a has the form $a = 2q + 1$ and is called *odd.* Now a^2 is either of the form $(2q)^2 = 4k$ or $(2q + 1)^2 = 4(q^2 + q) + 1 = 4k + 1$. The point to be made is that the square of an integer leaves the remainder 0 or 1 upon division by 4.

We can also show the following: The square of any odd integer is of the form $8k + 1$. For, by the Division Algorithm, any integer is representable as one of the four forms $4q, 4q + 1, 4q + 2, 4q + 3$. In this classification, only those integers of the forms $4q + 1$ and $4q + 3$ are odd. When the latter are squared, we find that

$$(4q + 1)^2 = 8(2q^2 + q) + 1 = 8k + 1$$

and similarly

$$(4q + 3)^2 = 8(2q^2 + 3q + 1) + 1 = 8k + 1.$$

As examples, the square of the odd integer 7 is $7^2 = 49 = 8 \cdot 6 + 1$, while the square of 13 is $13^2 = 169 = 8 \cdot 21 + 1$.

As these remarks indicate, the advantage of the Division Algorithm is that it allows us to prove assertions about all the integers by considering only a finite number of cases. Let us illustrate this with one final example.

Example 2-1

We propose to show that the expression $\dfrac{a(a^2 + 2)}{3}$ is an integer for all $a \geq 1$.

According to the Division Algorithm, every a is of the form $3q, 3q + 1$, or $3q + 2$. Assume the first of these cases. Then

$$a(a^2 + 2)/3 = q(9q^2 + 2),$$

which clearly is an integer. Similarly, if $a = 3q + 1$, then

$$(3q + 1)((3q + 1)^2 + 2)/3 = (3q + 1)(3q^2 + 2q + 1),$$

and $a(a^2 + 2)/3$ is an integer in this instance also. Finally, for $a = 3q + 2$, we get

$$(3q + 2)((3q + 2)^2 + 2)/3 = (3q + 2)(3q^2 + 4q + 2),$$

an integer once more. Consequently, our result is established in all cases.

PROBLEMS 2.1

1. Prove that if a and b are integers, with $b > 0$, then there exist unique integers q and r satisfying $a = qb + r$, where $2b \leq r < 3b$.

2. Show that any integer of the form $6k + 5$ is also of the form $3j + 2$, but not conversely.

3. Use the Division Algorithm to establish that
   (a) the square of any integer is of the form either $3k$ or $3k + 1$;
   (b) the cube of any integer has one of the forms $9k$, $9k + 1$ or $9k + 8$;
   (c) the fourth power of any integer is of the form either $5k$ or $5k + 1$.

4. Prove that $3a^2 - 1$ is never a perfect square. [*Hint:* Problem 3(a).]

5. For $n \geq 1$, prove that $n(n + 1)(2n + 1)/6$ is an integer. [*Hint:* By the Division Algorithm, n has one of the forms $6k, 6k + 1, \ldots, 6k + 5$; establish the result in each of these six cases.]

6. Show that the cube of any integer is of the form $7k$ or $7k \pm 1$.

7. Obtain the following version of the Division Algorithm: For integers a and b, with $b \neq 0$, *there exist unique integers q and r which satisfy $a = qb + r$, where $-\frac{1}{2}|b| < r \leq \frac{1}{2}|b|$.* [*Hint:* First write $a = q'b + r'$, where $0 \leq r' < |b|$. When $0 \leq r' \leq \frac{1}{2}|b|$, let $r = r'$ and $q = q'$; when $\frac{1}{2}|b| < r' < |b|$, let $r = r' - |b|$ and $q = q' + 1$ if $b > 0$ or $q = q' - 1$ if $b < 0$.]

8. Prove that no integer in the sequence

$$11, 111, 1111, 11111, \ldots$$

   is a perfect square. [*Hint:* A typical term $111 \cdots 111$ can be written as $111 \cdots 111 = 111 \cdots 108 + 3 = 4k + 3$.]

9. Verify that if an integer is simultaneously a square and a cube (as is the case with $64 = 8^2 = 4^3$), then it must be either of the form $7k$ or $7k + 1$.

10. For $n \geq 1$, establish that the integer $n(7n^2 + 5)$ is of the form $6k$.

11. If n is an odd integer, show that $n^4 + 4n^2 + 11$ is of the form $16k$.

2.2 THE GREATEST COMMON DIVISOR

Of special significance is the case in which the remainder in the Division Algorithm turns out to be zero. Let us look into this situation now.

DEFINITION 2-1.
An integer b is said to be *divisible* by an integer $a \neq 0$, in symbols $a \mid b$, if there exists some integer c such that $b = ac$. We write $a \nmid b$ to indicate that b is not divisible by a.

Thus, for example, -12 is divisible by 4, since $-12 = 4(-3)$. However, 10 is not divisible by 3; for there is no integer c which makes the statement $10 = 3c$ true.

There is other language for expressing the divisibility relation $a \mid b$. One could say that a is a *divisor* of b, that a is a *factor* of b or that b is a *multiple* of a. Notice that, in Definition 2-1, there is a restriction on the divisor a: whenever the notation $a \mid b$ is employed, it is understood that a is different from zero.

If a is a divisor of b, then b is also divisible by $-a$ (indeed, $b = ac$ implies that $b = (-a)(-c)$), so that the divisors of an integer always occur in pairs. In order to find all the divisors of a given integer, it is sufficient to obtain the positive divisors and then adjoin to them the corresponding negative integers. For this reason, we shall usually limit ourselves to a consideration of positive divisors.

It will be helpful to list some immediate consequences of Definition 2-1 (the reader is again reminded that, although not stated, divisors are assumed to be nonzero).

THEOREM 2-2.

For integers a, b, c, the following hold:
(1) $a \mid 0$, $1 \mid a$, $a \mid a$.
(2) $a \mid 1$ *if and only if* $a = \pm 1$.
(3) *If* $a \mid b$ *and* $c \mid d$, *then* $ac \mid bd$.
(4) *If* $a \mid b$ *and* $b \mid c$, *then* $a \mid c$.
(5) $a \mid b$ *and* $b \mid a$ *if and only if* $a = \pm b$.
(6) *If* $a \mid b$ *and* $b \neq 0$, *then* $|a| \leq |b|$.
(7) *If* $a \mid b$ *and* $a \mid c$, *then* $a \mid (bx + cy)$ *for arbitrary integers x and y.*

Proof: We shall prove assertions (6) and (7), leaving the other parts as an exercise. If $a \mid b$, then there exists an integer c such that $b = ac$; also, $b \neq 0$ implies that $c \neq 0$. Upon taking absolute values, we get $|b| = |ac| = |a||c|$. Since $c \neq 0$, it follows that $|c| \geq 1$, whence $|b| = |a||c| \geq |a|$.

As regards (7), the relations $a \mid b$ and $a \mid c$ ensure that $b = ar$ and $c = as$ for suitable integers r and s. But then

$$bx + cy = arx + asy = a(rx + sy)$$

whatever the choice of x and y. Since $rx + sy$ is an integer, this says that $a \mid (bx + cy)$, as desired.

It is worth pointing out that property (7) of the preceding theorem extends by induction to sums of more than two terms. That is, if $a \mid b_k$ for $k = 1, 2, \ldots, n$, then

$$a \mid (b_1x_1 + b_2x_2 + \cdots + b_nx_n)$$

for all integers $x_1, x_2, \ldots, x_n$. The few details needed for the proof are so straightforward that we omit them.

If a and b are arbitrary integers, then an integer d is said to be a *common divisor* of a and b if both $d \mid a$ and $d \mid b$. Since 1 is a divisor of every integer, 1 is a common divisor of a and b; hence, their set of positive common divisors is nonempty. Now every integer divides 0, so that if $a = b = 0$, then every integer serves as a common divisor of a and b. In this instance, the set of positive common divisors of a and b is infinite. However, when at least one of a or b is different from zero, there are only a finite number of positive common divisors. Among these, there is a largest one, called the greatest common divisor of a and b. Framed as a definition,

DEFINITION 2-2.

Let a and b be given integers, with at least one of them different from zero. The *greatest common divisor* of a and b, denoted by $\gcd(a, b)$, is the positive integer d satisfying

(1) $d \mid a$ and $d \mid b$,
(2) if $c \mid a$ and $c \mid b$, then $c \leq d$.

Example 2-2

The positive divisors of -12 are 1, 2, 3, 4, 6, 12, while those of 30 are 1, 2, 3, 5, 6, 10, 15, 30; hence, the positive common divisors of -12 and 30 are 1, 2, 3, 6. Since 6 is the largest of these integers, it follows that $\gcd(-12, 30) = 6$. In the same way, one can show that

$$\gcd(-5, 5) = 5, \quad \gcd(8, 17) = 1, \quad \text{and} \gcd(-8, -36) = 4.$$

The next theorem indicates that $\gcd(a, b)$ can be represented as a linear combination of a and b (by a *linear combination* of a and b, we mean an expression of the form $ax + by$, where x and y are integers). This is illustrated by, say

$$\gcd(-12, 30) = 6 = (-12)2 + 30 \cdot 1$$

or

$$\gcd(-8, -36) = 4 = (-8)4 + (-36)(-1).$$

Now for the theorem:

THEOREM 2-3.

Given integers a and b, not both of which are zero, there exist integers x and y such that

$$\gcd(a, b) = ax + by.$$

Proof: Consider the set S of all positive linear combinations of a and b:

$$S = \{au + bv \mid au + bv > 0; u, v \text{ integers}\}.$$

Notice first that S is not empty. For example, if $a \neq 0$, then the integer $|a| = au + b \cdot 0$ will lie in S, where we choose $u = 1$ or $u = -1$ according as a is positive or negative. By virtue of the Well-Ordering Principle, S must contain a smallest element d. Thus, from the very definition of S, there exist integers x and y for which $d = ax + by$. We claim that $d = \gcd(a, b)$.

Taking stock of the Division Algorithm, one can obtain integers q and r such that $a = qd + r$, where $0 \leq r < d$. Then r can be written in the form

$$r = a - qd = a - q(ax + by)$$
$$= a(1 - qx) + b(-qy).$$

Were $r > 0$, this representation would imply that r is a member of S, contradicting the fact that d is the least integer in S (recall that $r < d$). Therefore, $r = 0$ and so $a = qd$, or equivalently, $d \mid a$. By similar reasoning $d \mid b$, the effect of which is to make d a common divisor of both a and b.

Now if c is an arbitrary positive common divisor of the integers a and b, then part (7) of Theorem 2-2 allows us to conclude that $c \mid (ax + by)$; in other words,

$c \mid d$. By part (6) of the same theorem, $c = |c| \le |d| = d$, so that d is greater than every positive common divisor of a and b. Piecing the bits of information together, we see that $d = \gcd(a, b)$.

It should be noted that the foregoing argument is merely an "existence" proof and does not provide a practical method for finding the values of x and y; this will come later.

A perusal of the proof of Theorem 2-3 reveals that the greatest common divisor of a and b may be described as the smallest positive integer of the form $ax + by$. Consider the case in which $a = 6$ and $b = 15$. Here, the set S becomes

$$S = \{ 6(-2) + 15 \cdot 1, 6(-1) + 15 \cdot 1, 6 \cdot 1 + 15 \cdot 0, \cdots \}$$
$$= \{ 3, 9, 6, \cdots \}.$$

We observe that 3 is the smallest integer in S, whence $3 = \gcd(6, 15)$.

The nature of the members of S appearing in this illustration suggests another result:

COROLLARY.

If a and b are given integers, not both zero, then the set

$$T = \{ax + by \mid x, y \text{ are integers}\}$$

is precisely the set of all multiples of $d = \gcd(a, b)$.

Proof: Since $d \mid a$ and $d \mid b$, we know that $d \mid (ax + by)$ for all integers x, y. Thus, every member of T is a multiple of d. On the other hand, d may be written as $d = ax_0 + by_0$ for suitable integers x_0 and y_0, so that any multiple nd of d is of the form

$$nd = n(ax_0 + by_0) = a(nx_0) + b(ny_0).$$

Hence, nd is a linear combination of a and b, and, by definition, lies in T.

It may happen that 1 and -1 are the only common divisors of a given pair of integers a and b, whence $\gcd(a, b) = 1$. For example:

$$\gcd(2, 5) = \gcd(-9, 16) = \gcd(-27, -35) = 1.$$

This situation occurs often enough to prompt a definition.

DEFINITION 2-3.

Two integers a and b, not both of which are zero, are said to be *relatively prime* whenever $\gcd(a, b) = 1$.

The following theorem characterizes relatively prime integers in terms of linear combinations.

THEOREM 2-4.

Let a and b be integers, not both zero. Then a and b are relatively prime if and only if there exist integers x and y such that $1 = ax + by$.

Proof: If a and b are relatively prime so that $\gcd(a, b) = 1$, then Theorem 2-3 guarantees the existence of integers x and y satisfying $1 = ax + by$. As for the converse, suppose that $1 = ax + by$ for some choice of x and y, and that $d = \gcd(a, b)$. Since $d \mid a$ and $d \mid b$, Theorem 2-2 yields $d \mid (ax + by)$, or $d \mid 1$. Inasmuch as d is a positive integer, this last divisibility condition forces $d = 1$ (part (2) of Theorem 2-2 plays a role here) and the desired conclusion follows.

This result leads to an observation that is useful in certain situations; namely,

COROLLARY 1.

If $\gcd(a, b) = d$, then $\gcd(a/d, b/d) = 1$.

Proof: Before starting with the proof proper, we should observe that while a/d and b/d have the appearance of fractions, they are in fact integers since d is a divisor both of a and of b. Now, knowing that $\gcd(a, b) = d$, it is possible to find integers x and y such that $d = ax + by$. Upon dividing each side of this equation by d, one obtains the expression

$$1 = (a/d)x + (b/d)y.$$

Because a/d and b/d are integers, an appeal to the theorem is legitimate. The conclusion is that a/d and b/d are relatively prime.

For an illustration of the last corollary, let us observe that $\gcd(-12, 30) = 6$ and

$$\gcd(-12/6, 30/6) = \gcd(-2, 5) = 1,$$

as it should be.

It is not true, without adding an extra condition, that $a \mid c$ and $b \mid c$ together give $ab \mid c$. For instance, $6 \mid 24$ and $8 \mid 24$, but $6 \cdot 8 \nmid 24$. Were 6 and 8 relatively prime, of course, this situation would not arise. This brings us to

COROLLARY 2.

If $a \mid c$ and $b \mid c$, with $\gcd(a, b) = 1$, then $ab \mid c$.

Proof: Inasmuch as $a \mid c$ and $b \mid c$, integers r and s can be found such that $c = ar = bs$. Now the relation $\gcd(a, b) = 1$ allows us to write $1 = ax + by$ for some choice of integers x and y. Multiplying the last equation by c, it appears that

$$c = c \cdot 1 = c(ax + by) = acx + bcy.$$

If the appropriate substitutions are now made on the right-hand side, then

$$c = a(bs)x + b(ar)y = ab(sx + ry)$$

or, as a divisibility statement, $ab \mid c$.

Our next result seems mild enough, but it is of fundamental importance.

THEOREM 2-5 (Euclid's Lemma).

If $a \mid bc$, with $\gcd(a, b) = 1$, then $a \mid c$.

Proof: We start again from Theorem 2-3, writing $1 = ax + by$ where x and y are integers. Multiplication of this equation by c produces

$$c = 1 \cdot c = (ax + by)c = acx + bcy.$$

Since $a \mid ac$ and $a \mid bc$, it follows that $a \mid (acx + bcy)$, which can be recast as $a \mid c$.

If a and b are not relatively prime, then the conclusion of Euclid's Lemma may fail to hold. A specific example: $12 \mid 9 \cdot 8$, but $12 \nmid 9$ and $12 \nmid 8$.

The subsequent theorem often serves as a definition of $\gcd(a, b)$. The advantage of using it as a definition is that order relationship is not involved; thus it may be used in algebraic systems having no order relation.

THEOREM 2-6.

Let a, b be integers, not both zero. For a positive integer d, $d = \gcd(a, b)$ if and only if

(1)   *$d \mid a$ and $d \mid b$,*
(2)   *whenever $c \mid a$ and $c \mid b$, then $c \mid d$.*

Proof: To begin, suppose that $d = \gcd(a, b)$. Certainly, $d \mid a$ and $d \mid b$, so that (1) holds. In light of Theorem 2-3, d is expressible as $d = ax + by$ for some integers x, y. Thus, if $c \mid a$ and $c \mid b$, then $c \mid (ax + by)$, or rather $c \mid d$. In short, condition (2) holds. Conversely, let d be any positive integer satisfying the stated conditions. Given any common divisor c of a and b, we have $c \mid d$ from hypothesis (2). The implication is that $d \geq c$, and consequently d is the greatest common divisor of a and b.

Problems 2.2

1.   If $a \mid b$, show that $(-a) \mid b$, $a \mid (-b)$, and $(-a) \mid (-b)$.

2.   Given integers a, b, c, d, verify that
     (a)   if $a \mid b$, then $a \mid bc$;
     (b)   if $a \mid b$ and $a \mid c$, then $a^2 \mid bc$;
     (c) $a \mid b$ if and only if $ac \mid bc$, where $c \neq 0$;
     (d)   if $a \mid b$ and $c \mid d$, then $ac \mid bd$.

3.   Prove or disprove: if $a \mid (b + c)$, then either $a \mid b$ or $a \mid c$.

4.   For $n \geq 1$, use mathematical induction to establish each of the following divisibility statements:
     (a) $8 \mid 5^{2n} + 7$,
           [*Hint:* $5^{2(k+1)} + 7 = 5^2(5^{2k} + 7) + (7 - 5^2 \cdot 7).$]
     (b) $15 \mid 2^{4n} - 1$,
     (c) $5 \mid 3^{3n+1} + 2^{n+1}$,
     (d) $21 \mid 4^{n+1} + 5^{2n-1}$,
     (e) $24 \mid 2 \cdot 7^n + 3 \cdot 5^n - 5$.

5. Prove that for any integer a one of the integers a, $a + 2$, $a + 4$ is divisible by 3.

6. For an arbitrary integer a, verify that
   (a) $2 \mid a(a + 1)$, and $3 \mid a(a + 1)(a + 2)$;
   (b) $3 \mid a(2a^2 + 7)$;
   (c) if a is odd, then $32 \mid (a^2 + 3)(a^2 + 7)$.

7. Prove that if a and b are both odd integers, then $16 \mid a^4 + b^4 - 2$.

8. Prove that
   (a) the sum of the squares of two odd integers cannot be a perfect square;
   (b) the product of four consecutive integers is 1 less than a perfect square.

9. Establish that the difference of two consecutive cubes is never divisible by 2.

10. For a nonzero integer a, show that $\gcd(a, 0) = \mid a \mid$, $\gcd(a, a) = \mid a \mid$, and $\gcd(a, 1) = 1$.

11. If a and b are integers, not both of which are zero, verify that

$$\gcd(a, b) = \gcd(-a, b) = \gcd(a, -b) = \gcd(-a, -b).$$

12. Prove that, for a positive integer n and any integer a, $\gcd(a, a + n)$ divides n; hence, $\gcd(a, a + 1) = 1$.

13. Given integers a and b, prove that
    (a) there exist integers x and y for which $c = ax + by$ if and only if $\gcd(a, b) \mid c$;
    (b) if there exist integers x and y for which $ax + by = \gcd(a, b)$, then $\gcd(x, y) = 1$.

14. For any integer a, show that
    (a) $\gcd(2a + 1, 9a + 4) = 1$;
    (b) $\gcd(5a + 2, 7a + 3) = 1$;
    (c) if a is odd, then $\gcd(3a, 3a + 2) = 1$.

15. If a and b are integers, not both of which are zero, prove that $\gcd(2a - 3b, 4a - 5b)$ divides b; hence, $\gcd(2a + 3, 4a + 5) = 1$.

16. Given an odd integer a, establish that

$$a^2 + (a + 2)^2 + (a + 4)^2 + 1$$

is divisible by 12.

17. Prove that the expression $(3n)!/(3!)^n$ is an integer for all $n \geq 0$.

18. Prove: the product of any three consecutive integers is divisible by 6; the product of any four consecutive integers is divisible by 24; the product of any five consecutive integers is divisible by 120. [*Hint:* See Corollary 2 to Theorem 2-4.]

19. Establish each of the assertions below:
    (a) If a is an arbitrary integer, then $6 \mid a(a^2 + 11)$.
    (b) If a is an odd integer, then $24 \mid a(a^2 - 1)$. [*Hint:* The square of an odd integer is of the form $8k + 1$.]
    (c) If a and b are odd integers, then $8 \mid (a^2 - b^2)$.
    (d) If a is an integer not divisible by 2 or 3, then $24 \mid (a^2 + 23)$.
    (e) If a is an arbitrary integer, then $360 \mid a^2(a^2 - 1)(a^2 - 4)$.

20. Confirm the following properties of the greatest common divisor:

(a) If $\gcd(a, b) = 1$, and $\gcd(a, c) = 1$, then $\gcd(a, bc) = 1$.
    [*Hint:* Since $1 = ax + by = au + cv$ for some x, y, u, v, $1 = (ax + by)(au + cv)$
    $= a(aux + cvx + byu) + bc(yv)$.]

(b) If $\gcd(a, b) = 1$, and $c \mid a$, then $\gcd(b, c) = 1$.

(c) If $\gcd(a, b) = 1$, then $\gcd(ac, b) = \gcd(c, b)$.

(d) If $\gcd(a, b) = 1$, and $c \mid a + b$, then $\gcd(a, c) = \gcd(b, c) = 1$. [*Hint:* Let $d = \gcd(a, c)$. Then $d \mid a$, $d \mid c$ implies that $d \mid (a + b) - a$, or $d \mid b$.]

(e) If $\gcd(a, b) = 1$, $d \mid ac$, and $d \mid bc$, then $d \mid c$.

(f) If $\gcd(a, b) = 1$, then $\gcd(a^2, b^2) = 1$.
    [*Hint:* First show that $\gcd(a, b^2) = \gcd(a^2, b) = 1$.]

21. (a) Prove that if $d \mid n$, then $2^d - 1 \mid 2^n - 1$. [*Hint:* Use the identity

$$x^k - 1 = (x - 1)(x^{k-1} + x^{k-2} + \cdots + x + 1).]$$

(b) Verify that $2^{35} - 1$ is divisible by 31 and 127.

22. Let t_n denote the n'th triangular number. For what values of n does t_n divide the sum $t_1 + t_2 + \cdots + t_n$? [*Hint:* See Problem 1(c), Section 1.1.]

23. If $a \mid bc$, show that $a \mid \gcd(a, b) \gcd(a, c)$.

2.3 THE EUCLIDEAN ALGORITHM

The greatest common divisor of two integers can, of course, be found by listing all their positive divisors and picking out the largest one common to each; but this is cumbersome for large numbers. A more efficient process, involving repeated application of the Division Algorithm, is given in the seventh Book of the *Elements*. Although there is historical evidence that this method predates Euclid, it is today referred to as the Euclidean Algorithm.

The Euclidean Algorithm may be described as follows: Let a and b be two integers whose greatest common divisor is desired. Since $\gcd(\mid a \mid, \mid b \mid) = \gcd(a, b)$, there is no harm in assuming that $a \geq b > 0$. The first step is to apply the Division Algorithm to a and b to get

$$a = q_1 b + r_1, \qquad\qquad 0 \leq r_1 < b.$$

If it happens that $r_1 = 0$, then $b \mid a$ and $\gcd(a, b) = b$. When $r_1 \neq 0$, divide b by r_1 to produce integers q_2 and r_2 satisfying

$$b = q_2 r_1 + r_2, \qquad\qquad 0 \leq r_2 < r_1.$$

If $r_2 = 0$, then we stop; otherwise, proceed as before to obtain

$$r_1 = q_3 r_2 + r_3, \qquad\qquad 0 \leq r_3 < r_2.$$

This division process continues until some zero remainder appears, say at the $(n + 1)$th stage where r_{n-1} is divided by r_n (a zero remainder occurs sooner or later since the decreasing sequence $b > r_1 > r_2 > \cdots \geq 0$ cannot contain more than b integers).

The result is the following system of equations:

$$a = q_1 b + r_1, \qquad\qquad\qquad 0 < r_1 < b$$
$$b = q_2 r_1 + r_2, \qquad\qquad\qquad 0 < r_2 < r_1$$
$$r_1 = q_3 r_2 + r_3, \qquad\qquad\qquad 0 < r_3 < r_2$$

$$\vdots$$

$$r_{n-2} = q_n r_{n-1} + r_n, \qquad\qquad 0 < r_n < r_{n-1}$$
$$r_{n-1} = q_{n+1} r_n + 0.$$

We argue that r_n, the last nonzero remainder which appears in this manner, is equal to $\gcd(a, b)$. Our proof is based on the lemma below.

LEMMA.

If $a = qb + r$, then $\gcd(a, b) = \gcd(b, r)$.

Proof: If $d = \gcd(a, b)$, then the relations $d \mid a$ and $d \mid b$ together imply that $d \mid (a - qb)$, or $d \mid r$. Thus d is a common divisor of both b and r. On the other hand, if c is an arbitrary common divisor of b and r, then $c \mid (qb + r)$, whence $c \mid a$. This makes c a common divisor of a and b, so that $c \leq d$. It now follows from the definition of $\gcd(b, r)$ that $d = \gcd(b, r)$.

Using the result of this lemma, we simply work down the displayed system of equations obtaining

$$\gcd(a, b) = \gcd(b, r_1) = \cdots = \gcd(r_{n-1}, r_n) = \gcd(r_n, 0) = r_n,$$

as claimed.

Although Theorem 2-3 asserts that $\gcd(a, b)$ can be expressed in the form $ax + by$, the proof of the theorem gives no hint as to how to determine the integers x and y. For this, we fall back on the Euclidean Algorithm. Starting with the next-to-last equation arising from the algorithm, we write

$$r_n = r_{n-2} - q_n r_{n-1}.$$

Now solve the preceding equation in the algorithm for r_{n-1} and substitute to obtain

$$r_n = r_{n-2} - q_n(r_{n-3} - q_{n-1} r_{n-2})$$
$$= (1 + q_n q_{n-1}) r_{n-2} + (-q_n) r_{n-3}.$$

This represents r_n as a linear combination of r_{n-2} and r_{n-3}. Continuing backwards through the system of equations, we successively eliminate the remainders r_{n-1}, $r_{n-2}, \ldots, r_2, r_1$ until a stage is reached where $r_n = \gcd(a, b)$ is expressed as a linear combination of a and b.

Example 2-3

Let us see how the Euclidean Algorithm works in a concrete case by calculating, say, gcd(12378 , 3054). The appropriate applications of the Division Algorithm produce the equations

$$12378 = 4 \cdot 3054 + 162,$$
$$3054 = 18 \cdot 162 + 138,$$
$$162 = 1 \cdot 138 + 24,$$
$$138 = 5 \cdot 24 + 18,$$
$$24 = 1 \cdot 18 + 6,$$
$$18 = 3 \cdot 6 + 0.$$

Our previous discussion tells us that the last nonzero remainder appearing above, namely the integer 6, is the greatest common divisor of 12378 and 3054:

$$6 = \gcd(12378 , 3054).$$

To represent 6 as a linear combination of the integers 12378 and 3054, we start with the next-to-last of the displayed equations and successively eliminate the remainders 18, 24, 138, and 162:

$$\begin{aligned}
6 &= 24 - 18 \\
&= 24 - (138 - 5 \cdot 24) \\
&= 6 \cdot 24 - 138 \\
&= 6(162 - 138) - 138 \\
&= 6 \cdot 162 - 7 \cdot 138 \\
&= 6 \cdot 162 - 7(3054 - 18 \cdot 162) \\
&= 132 \cdot 162 - 7 \cdot 3054 \\
&= 132(12378 - 4 \cdot 3054) - 7 \cdot 3054 \\
&= 132 \cdot 12378 + (-535)3054.
\end{aligned}$$

Thus, we have

$$6 = \gcd(12378 , 3054) = 12378x + 3054y,$$

where $x = 132$ and $y = -535$. It might be well to record that this is not the only way to express the integer 6 as a linear combination of 12378 and 3054; among other possibilities, one could add and subtract $3054 \cdot 12378$ to get

$$\begin{aligned}
6 &= (132 + 3054)12378 + (-535 - 12378)3054 \\
&= 3186 \cdot 12378 + (-12913)3054.
\end{aligned}$$

The French mathematician Gabriel Lamé (1795–1870) proved that the number of steps required in the Euclidean Algorithm is at most five times the number of digits in the smaller integer. In Example 2-3, the smaller integer (namely 3054) has four digits, so that the total number of divisions cannot be greater than twenty; in actuality only six divisions were needed. Another observation of interest is that for each $n > 0$, it is possible to find integers a_n and b_n such that exactly n divisions are required in order to compute $\gcd(a_n, b_n)$ by the Euclidean Algorithm. We shall prove this fact in Chapter 13.

One more remark is necessary: The number of steps in the Euclidean Algorithm can usually be reduced by selecting remainders r_{k+1} such that $|r_{k+1}| < r_k/2$; that is, by working with least absolute remainders in the divisions. Thus, repeating Example 2-3, it would be more efficient to write

$$12378 = 4 \cdot 3054 + 162,$$
$$3054 = 19 \cdot 162 - 24,$$
$$162 = 7 \cdot 24 - 6,$$
$$24 = (-4)(-6) + 0.$$

As evidenced by the above set of equations, this scheme is apt to produce the negative of the value of the greatest common divisor of two integers (the last nonzero remainder being -6), rather than the greatest common divisor itself.

An important consequence of the Euclidean Algorithm is the following theorem.

THEOREM 2-7.

If $k > 0$, then $\gcd(ka, kb) = k \gcd(a, b)$.

Proof: If each of the equations appearing in the Euclidean Algorithm for a and b (see page 28) is multiplied by k, we obtain

$$ak = q_1(bk) + r_1k, \qquad\qquad 0 < r_1k < bk$$
$$bk = q_2(r_1k) + r_2k, \qquad\qquad 0 < r_2k < r_1k$$

$$\vdots$$

$$r_{n-2}k = q_n(r_{n-1}k) + r_nk, \qquad\qquad 0 < r_nk < r_{n-1}k$$
$$r_{n-1}k = q_{n+1}(r_nk) + 0.$$

But this is clearly the Euclidean Algorithm applied to the integers ak and bk, so that their greatest common divisor is the last nonzero remainder r_nk; that is,

$$\gcd(ka, kb) = r_nk = k \gcd(a, b),$$

as stated in the theorem.

COROLLARY.

For any integer $k \neq 0$, $\gcd(ka, kb) = |k| \gcd(a, b)$.

Proof: It suffices to consider the case in which $k < 0$. Then $-k = |k| > 0$ and, by Theorem 2-7,

$$\gcd(ak , bk) = \gcd(-ak , -bk)$$
$$= \gcd(a|k| , b|k|) = |k| \gcd(a , b).$$

An alternate proof of Theorem 2-7 runs very quickly as follows: $\gcd(ak , bk)$ is the smallest positive integer of the form $(ak)x + (bk)y$, which in its turn is equal to k times the smallest positive integer of the form $ax + by$; the latter value is equal to $k \gcd(a , b)$.

By way of illustrating Theorem 2-7, we see that

$$\gcd(12 , 30) = 3 \gcd(4 , 10) = 3 \cdot 2 \gcd(2 , 5) = 6 \cdot 1 = 6.$$

There is a concept parallel to that of the greatest common divisor of two integers, known as their least common multiple; but we shall not have much occasion to make use of it. An integer c is said to be a common multiple of two nonzero integers a and b whenever $a|c$ and $b|c$. Evidently, 0 is a common multiple of a and b. To see that common multiples which are not trivial do exist, just note that the products ab and $-(ab)$ are both common multiples of a and b, and one of these is positive. By the Well-Ordering Principle, the set of positive common multiples of a and b must contain a smallest integer; we call it the least common multiple of a and b.

For the record, here is the official definition.

DEFINITION 2-4.

The *least common multiple* of two nonzero integers a and b, denoted by $\text{lcm}(a , b)$, is the positive integer m satisfying

(1) $a|m$ and $b|m$,

(2)   if $a|c$ and $b|c$, with $c > 0$, then $m \le c$.

As an example, the positive common multiples of the integers -12 and 30 are 60, 120, 180, . . . ; hence, $\text{lcm}(-12 , 30) = 60$.

The following remark is clear from our discussion: Given nonzero integers a and b, $\text{lcm}(a , b)$ always exists and $\text{lcm}(a , b) \le |ab|$.

What we lack is a relationship between the ideas of greatest common divisor and least common multiple. This gap is filled by

THEOREM 2-8.

For positive integers a and b,

$$\gcd(a , b) \, \text{lcm}(a , b) = ab.$$

Proof: To begin, put $d = \gcd(a , b)$ and write $a = dr, b = ds$ for integers r and s. If $m = ab/d$, then $m = as = rb$, the effect of which is to make m a (positive) common multiple of a and b.

Now let c be any positive integer that is a common multiple of a and b; say for definiteness, $c = au = bv$. As we know, there exist integers x and y satisfying $d = ax + by$. In consequence,

$$\frac{c}{m} = \frac{cd}{ab} = \frac{c(ax + by)}{ab} = (c/b)x + (c/a)y = vx + uy.$$

This equation states that $m \mid c$, allowing us to conclude that $m \leq c$. Thus, in accordance with Definition 2-4, $m = \mathrm{lcm}(a , b)$; that is,

$$\mathrm{lcm}(a , b) = \frac{ab}{d} = \frac{ab}{\gcd(a , b)},$$

which is what we started out to prove.

Theorem 2-8 has a corollary that is worth a separate statement.

COROLLARY.

Given positive integers a and b, $\mathrm{lcm}(a , b) = ab$ if and only if $\gcd(a , b) = 1$.

Perhaps the chief virtue of Theorem 2-8 is that it makes the calculation of the least common multiple of two integers dependent on the value of their greatest common divisor—which in its turn can be calculated from the Euclidean Algorithm. When considering the integers 3054 and 12378, for instance, we found that $\gcd(3054 , 12378) = 6$; whence,

$$\mathrm{lcm}(3054 , 12378) = \frac{3054 \cdot 12378}{6} = 6{,}300{,}402.$$

Before moving on to other matters, let us observe that the notion of greatest common divisor can be extended to more than two integers in an obvious way. In the case of three integers, a, b, c, not all zero, $\gcd(a , b , c)$ is defined to be the positive integer d having the properties

(1) d is a divisor of each of $a, b, c,$
(2) if e divides the integers $a, b, c,$ then $e \leq d$.

To cite two examples, we have

$$\gcd(39 , 42 , 54) = 3 \text{ and } \gcd(49 , 210 , 350) = 7.$$

The reader is cautioned that it is possible for three integers to be relatively prime as a triple (in other words, $\gcd(a , b , c) = 1$), yet not relatively prime in pairs; this is brought out by the integers 6, 10, and 15.

PROBLEMS 2.3

1. Find $\gcd(143 , 227)$, $\gcd(306 , 657)$ and $\gcd(272 , 1479)$.

2. Use the Euclidean Algorithm to obtain integers x and y satisfying
    (a) $\gcd(56 , 72) = 56x + 72y$;
    (b) $\gcd(24 , 138) = 24x + 138y$;

(c) $\gcd(119 , 272) = 119x + 272y;$

(d) $\gcd(1769 , 2378) = 1769x + 2378y.$

3. Prove that if d is a common divisor of a and b, then $d = \gcd(a , b)$ if and only if $\gcd(a/d , b/d) = 1$. [*Hint:* Use Theorem 2-7.]

4. Assuming that $\gcd(a , b) = 1$, prove the following:
    (a) $\gcd(a + b , a - b) = 1$ or 2.
         [*Hint:* Let $d = \gcd(a + b , a - b)$ and show that $d \mid 2a$, $d \mid 2b;$ thus, that $d \le \gcd(2a , 2b) = 2 \gcd(a , b)$.]
    (b) $\gcd(2a + b , a + 2b) = 1$ or 3.
    (c) $\gcd(a + b , a^2 + b^2) = 1$ or 2.
         [*Hint:* $a^2 + b^2 = (a + b)(a - b) + 2b^2$.]
    (d) $\gcd(a + b , a^2 - ab + b^2) = 1$ or 3.
         [*Hint:* $a^2 - ab + b^2 = (a + b)^2 - 3ab$.]

5. For positive integers a, b and $n \ge 1$, show that
    (a) If $\gcd(a , b) = 1$, then $\gcd(a^n , b^n) = 1$. [*Hint:* See Problem 20(a), Section 2-2.]
    (b) The relation $a^n \mid b^n$ implies that $a \mid b$. [*Hint:* Put $d = \gcd(a , b)$ and write $a = rd$, $b = sd$, where $\gcd(r , s) = 1$. By part (a), $\gcd(r^n, s^n) = 1$. Show that $r = 1$, whence $a = d$.]

6. Prove that if $\gcd(a , b) = 1$, then $\gcd(a + b , ab) = 1$.

7. For nonzero integers a and b, verify that the following conditions are equivalent:
    (a) $a \mid b$ (b) $\gcd(a , b) = |a|$ (c) $\text{lcm}(a , b) = |b|$.

8. Find $\text{lcm}(143 , 227)$, $\text{lcm}(306 , 657)$ and $\text{lcm}(272 , 1479)$.

9. Prove that the greatest common divisor of two positive integers divides their least common multiple.

10. Given nonzero integers a and b, establish the following facts concerning $\text{lcm}(a , b)$:
    (a) $\gcd(a , b) = \text{lcm}(a , b)$ if and only if $a = \pm b$.
    (b) If $k > 0$, then $\text{lcm}(ka , kb) = k \text{lcm}(a , b)$.
    (c) If m is any common multiple of a and b, then $\text{lcm}(a , b) \mid m$.
         [*Hint:* Put $t = \text{lcm}(a , b)$ and use the Division Algorithm to write $m = qt + r$, where $0 \le r < t$. Show that r is a common multiple of a and b.]

11. Let a, b, c be integers, no two of which are zero, and $d = \gcd(a , b , c)$. Show that

$$d = \gcd(\gcd(a , b), c) = \gcd(a , \gcd(b , c)) = \gcd(\gcd(a , c), b).$$

12. Find integers x, y, z satisfying

$$\gcd(198 , 288 , 512) = 198x + 288y + 512z.$$

[*Hint:* Put $d = \gcd(198 , 288)$. Since $\gcd(198 , 288 , 512) = \gcd(d , 512)$, first find integers u and v for which $\gcd(d , 512) = du + 512v$.]

2.4 THE DIOPHANTINE EQUATION $ax + by = c$

We now change focus somewhat and take up the study of Diophantine equations. The name honors the mathematician Diophantus, who initiated the study of such equations. Practically nothing is known of Diophantus as an individual, save that he lived in Alexandria sometime around 250 A.D. The only positive evidence as to the date of his activity is that the Bishop of Laodicea, who began his episcopate in 270, dedicated a book on Egyptian computation to his friend Diophantus. While Diophantus' works were written in Greek and he displayed the Greek genius for theoretical abstraction, he was most likely a Hellenized Babylonian. The only personal particulars we have of his career come from the wording of an epigram-problem (apparently dating from the 4th century) to the effect: his boyhood lasted 1/6 of his life; his beard grew after 1/12 more; after 1/7 more he married, and his son was born 5 years later; the son lived to half his father's age and the father died four years after his son. If x was the age at which Diophantus died, these data lead to the equation

$$\frac{1}{6}x + \frac{1}{12}x + \frac{1}{7}x + 5 + \frac{1}{2}x + 4 = x,$$

with solution $x = 84$. Thus he must have reached an age of 84, but in what year or even in what century is not certain.

The great work upon which the reputation of Diophantus rests is his *Arithmetica,* which may be described as the earliest treatise on algebra. Only six Books of the original thirteen have been preserved. It is in the *Arithmetica* that we find the first systematic use of mathematical notation, although the signs employed are of the nature of abbreviations for words rather than algebraic symbols in our sense. Special symbols are introduced to represent frequently occurring concepts, such as the unknown quantity in an equation and the different powers of the unknown up to the sixth power; Diophantus also had a symbol to express subtraction, and another for equality.

It is customary to apply the term *Diophantine equation* to any equation in one or more unknowns which is to be solved in the integers. The simplest type of Diophantine equation that we shall consider is the linear Diophantine equation in two unknowns:

$$ax + by = c,$$

where a, b, c are given integers and a, b not both zero. A solution of this equation is a pair of integers x_0, y_0 which, when substituted into the equation, satisfy it; that is, we ask that $ax_0 + by_0 = c$. Curiously enough, the linear equation does not appear in the extant works of Diophantus (the theory required for its solution is to be found in Euclid's *Elements*), possibly because he viewed it as trivial; most of his problems dealt with finding squares or cubes with certain properties.

A given linear Diophantine equation can have a number of solutions, as is the case with $3x + 6y = 18$, where

$$3 \cdot 4 + 6 \cdot 1 = 18,$$

$$3(-6) + 6 \cdot 6 = 18,$$

$$3 \cdot 10 + 6(-2) = 18.$$

By contrast, there is no solution to the equation $2x + 10y = 17$. Indeed, the left-hand side is an even integer whatever the choice of x and y, while the right-hand side is not. Faced with this, it is reasonable to inquire about the circumstances under which a solution is possible and, when a solution does exist, whether we can determine all solutions explicitly.

The condition for solvability is easy to state: The Diophantine equation $ax + by = c$ admits a solution if and only if $d \mid c$, where $d = \gcd(a, b)$. We know that there are integers r and s for which $a = dr$ and $b = ds$. If a solution of $ax + by = c$ exists, so that $ax_0 + by_0 = c$ for suitable x_0 and y_0, then

$$c = ax_0 + by_0 = drx_0 + dsy_0 = d(rx_0 + sy_0),$$

which simply says that $d \mid c$. Conversely, assume that $d \mid c$, say $c = dt$. Using Theorem 2-3, integers x_0 and y_0 can be found satisfying $d = ax_0 + by_0$. When this relation is multiplied by t, we get

$$c = dt = (ax_0 + by_0)t = a(tx_0) + b(ty_0).$$

Hence, the Diophantine equation $ax + by = c$ has $x = tx_0$ and $y = ty_0$ as a particular solution. This proves part of our next theorem.

THEOREM 2-9.

The linear Diophantine equation $ax + by = c$ has a solution if and only if $d \mid c$, where $d = \gcd(a, b)$. If x_0, y_0 is any particular solution of this equation, then all other solutions are given by

$$x = x_0 + (b/d)t, \qquad y = y_0 - (a/d)t$$

for varying integers t.

Proof: To establish the second assertion of the theorem, let us suppose that a solution x_0, y_0 of the given equation is known. If x', y' is any other solution, then

$$ax_0 + by_0 = c = ax' + by',$$

which is equivalent to

$$a(x' - x_0) = b(y_0 - y').$$

By the Corollary to Theorem 2-4, there exist relatively prime integers r and s such that $a = dr$, $b = ds$. Substituting these values into the last-written equation and cancelling the common factor d, we find that

$$r(x' - x_0) = s(y_0 - y').$$

The situation is now this: $r \mid s(y_0 - y')$, with $\gcd(r, s) = 1$. Using Euclid's Lemma, it must be the case that $r \mid (y_0 - y')$; or, in other words, $y_0 - y' = rt$ for some integer t. Substituting, we obtain

$$x' - x_0 = st.$$

This leads us to the formulas

$$x' = x_0 + st = x_0 + (b/d)t,$$
$$y' = y_0 - rt = y_0 - (a/d)t.$$

It is easy to see that these values satisfy the Diophantine equation, regardless of the choice of the integer t; for,

$$ax' + by' = a\,[x_0 + (b/d)t] + b\,[y_0 - (a/d)t]$$
$$= (ax_0 + by_0) + (ab/d - ab/d)t$$
$$= c + 0 \cdot t = c.$$

Thus there are an infinite number of solutions of the given equation, one for each value of t.

Example 2-4

Consider the linear Diophantine equation

$$172x + 20y = 1000.$$

Applying Euclid's Algorithm to the evaluation of $\gcd(172\,,\,20)$, we find that

$$172 = 8 \cdot 20 + 12,$$
$$20 = 1 \cdot 12 + 8,$$
$$12 = 1 \cdot 8 + 4,$$
$$8 = 2 \cdot 4,$$

whence $\gcd(172\,,\,20) = 4$. Since $4\,|\,1000$, a solution to this equation exists. To obtain the integer 4 as a linear combination of 172 and 20, we work backwards through the above calculations, as follows:

$$4 = 12 - 8$$
$$= 12 - (20 - 12)$$
$$= 2 \cdot 12 - 20$$
$$= 2(172 - 8 \cdot 20) - 20$$
$$= 2 \cdot 172 + (-17)20.$$

Upon multiplying this relation by 250, one arrives at

$$1000 = 250 \cdot 4 = 250[2 \cdot 172 + (-17)20]$$
$$= 500 \cdot 172 + (-4250)20,$$

so that $x = 500$ and $y = -4250$ provides one solution to the Diophantine equation in question. All other solutions are expressed by

$$x = 500 + (20/4)t = 500 + 5t,$$
$$y = -4250 - (172/4)t = -4250 - 43t$$

for some integer t.

A little further effort produces the solutions in the positive integers, if any happen to exist. For this, t must be chosen so as to satisfy simultaneously the inequalities

$$5t + 500 > 0, \qquad -43t - 4250 > 0$$

or, what amounts to the same thing,

$$-98\frac{36}{43} > t > -100.$$

Since t must be an integer, we are forced to conclude that $t = -99$. Thus our Diophantine equation has a unique positive solution $x = 5, y = 7$ corresponding to the value $t = -99$.

It might be helpful to record the form that Theorem 2-9 takes when the coefficients are relatively prime integers.

COROLLARY.

If $\gcd(a, b) = 1$ *and if* x_0, y_0 *is a particular solution of the linear Diophantine equation* $ax + by = c$, *then all solutions are given by*

$$x = x_0 + bt, \, y = y_0 - at$$

for integral values of t.

For example: The equation $5x + 22y = 18$ has $x_0 = 8, y_0 = -1$ as one solution; from the Corollary, a complete solution is given by $x = 8 + 22t, y = -1 - 5t$ for arbitrary t.

Diophantine equations frequently arise in the solving of certain types of traditional "word problems," as evidenced by our next example.

Example 2-5

A customer bought a dozen pieces of fruit, apples and oranges, for $1.32. If an apple costs 3 cents more than an orange and more apples than oranges were purchased, how many pieces of each kind were bought?

To set up this problem as a Diophantine equation, let x be the number of apples and y the number of oranges purchased; also, let z represent the cost (in cents) of an orange. Then the conditions of the problem lead to

$$(z + 3)x + zy = 132$$

or equivalently

$$3x + (x + y)z = 132.$$

Since $x + y = 12$, the above equation may be replaced by

$$3x + 12z = 132,$$

which in turn simplifies to $x + 4z = 44$.

Stripped of inessentials, the object is to find integers x and z satisfying the Diophantine equation

(1) $x + 4z = 44.$

Inasmuch as gcd $(1 , 4) = 1$ is a divisor of 44, there is a solution to this equation. Upon multiplying the relation $1 = 1(-3) + 4 \cdot 1$ by 44 to get

$$44 = 1(-132) + 4 \cdot 44,$$

it follows that $x_0 = -132, z_0 = 44$ serves as one solution. All other solutions of (1) are of the form

$$x = -132 + 4t,$$
$$z = 44 - t,$$

where t is an integer.

Not all of the infinite set of values of t furnish solutions to the original problem. Only values of t should be considered which will ensure that $12 \geq x > 6$. This requires obtaining those t such that

$$12 \geq -132 + 4t > 6.$$

Now, $12 \geq -132 + 4t$ implies that $t \leq 36$, while $-132 + 4t > 6$ gives $t > 34 \frac{1}{2}$. The only integral values of t to satisfy both inequalities are $t = 35$ and $t = 36$. Thus there are two possible purchases: a dozen apples costing 11 cents apiece (the case where $t = 36$), or else 8 apples at 12 cents each and 4 oranges at 9 cents each (the case where $t = 35$).

Linear indeterminate problems such as these have a long history, occurring as early as the first century in the Chinese mathematical literature. Owing to a lack of algebraic symbolism, they often appeared in the guise of rhetorical puzzles or riddles. The contents of the *Mathematical Classic* of Chang Ch' iu-chien (sixth century) attest to the algebraic abilities of the Chinese scholars. This elaborate treatise contains one of the most famous problems in indeterminate equations, in the sense of transmission to other societies—the problem of the "hundred fowls." The problem states:

If a cock is worth 5 coins, a hen 3 coins, and three chicks together 1 coin, how many cocks, hens and chicks, totaling 100, can be bought for 100 coins?

In terms of equations, the problem would be written (if x equals the number of cocks, y the number of hens, z the number of chickens):

$$5x + 3y + \frac{1}{3} z = 100, \quad x + y + z = 100.$$

Eliminating one of the unknowns, we are left with a linear Diophantine equation in the two other unknowns. Specifically, since the quantity $z = 100 - x - y$, we have $5x + 3y + \frac{1}{3}(100 - x - y) = 100$, or

$$7x + 4y = 100.$$

This equation has the general solution $x = 4t$, $y = 25 - 7t$, so that $z = 75 + 3t$, where t is an arbitrary integer. Chang himself gave several answers:

$$x = 4, y = 18, z = 78;$$

$$x = 8, y = 11, z = 81;$$

$$x = 12, y = 4, z = 84.$$

A little further effort produces all solutions in the positive integers. For this, t must be chosen to satisfy simultaneously the inequalities

$$4t > 0, \quad 25 - 7t > 0, \quad 75 + 3t > 0.$$

The last two of these are equivalent to the requirement $-25 < t < 3\frac{4}{7}$. Since t must have a positive value, we conclude that $t = 1, 2, 3$, leading to precisely the values Chang obtained.

PROBLEMS 2.4

1. Which of the following Diophantine equations cannot be solved?
    (a) $6x + 51y = 22$;
    (b) $33x + 14y = 115$;
    (c) $14x + 35y = 93$.

2. Determine all solutions in the integers of the following Diophantine equations:
    (a) $56x + 72y = 40$;
    (b) $24x + 138y = 18$;
    (c) $221x + 35y = 11$.

3. Determine all solutions in the positive integers of the following Diophantine equations:
    (a) $18x + 5y = 48$;
    (b) $54x + 21y = 906$;
    (c) $123x + 360y = 99$;
    (d) $158x - 57y = 7$.

4. If a and b are relatively prime positive integers, prove that the Diophantine equation $ax - by = c$ has infinitely many solutions in the positive integers.
    [*Hint:* There exist integers x_0 and y_0 such that $ax_0 + by_0 = c$. For any integer t, which is larger than both $|x_0|/b$ and $|y_0|/a$, $x = x_0 + bt$ and $y = -(y_0 - at)$ are a positive solution of the given equation.]

5. (a) Prove that the Diophantine equation $ax + by + cz = d$ is solvable in the integers if and only if $\gcd(a, b, c)$ divides d.
    (b) Find all solutions in the integers of $15x + 12y + 30z = 24$.
        [*Hint:* Put $y = 3s - 5t$ and $z = -s + 2t$.]

6. (a) A man has \$4.55 in change composed entirely of dimes and quarters. What are the maximum and minimum number of coins that he can have? Is it possible for the number of dimes to equal the number of quarters?
    (b) The neighborhood theater charges \$1.80 for adult admissions and 75 cents for children. On a particular evening the total receipts were \$90. Assuming that more adults than children were present, how many people attended?

(c) A certain number of sixes and nines are added to give a sum of 126; if the number of sixes and nines are interchanged, the new sum is 114. How many of each were there originally?

7. A farmer purchased one hundred head of livestock for a total cost of $4000. Prices were as follow: calves, $120 each; lambs, $50 each; piglets, $25 each. If the farmer obtained at least one animal of each type, how many did he buy?

8. When Mr. Smith cashed a check at his bank, the teller mistook the number of cents for the number of dollars and vice versa. Unaware of this, Mr. Smith spent 68 cents and then noticed to his surprise that he had twice the amount of the original check. Determine the smallest value for which the check could have been written.
[*Hint:* If x is the number of dollars and y the number of cents in the check, then $100y + x - 68 = 2(100x + y)$.]

9. Solve each of the puzzle-problems below:
   (a) Alcuin of York, 775. A hundred bushels of grain are distributed among 100 persons in such a way that each man receives 3 bushels, each woman 2 bushels, and each child ½ bushel. How many men, women, and children are there?
   (b) Mahaviracarya, 850. There were 63 equal piles of plantain fruit put together and 7 single fruits. They were divided evenly among 23 travelers. What is the number of fruits in each pile?
   [*Hint:* Consider the Diophantine equation $63x + 7 = 23y$.]
   (c) Yen Kung, 1372. We have an unknown number of coins. If you make 77 strings of them, you are 50 coins short; but if you make 78 strings, it is exact. How many coins are there? [*Hint:* If N is the number of coins, then $N = 77x + 27 = 78y$ for integers x and y.]
   (d) Christoff Rudolff, 1526. Find the number of men, women and children in a company of 20 persons if together they pay 20 coins, each man paying 3, each woman 2, and each child ½.
   (e) Euler, 1770. Divide 100 into two summands such that one is divisible by 7 and the other by 11.

3

Primes and their Distribution

"Mighty are numbers, joined with art resistless."
EURIPIDES

3.1 THE FUNDAMENTAL THEOREM OF ARITHMETIC

Essential to everything discussed herein—in fact, essential to every aspect of number theory—is the notion of a prime number. We have previously observed that any integer $a > 1$ is divisible by ± 1 and $\pm a$; if these exhaust the divisors of a, then it is said to be a prime number. Put somewhat differently:

DEFINITION 3-1.
An integer $p > 1$ is called a *prime number,* or simply a *prime,* if its only positive divisors are 1 and p. An integer greater than 1 which is not a prime is termed *composite.*

Among the first ten positive integers 2, 3, 5, 7 are all primes, while 4, 6, 8, 9, 10 are composite numbers. Note that the integer 2 is the only even prime, and according to our definition the integer 1 plays a special role, being neither prime nor composite.

For the rest of the book, the letters p and q will be reserved, so far as is possible, for primes.

Proposition 14 of Book IX of Euclid's *Elements* embodies the result which later became known as the Fundamental Theorem of Arithmetic, namely, that every integer greater than 1 can, except for the order of the factors, be represented as a product of primes in one and only one way. To quote the proposition itself: "If a number be the least that is measured by prime numbers, it will not be measured by any other prime except those originally measuring it." Since every number is either a prime or, by the Fundamental Theorem, can be broken down into unique prime factors and no further, the primes serve as the "building blocks" from which all other integers can be made. Accordingly, the prime numbers have intrigued mathematicians through the ages, and while a number of remarkable theorems relating to their distribution in the sequence of positive integers have been proved, even more remarkable is what remains unproved. The open questions can be counted among the outstanding unsolved problems of all mathematics.

To begin on a simpler note, we observe that the prime 3 divides the integer 36, where 36 may be written as any one of the products

$$6 \cdot 6 = 9 \cdot 4 = 12 \cdot 3 = 18 \cdot 2.$$

In each instance, 3 divides at least one of the factors involved in the product. This is typical of the general situation, the precise result being:

THEOREM 3-1.

> *If p is a prime and $p \mid ab$, then $p \mid a$ or $p \mid b$.*
>
> *Proof:* If $p \mid a$, then we need go no further, so let us assume that $p \nmid a$. Since the only positive divisors of p are 1 and p itself, this implies that $\gcd(p, a) = 1$. (In general, $\gcd(p, a) = p$ or $\gcd(p, a) = 1$ according as $p \mid a$ or $p \nmid a$.) Hence, citing Euclid's Lemma, we get $p \mid b$.

This theorem easily extends to products of more than two terms.

COROLLARY 1.

> *If p is a prime and $p \mid a_1 a_2 \ldots a_n$, then $p \mid a_k$ for some k, where $1 \leq k \leq n$.*
>
> *Proof:* We proceed by induction on n, the number of factors. When $n = 1$, the stated conclusion obviously holds, while for $n = 2$ the result is the content of Theorem 3-1. Suppose, as the induction hypothesis, that $n > 2$ and that whenever p divides a product of less than n factors, then it divides at least one of the factors. Now, let $p \mid a_1 a_2 \ldots a_n$. From Theorem 3-1, either $p \mid a_n$ or else $p \mid a_1 a_2 \ldots a_{n-1}$. If $p \mid a_n$, then we are through. As regards the case where $p \mid a_1 a_2 \ldots a_{n-1}$, the induction hypothesis ensures that $p \mid a_k$ for some choice of k, with $1 \leq k \leq n - 1$. In any event, p divides one of the integers $a_1, a_2, \ldots, a_n$.

COROLLARY 2.

> *If $p, q_1, q_2, \ldots, q_n$ are all primes and $p \mid q_1 q_2 \ldots q_n$, then $p = q_k$ for some k, where $1 \leq k \leq n$.*
>
> *Proof:* By virtue of Corollary 1, we know that $p \mid q_k$ for some k, with $1 \leq k \leq n$. Being a prime, q_k is not divisible by any positive integer other than 1 or q_k itself. Since $p > 1$, we are forced to conclude that $p = q_k$.

With this preparation out of the way, we arrive at one of the cornerstones of our development, the Fundamental Theorem of Arithmetic. As indicated earlier, this theorem asserts that every integer greater than 1 can be factored into primes in essentially one way; the linguistic ambiguity "essentially" means that $2 \cdot 3 \cdot 2$ is not considered as being a different factorization of 12 from $2 \cdot 2 \cdot 3$. Stated precisely:

THEOREM 3-2 (Fundamental Theorem of Arithmetic).

> *Every positive integer $n > 1$ can be expressed as a product of primes; this representation is unique, apart from the order in which the factors occur.*
>
> *Proof:* Either n is a prime or it is composite; in the former case, there is nothing more to prove. If n is composite, then there exists an integer d satisfying $d \mid n$ and

$1 < d < n$. Among all such integers d choose p_1 to be the smallest (this is possible by the Well-Ordering Principle). Then p_1 must be a prime number. Otherwise, it too would have a divisor q with $1 < q < p_1$; but then $q \mid p_1$ and $p_1 \mid n$ imply that $q \mid n$, which contradicts the choice of p_1 as the smallest positive divisor, not equal to 1, of n.

We may therefore write $n = p_1 n_1$, where p_1 is prime and $1 < n_1 < n$. If n_1 happens to be a prime, then we have our representation. In the contrary case, the argument is repeated to produce a second prime number p_2 such that $n_1 = p_2 n_2$; that is,

$$n = p_1 p_2 n_2, \qquad\qquad 1 < n_2 < n_1.$$

If n is a prime, then it is not necessary to go further. Otherwise, write $n_2 = p_3 n_3$, with p_3 a prime:

$$n = p_1 p_2 p_3 n_3, \qquad\qquad 1 < n_3 < n_2.$$

The decreasing sequence

$$n > n_1 > n_2 > \cdots > 1$$

cannot continue indefinitely, so that after a finite number of steps n_{k-1} is a prime, say p_k. This leads to the prime factorization

$$n = p_1 p_2 \cdots p_k .$$

To establish the second part of the proof—the uniqueness of the prime factorization—let us suppose that the integer n can be represented as a product of primes in two ways; say

$$n = p_1 p_2 \cdots p_r = q_1 q_2 \cdots q_s, \qquad\qquad r \le s$$

where the p_i and q_j are all primes, written in increasing magnitude so that

$$p_1 \le p_2 \le \cdots \le p_r, q_1 \le q_2 \le \cdots \le q_s .$$

Since $p_1 \mid q_1 q_2 \cdots q_s$, Corollary 2 of Theorem 3-1 tells us that $p_1 = q_k$ for some k; but then $p_1 \ge q_1$. Similar reasoning gives $q_1 \ge p_1$, whence $p_1 = q_1$. We may cancel this common factor and obtain

$$p_2 p_3 \cdots p_r = q_2 q_3 \cdots q_s .$$

Now repeat the process to get $p_2 = q_2$ and, in its turn,

$$p_3 p_4 \cdots p_r = q_3 q_4 \cdots q_s .$$

Continue in this fashion. If the inequality $r < s$ held, we would eventually arrive at

$$1 = q_{r+1} q_{r+2} \cdots q_s$$

which is absurd, since each $q_1 > 1$. Hence $r = s$ and

$$p_1 = q_1, p_2 = q_2, \ldots, p_r = q_r ,$$

making the two factorizations of n identical. The proof is now complete.

Of course, several of the primes which appear in the factorization of a given positive integer may be repeated as is the case with $360 = 2 \cdot 2 \cdot 2 \cdot 3 \cdot 3 \cdot 5$. By collecting like primes and replacing them by a single factor, we could rephrase Theorem 3–2 as

COROLLARY.

Any positive integer $n > 1$ can be written uniquely in a canonical form

$$n = p_1^{k_1} p_2^{k_2} \ldots p_r^{k_r},$$

where, for $i = 1, 2, \ldots, r$, each k_i is a positive integer and each p_i is a prime, with $p_1 < p_2 < \cdots < p_r$.

To illustrate: the canonical form of the integer 360 is $360 = 2^3 \cdot 3^2 \cdot 5$. As further examples we cite

$$4725 = 3^3 \cdot 5^2 \cdot 7, \text{ and } 17460 = 2^3 \cdot 3^2 \cdot 5 \cdot 7^2.$$

Theorem 3–2 should not be taken lightly, for there do exist number systems in which the factorization into "primes" is not unique. Perhaps the most elemental example is the set E of all positive even integers. Let us agree to call an even integer an *e-prime* if it is not the product of two other even integers. Thus, 2, 6, 10, 14, . . . are all *e*-primes while 4, 8, 12, 16, . . . are not. It is not difficult to see that the integer 60 can be factored into *e*-primes in two distinct ways; namely,

$$60 = 2 \cdot 30 = 6 \cdot 10.$$

Part of the trouble arises from the fact that Theorem 3–1 is lacking in the set E: $6 \mid 2 \cdot 30$, but $6 \nmid 2$ and $6 \nmid 30$.

This is an opportune moment to insert a famous result of Pythagoras. Mathematics as a science began with Pythagoras (569–500 B.C.), and much of the content of Euclid's *Elements* is due to Pythagoras and his School. The Pythagoreans deserve the credit for being the first to classify numbers into odd and even, prime and composite.

THEOREM 3-3 (Pythagoras).

The number $\sqrt{2}$ is irrational.

Proof: Suppose to the contrary that $\sqrt{2}$ is a rational number; say, $\sqrt{2} = a/b$, where a and b are both integers with $\gcd(a, b) = 1$. Squaring, we get $a^2 = 2b^2$, so that $b \mid a^2$. If $b > 1$, then the Fundamental Theorem of Arithmetic guarantees the existence of a prime p such that $p \mid b$. It follows that $p \mid a^2$ and, by Theorem 3–1, that $p \mid a$; hence, $\gcd(a, b) \geq p$. We therefore arrive at a contradiction, unless $b = 1$. But if this happens, then $a^2 = 2$, which is impossible (we assume that the reader is willing to grant that no integer can be multiplied by itself to give 2). Our supposition that $\sqrt{2}$ is a rational number is untenable and so $\sqrt{2}$ must be irrational.

There is an interesting variation on the proof of Theorem 3-3: Since $\sqrt{2} = a/b$ with $\gcd(a,b) = 1$, there must exist integers r and s satisfying $ar + bs = 1$. As a result,

$$\sqrt{2} = \sqrt{2}(ar + bs) = (\sqrt{2}a)r + (\sqrt{2}b)s = 2br + as.$$

This representation leads us to conclude that $\sqrt{2}$ is an integer, an obvious impossibility.

PROBLEMS 3.1

1. It has been conjectured that there are infinitely many primes of the form $n^2 - 2$. Exhibit five such primes.

2. Give an example to show that the following conjecture is not true: Every positive integer can be written in the form $p + a^2$, where p is either a prime or 1, and $a \geq 0$.

3. Prove each of the assertions below:
    (a) Any prime of the form $3n + 1$ is also of the form $6m + 1$.
    (b) Each integer of the form $3n + 2$ has a prime factor of this form.
    (c) The only prime of the form $n^3 - 1$ is 7.
         [*Hint:* Write $n^3 - 1$ as $(n - 1)(n^2 + n + 1)$.]
    (d) The only prime p for which $3p + 1$ is a perfect square is $p = 5$.
    (e) The only prime of the form $n^2 - 4$ is 5.

4. If $p \geq 5$ is a prime number, show that $p^2 + 2$ is composite. [*Hint:* p takes one of the forms $6k + 1$ or $6k + 5$.]

5. (a) Given that p is a prime and $p \mid a^n$, prove that $p^n \mid a^n$.
    (b) If $\gcd(a , b) = p$, a prime, what are the possible values of $\gcd(a^2 , b^2)$, $\gcd(a^2 , b)$ and $\gcd(a^3 , b^2)$?

6. Establish each of the following statements:
    (a) Every integer of the form $n^4 + 4$, with $n > 1$, is composite.
         [*Hint:* Write $n^4 + 4$ as a product of two quadratic factors.]
    (b) If $n > 4$ is composite, then n divides $(n - 1)!$.
    (c) Any integer of the form $8^n + 1$, where $n \geq 1$, is composite.
         [*Hint:* $2^n + 1 \mid 2^{3n} + 1$.]
    (d) Each integer $n > 11$ can be written as the sum of two composite numbers. [*Hint:* If n is even, say $n = 2k$, then $n - 6 = 2(k - 3)$; for n odd, consider the integer $n - 9$.]

7. Find all prime numbers that divide 50!.

8. If $p \geq q \geq 5$ and p and q are both primes, prove that $24 \mid p^2 - q^2$.

9. (a) An unanswered question is whether there are infinitely many primes which are 1 more than a power of 2, such as $5 = 2^2 + 1$. Find two more of these primes.
    (b) A more general conjecture is that there exist infinitely many primes of the form $n^2 + 1$; for example, $257 = 16^2 + 1$. Exhibit five more primes of this type.

10. If $p \neq 5$ is an odd prime, prove that either $p^2 - 1$ or $p^2 + 1$ is divisible by 10.

11. Another unproven conjecture is that there are an infinitude of primes which are 1 less than a power of 2, such as $3 = 2^2 - 1$.
     (a) Find four more of these primes.
     (b) If $p = 2^k - 1$ is prime, show that k is an odd integer, except when $k = 2$. [*Hint:* $3 \mid 4^n - 1$ for all $n \geq 1$.]

12. Find the prime factorization of the integers 1234, 10140, and 36000.

13. If $n > 1$ is an integer not of the form $6k + 3$, prove that $n^2 + 2^n$ is composite. [*Hint:* Show that either 2 or 3 divides $n^2 + 2^n$.]

14. It has been conjectured that every even integer can be written as the difference of two consecutive primes in infinitely many ways. For example,

$$6 = 29 - 23 = 137 - 131 = 599 - 593 = 1019 - 1013 = \cdots .$$

Express the integer 10 as the difference of two consecutive primes in fifteen ways.

15. Prove that a positive integer $a > 1$ is a square if and only if in the canonical form of a all the exponents of the primes are even integers.

16. An integer is said to be *square-free* if it is not divisible by the square of any integer greater than 1. Prove that
    (a) an integer $n > 1$ is square-free if and only if n can be factored into a product of distinct primes;
    (b) every integer $n > 1$ is the product of a square-free integer and a perfect square. [*Hint:* If $n = p_1^{k_1} p_2^{k_2} \cdots p_s^{k_s}$ is the canonical factorization of n, then write $k_i = 2q_i + r_i$ where $r_i = 0$ or 1 according as k_i is even or odd.]

17. Verify that any integer n can be expressed as $n = 2^k m$, where $k \geq 0$ and m is an odd integer.

18. Numerical evidence makes it plausible that there are infinitely many primes p such that $p + 50$ is also prime. List fifteen of these primes.

19. A positive integer n is called *square-full*, or *powerful*, if $p^2 \mid n$ for every prime factor p of n (there are 997 square-full numbers less than 250,000). If n is square-full, show that it can be written in the form $n = a^2 b^3$, with a and b positive integers.

3.2 THE SIEVE OF ERATOSTHENES

Given a particular integer, how can we determine whether it is prime or composite and, in the latter case, how can we actually find a nontrivial divisor? The most obvious approach consists of successively dividing the integer in question by each of the numbers preceding it; if none of them (except 1) serves as a divisor, then the integer must be prime. Although this method is very simple to describe, it cannot be regarded as useful in practice. For even if one is undaunted by large calculations, the amount of time and work involved may be prohibitive.

There is a property of composite numbers which allows us to reduce materially the necessary computations—but still the process remains cumbersome. If an integer $a > 1$ is composite, then it may be written as $a = bc$, where $1 < b < a$ and $1 < c < a$. Assuming that $b \leq c$, we get $b^2 \leq bc = a$ and so $b \leq \sqrt{a}$. Since $b > 1$, Theorem 3-2 ensures that b has at least one prime factor p. Then $p \leq b \leq \sqrt{a}$; furthermore, because $p \mid b$ and $b \mid a$, it follows that $p \mid a$. The point is simply this: a composite number a will always possess a prime divisor p satisfying $p \leq \sqrt{a}$.

In testing the primality of a specific integer $a > 1$, it therefore suffices to divide a by those primes not exceeding $\sqrt{a}$ (presuming, of course, the availability of a list of primes up to $\sqrt{a}$). This may be clarified by considering the integer $a = 509$. Inasmuch as $22 < \sqrt{509} < 23$, we need only try out the primes which are not larger than 22 as possible divisors; namely, the primes 2, 3, 5, 7, 11, 13, 17, 19. Dividing 509 by each of these in turn, we find that none serves as a divisor of 509. The conclusion is that 509 must be a prime number.

Example 3-1

The foregoing technique provides a practical means for determining the canonical form of an integer, say $a = 2093$. Since $45 < \sqrt{2093} < 46$, it is enough to examine the primes 2, 3, 5, 7, 11, 13, 17, 19, 23, 29, 31, 37, 41, 43. By trial, the first of these to divide 2093 is 7 and $2093 = 7 \cdot 299$. As regards the integer 299, the seven primes which are less than 18 (note that $17 < \sqrt{299} < 18$) are 2, 3,

5, 7, 11, 13, 17. The first prime divisor of 299 is 13 and, carrying out the required division, we obtain $299 = 13 \cdot 23$. But 23 is itself a prime, whence 2093 has exactly three prime factors, 7, 13, and 23:

$$2093 = 7 \cdot 13 \cdot 23.$$

Another Greek mathematician whose work in number theory remains significant is Eratosthenes of Cyrene (276–194 B.C.). While posterity remembers him mainly as the director of the world-famous library at Alexandria, Eratosthenes was gifted in all branches of learning, if not of first rank in any; in his own day, he was nicknamed "Beta" because, it was said, he stood at least second in every field. Perhaps the most impressive feat of Eratosthenes was the accurate measurement of the earth's circumference by a simple application of Euclidean geometry.

We have seen that if an integer $a > 1$ is not divisible by a prime $p \leq \sqrt{a}$, then a is of necessity a prime. Eratosthenes used this fact as the basis of a clever technique, called the "Sieve of Eratosthenes," for finding all primes below a given integer n. The scheme calls for writing down the integers from 2 to n in their natural order and then systematically eliminating all the composite numbers by striking out all multiples $2p$, $3p$, $4p$, $5p$, . . . of the primes $p \leq \sqrt{n}$. The integers that are left on the list—those that do not fall through the "sieve"—are primes.

To see an example of how this works, suppose that we wish to find all primes not exceeding 100. Consider the sequence of consecutive integers 2, 3, 4, . . . ,100. Recognizing that 2 is a prime, we begin by crossing out all even integers from our listing, except 2 itself. The first of the remaining integers is 3, which must be a prime. We keep 3, but strike out all higher multiples of 3, so that 9, 15, 21, . . . are now removed (the even multiples of 3 having been removed in the previous step). The smallest integer after 3 which has not yet been deleted is 5. It is not divisible by either 2 or 3—otherwise it would have been crossed out—hence it is also a prime. All proper multiples of 5 being composite numbers, we next remove 10, 15, 20, . . . (some of these are, of course, already missing), while retaining 5 itself. The first surviving integer 7 is a prime, for it is not divisible by 2, 3, or 5, the only primes that precede it. After eliminating the proper multiples of 7, the largest prime less than $\sqrt{100} = 10$, all composite integers in the sequence 2, 3, 4, . . . ,100 have fallen through the sieve. The positive integers which remain, to wit, 2, 3, 5, 7, 11, 13, 17, 19, 23, 29, 31, 37, 41, 43, 47, 53, 59, 61, 67, 71, 73, 79, 83, 89, 97, are all of the primes less than 100.

The table below represents the result of the completed sieve. The multiples of 2 are crossed out by \ ; the multiples of 3 are crossed out by / ; the multiples of 5 are crossed out by —; the multiples of 7 are crossed out by $\sim$.

2	3	4	5	6	7	8	9	10
	2	3	~~4~~	5	~~6~~	7	~~8~~	~~9~~	~~10~~
11	~~12~~	13	~~14~~	~~15~~	~~16~~	17	~~18~~	19	~~20~~
~~21~~	~~22~~	23	~~24~~	~~25~~	~~26~~	~~27~~	~~28~~	29	~~30~~
31	~~32~~	~~33~~	~~34~~	~~35~~	~~36~~	37	~~38~~	~~39~~	~~40~~
41	~~42~~	43	~~44~~	~~45~~	~~46~~	47	~~48~~	~~49~~	~~50~~
~~51~~	~~52~~	53	~~54~~	~~55~~	~~56~~	~~57~~	~~58~~	59	~~60~~
61	~~62~~	~~63~~	~~64~~	~~65~~	~~66~~	67	68	~~69~~	~~70~~
71	~~72~~	73	~~74~~	~~75~~	~~76~~	~~77~~	~~78~~	79	~~80~~
~~81~~	82	83	~~84~~	~~85~~	~~86~~	~~87~~	88	89	~~90~~
~~91~~	92	~~93~~	~~94~~	~~95~~	~~96~~	97	~~98~~	~~99~~	~~100~~

By this point, an obvious question must have occurred to the reader. Is there a largest prime number, or do the primes go on forever? The answer is to be found in a remarkably simple proof given by Euclid in Book IX of his *Elements*. Euclid's argument is universally regarded as a model of mathematical elegance. Loosely speaking, it goes like this: Given any finite list of prime numbers, one can always find a prime not on the list; hence, the number of primes is infinite. The actual details appear below.

THEOREM 3-4 (Euclid).

There are an infinite number of primes.

Proof: Euclid's proof is by contradiction. Let $p_1 = 2$, $p_2 = 3$, $p_3 = 5$, $p_4 = 7$, . . . be the primes in ascending order, and suppose that there is a last prime; called p_n. Now consider the positive integer

$$P = p_1 p_2 \cdots p_n + 1.$$

Since $P > 1$, we may put Theorem 3–2 to work once again and conclude that P is divisible by some prime p. But $p_1, p_2, \ldots, p_n$ are the only prime numbers, so that p must be equal to one of $p_1, p_2, \ldots, p_n$. Combining the divisibility relation $p \mid p_1 p_2 \cdots p_n$ with $p \mid P$, we arrive at $p \mid P - p_1 p_2 \cdots p_n$ or, equivalently, $p \mid 1$. The only positive divisor of the integer 1 is 1 itself and, since $p > 1$, a contradiction arises. Thus no finite list of primes is complete, whence the number of primes is infinite.

For a prime p, define $p^\#$ to be the product of all primes that are less than or equal to p. Numbers of the form $p^\# + 1$ might be dubbed "Euclidean numbers," since they appear in Euclid's scheme for proving the infinitude of primes. It is interesting to note that in forming these integers, the first five, namely

$$2^\# + 1 = 2 + 1 = 3,$$

$$3^\# + 1 = 2 \cdot 3 + 1 = 7,$$

$$5^\# + 1 = 2 \cdot 3 \cdot 5 + 1 = 31,$$

$$7^\# + 1 = 2 \cdot 3 \cdot 5 \cdot 7 + 1 = 211,$$

$$11^\# + 1 = 2 \cdot 3 \cdot 5 \cdot 7 \cdot 11 + 1 = 2311,$$

are all prime numbers. However,

$$13^\# + 1 = 59 \cdot 509,$$

$$17^\# + 1 = 19 \cdot 97 \cdot 277,$$

$$19^\# + 1 = 347 \cdot 27953$$

are not prime. A question whose answer is not known is whether there are infinitely many primes p for which $p^\# + 1$ is also prime. For that matter, are there infinitely many composite $p^\# + 1$?

At present, sixteen primes of the form $p^{\#} + 1$ have been identified. These correspond to the values $p = 2, 3, 5, 7, 11, 31, 379, 1019, 1021, 2657, 3229, 4547, 4787, 11549,$ and $13649, 18523$; the largest of these, a number consisting of 8002 digits, was discovered in 1989. The integer $p^{\#} + 1$ is composite for all other $p \leq 19051$.

Euclid's Theorem is too important for us to be content with a single proof. Here is a variation in the reasoning: Form the infinite sequence of positive integers

$$n_1 = 2,$$

$$n_2 = n_1 + 1,$$

$$n_3 = n_1 n_2 + 1,$$

$$n_4 = n_1 n_2 n_3 + 1,$$

$$\vdots$$

$$n_k = n_1 n_2 \cdots n_{k-1} + 1,$$

$$\vdots$$

Since each $n_k > 1$, each of these integers is divisible by a prime. But no two n_k can have the same prime divisor. To see this, let $d = \gcd(n_i, n_k)$ and suppose that $i < k$. Then d divides n_i, hence must divide $n_1 n_2 \cdots n_{k-1}$. Since $d \mid n_k$, Theorem 2-2 (7) tells us that $d \mid n_k - n_1 n_2 \cdots n_{k-1}$ or $d \mid 1$. The implication is that $d = 1$ and so the integers n_k $(k = 1, 2, \ldots)$ are pairwise relatively prime. The point which we wish to make is that there are as many distinct primes as there are integers n_k, namely, infinitely many of them.

Let p_n denote the nth of the prime numbers in their natural order. Euclid's proof shows that the expression $p_1 p_2 \cdots p_n + 1$ is divisible by at least one prime. If there are several, then the smallest will be p_{n+1} so that $p_{n+1} \leq p_1 p_2 \cdots p_n + 1$ for $n \geq 1$. Another way of saying the same thing is that

$$p_n \leq p_1 p_2 \cdots p_{n-1} + 1, \qquad\qquad n \geq 2.$$

With a slight modification of Euclid's reasoning this inequality can be improved to give

$$p_n \leq p_1 p_2 \cdots p_{n-1} - 1, \qquad\qquad n \geq 3.$$

For instance, when $n = 5$, this tells us that

$$11 = p_5 \leq 2 \cdot 3 \cdot 5 \cdot 7 - 1 = 209.$$

One can see that the estimate is rather extravagant. A sharper limitation on the size of p_n is given by *Bonse's inequality,* which states that

$$p_n^2 < p_1 p_2 \cdots p_{n-1}, \qquad\qquad n \geq 5.$$

This inequality yields $p_5^2 < 210$, or $p_5 \leq 14$. A somewhat better size-estimate for p_5 comes from the inequality

$$p_{2n} \leq p_2 p_3 \cdots p_n - 2, \qquad\qquad n \geq 3.$$

Here, we obtain

$$p_5 < p_6 \leq p_2 p_3 - 2 = 3 \cdot 5 - 2 = 13.$$

To approximate the size of p_n from these formulas, it is necessary to know the values of $p_1, p_2, \cdots, p_{n-1}$. For a bound in which the preceding primes do not enter the picture, we have the following theorem:

THEOREM 3-5.

If p_n is the nth prime number, then $p_n \leq 2^{2^n - 1}$.

Proof: Let us proceed by induction on n, the asserted inequality being clearly true when $n = 1$. As hypothesis of the induction, we assume that $n > 1$ and that the result holds for all integers up to n. Then

$$p_{n+1} \leq p_1 p_2 \cdots p_n + 1$$

$$\leq 2 \cdot 2^2 \cdots 2^{2^{n-1}} + 1 = 2^{1 + 2 + 2^2 + \cdots + 2^{n-1}} + 1.$$

Recalling the identity $1 + 2 + 2^2 + \cdots + 2^{n-1} = 2^n - 1$, we obtain

$$p_{n+1} \leq 2^{2^n - 1} + 1.$$

But $1 \leq 2^{2^n - 1}$ for all n; whence

$$p_{n+1} \leq 2^{2^n - 1} + 2^{2^n - 1}$$

$$= 2 \cdot 2^{2^n - 1} = 2^{2^n},$$

completing the induction step, and the argument.

There is a corollary to Theorem 3–5 which is of interest.

COROLLARY.

For $n \geq 1$, there are at least $n + 1$ primes less than 2^{2^n}.

Proof: From the theorem, we know that $p_1, p_2, \ldots, p_{n+1}$ are all less than 2^{2^n}.

We can do considerably better than is indicated by Theorem 3–5. In 1845, Joseph Bertrand conjectured that the prime numbers are well-distributed in the sense that between $n \geq 2$ and $2n$ there is at least one prime. He was unable to establish his conjecture, but verified it for all $n \leq 3000000$. (One way of achieving this is to consider a sequence of primes 3, 5, 7, 13, 23, 43, 83, 163, 371, 631, 1259, 2503, 4999, 9973, 19937, 39869, 79699, 159389, $\cdots$, each of which is less than twice the preceding.) Since it takes some real effort to substantiate this famous conjecture, let us content ourselves with saying that the first proof was carried out by the Russian mathematician P. L. Tchebychef in 1852. Granting the result, it is not difficult to show that

$$p_n < 2^n, \qquad\qquad n \geq 2,$$

and as a direct consequence, $p_{n+1} < 2p_n$ for $n \geq 3$. In particular,

$$11 = p_5 < 2^4 = 16.$$

To see that $p_n < 2^n$, we argue by induction on n. Clearly, $p_2 = 3 < 2^2$, so that the inequality is true here. Now assume that the inequality holds for an integer n, whence $p_n < 2^n$. Invoking Bertrand's conjecture, there exists a prime number p satisfying $2^n < p < 2^{n+1}$; that is, $p_n < p$. This immediately leads to the conclusion that $p_{n+1} \leq p < 2^{n+1}$, which completes the induction and the proof.

Primes of special form have been of perennial interest. Among these, the repunit primes are outstanding in their simplicity. A *repunit* is an integer written (in decimal notation) as a string of 1s, such as 11, 111, or 1111. Each such integer must have the form $(10^n - 1)/9$. We use the symbol R_n to denote the repunit consisting of n consecutive 1s. One peculiar feature of these numbers is the apparent scarcity of primes among them. So far, only

$$R_2, R_{19}, R_{23}, R_{317}, \text{ and } R_{1031}$$

have been identified as primes (the last one in 1985). It is known that the only possible repunit primes R_n for all $n \leq 10000$ are the five numbers just indicated. No conjecture has been made as to the existence of any others. For a repunit R_n to be prime, the subscript n must be a prime; that this is not a sufficient condition is shown by

$$R_5 = 11111 = 41 \cdot 271, R_7 = 1111111 = 239 \cdot 4649.$$

PROBLEMS 3.2

1. Determine whether the integer 701 is prime by testing all primes $p \leq \sqrt{701}$ as possible divisors. Do the same for the integer 1009.

2. Employing the Sieve of Eratosthenes, obtain all the primes between 100 and 200.

3. Given that $p \nmid n$ for all primes $p \leq \sqrt[3]{n}$, show that $n > 1$ is either a prime or the product of two primes. [*Hint:* Assume to the contrary that n contains at least three prime factors.]

4. Establish the following facts:
   (a) $\sqrt{p}$ is irrational for any prime p.
   (b) If $a > 0$ and $\sqrt[n]{a}$ is rational, then $\sqrt[n]{a}$ must be an integer.
   (c) For $n \geq 2$, $\sqrt[n]{n}$ is irrational. [*Hint:* Use the fact that $2^n > n$.]

5. Show that any composite three-digit number must have a prime factor less than or equal to 31.

6. Fill in any missing details in this sketch of a proof of the infinitude of primes: Assume that there are only finitely many primes, say $p_1, p_2, \ldots, p_n$. Let A be the product of any r of these primes and put $B = p_1 p_2 \cdots p_n/A$. Then each p_k divides either A or B, but not both. Since $A + B > 1$, $A + B$ has a prime divisor different from any of the p_k, which is a contradiction.

7. Modify Euclid's proof that there are infinitely many primes by assuming the existence of a largest prime p and using the integer $N = p! + 1$ to arrive at a contradiction.

8. Give another proof of the infinitude of primes by assuming that there are only finitely many primes, say $p_1, p_2, \ldots, p_n$, and using the integer

$$N = p_2 p_3 \cdots p_n + p_1 p_3 \cdots p_n + \cdots + p_1 p_2 \cdots p_{n-1}$$

to arrive at a contradiction.

9. (a) Prove that if $n > 2$, then there exists a prime p satisfying $n < p < n!$. [*Hint:* If $n! - 1$ is not prime, then it has a prime divisor p; and $p \le n$ implies $p \mid n!$, leading to a contradiction.]
   (b) For $n > 1$, show that every prime divisor of $n! + 1$ is an odd integer which is greater than n.

10. Let q_n be the smallest prime which is strictly greater than $P_n = p_1 p_2 \cdots p_n + 1$. It has been conjectured that the difference $(p_1 p_2 \cdots p_n) - q_n$ is always a prime. Confirm this for the first five values of n.

11. If p_n denotes the nth prime number, put $d_n = p_{n+1} - p_n$. An open question is whether the equation $d_n = d_{n+1}$ has infinitely many solutions; give five solutions.

12. Assuming that p_n is the nth prime number, establish each of the following statements:

   (a) $p_n > 2n - 1$ for $n \ge 5$.
   (b) None of the integers $P_n = p_1 p_2 \cdots p_n + 1$ is a perfect square. [*Hint:* Each P_n is of the form $4k + 3$ for $n > 1$.]
   (c) The sum

$$\frac{1}{p_1} + \frac{1}{p_2} + \cdots + \frac{1}{p_n}$$

   is never an integer.

13. For the repunits R_n, verify the assertions below:
   (a) If $n \mid m$, then $R_n \mid R_m$. [*Hint:* If $n = km$, consider the identity

$$x^m - 1 = (x^n - 1)(x^{(k-1)n} + x^{(k-2)n} + \cdots + x^n + 1).]$$

   (b) If $d \mid R_n$ and $d \mid R_m$, then $d \mid R_{n+m}$. [*Hint:* Show that $R_{m+n} = R_n 10^m + R_m$.]
   (c) If $\gcd(n, m) = 1$, then $\gcd(R_n, R_m) = 1$.

14. Use the previous problem to obtain the prime factors of the repunit R_{10}.

3.3 THE GOLDBACH CONJECTURE

While there is an infinitude of primes, their distribution within the positive integers is most mystifying. Repeatedly in their distribution one finds hints or, as it were, shadows of a pattern; yet an actual pattern amenable to precise description remains unfound. The difference between consecutive primes can be small as with the pairs 11 and 13, 17, and 19, or for that matter 1,000,000,000,061 and 1,000,000,000,063. At the same time there exist arbitrarily long intervals in the sequence of integers which are totally devoid of any primes.

It is an unanswered question whether there are infinitely many pairs of *twin primes;* that is, pairs of successive odd integers p and $p + 2$ which are both primes. Numerical evidence leads us to suspect an affirmative conclusion. Electronic computers have discovered 152,892 pairs of twin primes less than 30,000,000 and twenty pairs between 10^{12} and $10^{12} + 10,000$, which hints at their growing scarcity as the positive integers increase in magnitude. Many immense examples of twins are known. The largest to date is

$$1706595 \cdot 2^{11235} \pm 1,$$

each 3389 decimal digits long, were discovered in 1989.

Consecutive primes can not only be close together, but also be far apart; that is, arbitrarily large gaps can occur between consecutive primes. Stated precisely: Given any positive integer n, there exist n consecutive integers, all of which are composite. To prove this, we need simply consider the integers

$$(n + 1)! + 2, (n + 1)! + 3, \ldots, (n + 1)! + (n + 1),$$

where $(n + 1)! = (n + 1) \cdot n \cdots 3 \cdot 2 \cdot 1$. Clearly there are n integers listed and they are consecutive. What is important is that each integer is composite; for, $(n + 1)! + 2$ is divisible by 2, $(n + 1)! + 3$ is divisible by 3, and so on.

For instance, if a sequence of four consecutive composite integers is desired, then the argument above produces 122, 123, 124, and 125:

$$5! + 2 = 122 = 2 \cdot 61,$$
$$5! + 3 = 123 = 3 \cdot 41,$$
$$5! + 4 = 124 = 4 \cdot 31,$$
$$5! + 5 = 125 = 5 \cdot 25.$$

Of course, one can find other sets of four consecutive composites, such as 24, 25, 26, 27 or 32, 33, 34, 35.

As this example suggests, our procedure for constructing gaps between two consecutive primes gives a gross overestimate of where they occur among the integers. The first occurrences of prime gaps of specific lengths, where all the intervening integers are composite, have been sought through computer searches. A large gap of width 778 (that is, $p_{n+1} - p_n = 778$) was disclosed recently following the prime 42,842,283,925,351. There is no gap of this size between smaller primes. The largest identifiable gap between consecutive prime numbers has length 784, a string of 783 composites immediately after the prime 2,500,107,922,440,823, but it is not known if this is a first occurrence. Interestingly, the computer specialists have not found examples of prime gaps of every possible width up to 784; the smallest missing gap size is 676.

This brings us to another unsolved problem concerning primes, the Goldbach Conjecture. In a letter to Leonhard Euler (1742), Christian Goldbach hazarded the guess that every even integer is the sum of two numbers that are either primes or 1. A somewhat more general formulation is that every even integer greater than 4 can be written as a sum of two odd prime numbers. This is easy to confirm for the first few even integers:

$$2 = 1 + 1$$
$$4 = 2 + 2 = 1 + 3$$
$$6 = 3 + 3 = 1 + 5$$
$$8 = 3 + 5 = 1 + 7$$
$$10 = 3 + 7 = 5 + 5$$
$$12 = 5 + 7 = 1 + 11$$
$$14 = 3 + 11 = 7 + 7 = 1 + 13$$
$$16 = 3 + 13 = 5 + 11$$
$$18 = 5 + 13 = 7 + 11 = 1 + 17$$
$$20 = 3 + 17 = 7 + 13 = 1 + 19$$
$$22 = 3 + 19 = 5 + 17 = 11 + 11$$
$$24 = 5 + 19 = 7 + 17 = 11 + 13 = 1 + 23$$
$$26 = 3 + 23 = 7 + 19 = 13 + 13$$
$$28 = 5 + 23 = 11 + 17$$
$$30 = 7 + 23 = 11 + 19 = 13 + 17 = 1 + 29.$$

It seems that Euler never tried to prove the result, but, writing to Goldbach at a later date he countered with a conjecture of his own: any even integer (≥ 6) of the form $4n + 2$ is a sum of two numbers each being either primes of the form $4n + 1$ or 1.

The numerical data suggesting the truth of Goldbach's conjecture is overwhelming. It has been verified by direct computation for all even integers less than 10^{10}. Most even numbers $2m$, where $0 < 2m < 10^{10}$, were found to be sums of a "small" prime (*small* meaning one of the first 150 primes) and a prime rather close to $2m$. Although this supports the feeling that Goldbach was correct in his conjecture, it is far from a mathematical proof, and all attempts to obtain a proof have been completely unsuccessful. One of the most famous number theorists of this century, G. H. Hardy, in his address to the Mathematical Society of Copenhagen in 1921, stated that the Goldbach conjecture appeared ". . . probably as difficult as any of the unsolved problems in mathematics."

We remark that if the conjecture of Goldbach is true, then each odd number larger than 7 must be the sum of three odd primes. For, take n to be an odd integer greater than 7, so that $n - 3$ is even and greater than 4; if $n - 3$ could be expressed as the sum of two odd primes, then n would be the sum of three.

The first real progress on the conjecture in nearly 200 years was made by Hardy and Littlewood in 1922. On the basis of a certain unproved hypothesis, the so-called generalized Riemann hypothesis, they showed that every sufficiently large odd number is the sum of three odd primes. In 1937, the Russian mathematician I. M. Vinogradov was able to remove the dependence on the generalized Riemann hypothesis, thereby giving an unconditional proof of this result; that is to say, he established that all odd integers greater than some effectively computable n_0 can be written as the sum of three odd primes.

$$n = p_1 + p_2 + p_3 \qquad \text{(n odd, n sufficiently large).}$$

Vinogradov was unable to decide how large n_0 should be, but Borozdkin (1956) proved that $n_0 < 3^{3^{15}}$. It follows immediately that every even integer from some point on is the sum of either two or four primes. Thus, it is enough to answer the question for every odd integer n in the range $9 \leq n \leq n_0$, which, for a given integer, becomes a matter of tedious computation (unfortunately, n_0 is so large that this exceeds the capabilities of the most modern electronic computers).

Another problem closely connected with the conjecture of Goldbach is whether every even number is the sum of two "almost primes"; that is, the sum of two integers each having not more than a certain number of prime factors. The smaller the number of factors, the better the result. The first theorem of this kind was obtained by Brun (1920), who showed that every sufficiently large even number can be written as the sum of two terms, where each has at most 9 prime factors. Later, Buchstab (1940) improved the result to 4 prime factors.

In 1948, the Hungarian mathematician Renyi established that every large even integer n is the sum of a prime and an "almost prime":

$$n = p + p_1 p_2 \cdots p_r \qquad \text{(n even, n sufficiently large).}$$

In Renyi's proof, r is very large. If it could be shown that $r = 1$, this would prove Goldbach's Conjecture for all large n. The subsequent work of Wang (1959) enabled one to take $r \leq 4$, while A. I. Vinogradov (1965) further reduced the estimate to $r \leq 3$. The closest anyone has come to settling the conjecture by this approach is the 1966 result of Chen Jing-Run, which says $r \leq 2$; in other words, from some point on, every even integer is the sum of a prime and a product of at most two primes. Chen's original proof was very long, but in 1973 he improved the argument and reduced its length to 20 pages.

Because of the strong evidence in favor of Goldbach's famous conjecture, one readily becomes convinced that it is true. It nevertheless might be false. Vinogradov showed that if $A(x)$ is the number of even integers $n \leq x$ which are not the sum of two primes, then

$$\lim_{x \to \infty} A(x)/x = 0.$$

This allows us to say that "almost all" even integers satisfy the conjecture. As Edmund Landau so aptly put it, "The Goldbach Conjecture is false for at most 0% of all even integers; this *at most 0%* does not exclude, of course, the possibility that there are infinitely many exceptions."

Having digressed somewhat, let us observe that according to the Division Algorithm, every positive integer can be written uniquely in one of the forms

$$4n, \ 4n + 1, \ 4n + 2, \ 4n + 3$$

for some suitable $n \geq 0$. Clearly the integers $4n$ and $4n + 2 = 2(2n + 1)$ are both even. Thus, all odd integers fall into two progressions: one containing integers of the form $4n + 1$,

$$1, 5, 9, 13, 17, 21, \ . \ . \ .$$

and the other containing integers of the form $4n + 3$,

$$3, 7, 11, 15, 19, 23, \ . \ . \ . \ .$$

While each of these progressions includes some obviously prime numbers, the question arises as to whether each of them contains infinitely many primes. This provides a pleasant opportunity for a repeat performance of Euclid's method for proving the existence of an infinitude of primes. A slight modification of his argument reveals that there are an infinite number of primes of the form $4n + 3$. We approach the proof through a simple lemma.

LEMMA.

The product of two or more integers of the form $4n + 1$ is of the same form.

Proof: It is sufficient to consider the product of just two integers. Let us take $k = 4n + 1$ and $k' = 4m + 1$. Multiplying these together, we obtain

$$kk' = (4n + 1)(4m + 1)$$
$$= 16nm + 4n + 4m + 1 = 4(4nm + n + m) + 1,$$

which is of the desired form.

This paves the way for:

THEOREM 3-6.

There is an infinite number of primes of the form $4n + 3$.

Proof: In anticipation of a contradiction, let us assume that there exist only finitely many primes of the form $4n + 3$; call them $q_1, q_2, \ . \ . \ ., q_s$. Consider the positive integer

$$N = 4q_1q_2 \cdots q_s - 1 = 4(q_1q_2 \cdots q_s - 1) + 3$$

and let $N = r_1 r_2 \cdots r_t$ be its prime factorization. Since N is an odd integer, we have $r_k \neq 2$ for all k, so that each r_k is either of the form $4n + 1$ or $4n + 3$. By the Lemma, the product of any number of primes of the form $4n + 1$ is again an integer of this type. For N to take the form $4n + 3$, as it clearly does, N must contain at least one prime factor r_i of the form $4n + 3$. But r_i cannot be found among the listing $q_1, q_2, \ldots, q_s$, for this would lead to the contradiction that $r_i | 1$. The only possible conclusion is that there are infinitely many primes of the form $4n + 3$.

Having just seen that there are infinitely many primes of the form $4n + 3$, one might reasonably ask: Is the number of primes of the form $4n + 1$ also infinite? This answer is likewise in the affirmative, but a demonstration must await the development of the necessary mathematical machinery. Both these results are special cases of a remarkable theorem by P. G. L. Dirichlet on primes in arithmetic progressions, established in 1837. The proof is much too difficult for inclusion here, so that we content ourselves with the mere statement.

THEOREM 3-7 (Dirichlet).

If a and b are relatively prime positive integers, then the arithmetic progression

$$a, a + b, a + 2b, a + 3b, \ldots$$

contains infinitely many primes.

Dirichlet's Theorem tells us, for instance, that there are infinitely many prime numbers ending in 999, such as 1999, 100999, 1000999, . . . , for these appear in the arithmetic progression determined by $1000n + 999$, where $\gcd(1000, 999) = 1$.

There is no arithmetic progression $a, a + b, a + 2b, \ldots$ that consists solely of prime numbers. To see this, suppose that $a + nb = p$, where p is a prime. If we put $n_k = n + kp$ for $k = 1, 2, 3, \ldots$, then the n_kth term in the progression is

$$a + n_k b = a + (n + kp)b = (a + nb) + kpb = p + kpb.$$

Since each term on the right-hand side is divisible by p, so is $a + n_k b$. In other words, the progression must contain infinitely many composite numbers.

It is an old, but still unsolved question whether there exist arbitrarily long but finite arithmetic progressions consisting only of prime numbers (not necessarily consecutive primes). The longest progression found to date is composed of the 21 primes

$$142072321123 + 1419763024n \qquad (0 \leq n \leq 20).$$

The prime factorization of the common difference between the terms is

$$2^3 \cdot 3 \cdot 5 \cdot 7 \cdot 11 \cdot 13 \cdot 17 \cdot 19 \cdot 23 \cdot 37 \cdot 43,$$

which is divisible by 9699690, the product of the primes less than 21. This takes place according to the following theorem:

THEOREM 3-8.

If the $n > 2$ terms of the arithmetic progression

$$p, p + d, p + 2d, \ldots, p + (n - 1)d$$

are all prime numbers, then the common difference d is divisible by every prime $q < n$.

Proof: Consider a prime number $q < n$, and assume to the contrary that $q \nmid d$. We claim that the first q terms of the progression,

(1) $p, p + d, p + 2d, \ldots, p + (q - 1)d,$

will leave different remainders when divided by q. Otherwise, there exist integers j and k with $0 \leq j < k \leq q - 1$ such that the numbers $p + jd$ and $p + kd$ yield the same remainder upon division by q. Then q divides their difference $(k - j)d$. But $\gcd(q , d) = 1$, and so Euclid's Lemma leads to $q \mid k - j$, which is nonsense in light of the inequality $k - j \leq q - 1$.

Since the q different remainders produced from (1) are drawn from the q integers $0, 1, \ldots, q - 1$, one of these remainders must be zero. This means that $q \mid p + td$ for some t satisfying $0 \leq t \leq q - 1$. Because of the inequality $q < n \leq p \leq p + td$, we are therefore forced to conclude that $p + td$ is composite. (If $p < n$, one of the terms of the progression would be $p + pd = p(1 + d)$.) With this contradiction, the proof that $q \mid d$ is complete.

It has been conjectured that there exist arithmetic progressions of finite (but otherwise arbitrary) length, composed of consecutive prime numbers. Examples of such progressions consisting of three and four primes, respectively, are 41, 47, 53 and 251, 257, 263, 269. Not long ago, a computer search revealed progressions which are composed of five and six consecutive primes, the terms having a common difference of 30; these begin with the primes

$$9,843,019 \qquad \text{and} \qquad 121,174,811.$$

We are not able to discover, at least for the time being, an arithmetic progression consisting of seven consecutive primes. When the restriction that the prime numbers involved be consecutive is removed, then it is possible to find infinitely many sets of seven primes in an arithmetic progression; one such is 7, 157, 307, 457, 607, 757, 907.

In interests of completeness, we might mention another famous problem that so far has resisted the most determined attack. For centuries, mathematicians have sought a simple formula that would yield every prime number or, failing this, a formula that would produce nothing but primes. At first glance, the request seems modest enough: find a function $f(n)$ whose domain is, say, the nonnegative integers and whose range is some infinite subset of the set of all primes. It was widely believed in the Middle Ages that the quadratic polynomial

$$f(n) = n^2 + n + 41$$

assumed only prime values. As evidenced by the following table, the claim is a correct one for $n = 0, 1, 2, \ldots, 39$.

n	$f(n)$	n	$f(n)$	n	$f(n)$
0	41	14	251	28	853
1	43	15	281	29	911
2	47	16	313	30	971
3	53	17	347	31	1033
4	61	18	383	32	1097
5	71	19	421	33	1163
6	83	20	461	34	1231
7	97	21	503	35	1301
8	113	22	547	36	1373
9	131	23	593	37	1447
10	151	24	641	38	1523
11	173	25	691	39	1601
12	197	26	743		
13	223	27	797		

However, this provocative conjecture is shattered in the cases $n = 40$ and $n = 41$, where there is a factor of 41:

$$f(40) = 40 \cdot 41 + 41 = 41^2$$

and

$$f(41) = 41 \cdot 42 + 41 = 41 \cdot 43.$$

The next value $f(42) = 1747$ turns out to be prime once again. It is not presently known whether $f(n) = n^2 + n + 41$ assumes infinitely many prime values for integral n.

Recent calculations have recently shown that the polynomial

$$g(n) = 103n^2 - 3945n + 34381$$

yields prime numbers for $n = 0,1,2, \ldots , 44$. This surpasses by 5 Euler's record for the longest string of prime values taken on by a quadratic polynomial $an^2 + bn + c$ at $n = 0,1,2, \ldots$.

The failure of the above function to be prime-producing is no accident, for it is easy to prove that there is no nonconstant polynomial $f(n)$ with integral coefficients which takes on just prime values for integral n. We assume that such a polynomial $f(n)$ actually does exist and argue until a contradiction is reached. Let

$$f(n) = a_k n^k + a_{k-1} n^{k-1} + \cdots + a_2 n^2 + a_1 n + a_0,$$

where the coefficients $a_0, a_1, \ldots , a_k$ are all integers and $a_k \neq 0$. For a fixed value of n, say $n = n_0$, $p = f(n_0)$ is a prime number. Now, for any integer t, we consider the expression $f(n_0 + tp)$:

$$f(n_0 + tp) = a_k(n_0 + tp)^k + \cdots + a_1(n_0 + tp) + a_0$$

$$= (a_k n_0^k + \cdots + a_1 n_0 + a_0) + pQ(t)$$

$$= f(n_0) + pQ(t)$$

$$= p + pQ(t) = p(1 + Q(t)),$$

where $Q(t)$ is a polynomial in t having integral coefficients. Our reasoning shows that $p \mid f(n_0 + tp)$; hence, from our own assumption that $f(n)$ takes on only prime values, $f(n_0 + tp) = p$ for any integer t. Since a polynomial of degree k cannot assume the same value more than k times, we have obtained the required contradiction.

Recent years have seen a measure of success in the search for prime-producing functions. W. H. Mills proved (1947) that there exists a positive real number r such that the expression $f(n) = [r^{3^n}]$ is prime for $n = 1, 2, 3, \ldots$ (the bracket indicates the greatest integer function). Needless to say, this is strictly an existence theorem and nothing is known about the actual value of r. Mills's function does not produce all the primes.

PROBLEMS 3.3

1. Verify that the integers 1949 and 1951 are twin primes.

2. (a) If 1 is added to a product of twin primes, prove that a perfect square is always obtained.
   (b) Show that the sum of twin primes p and $p + 2$ is divisible by 12, provided that $p > 3$.

3. Find all pairs of primes p and q satisfying $p - q = 3$.

4. Sylvester (1896) rephrased Goldbach's Conjecture so as to read: Every even integer $2n$ greater than 4 is the sum of two primes, one larger than $n/2$ and the other less than $3n/2$. Verify this version of the conjecture for all even integers between 6 and 76.

5. In 1752, Goldbach submitted the following conjecture to Euler: Every odd integer can be written in the form $p + 2a^2$, where p is either a prime or 1 and $a \geq 0$. Show that the integer 5777 refutes this conjecture.

6. Prove that Goldbach's Conjecture that every even integer greater than 2 is the sum of two primes is equivalent to the statement that every integer greater than 5 is the sum of three primes. [*Hint:* If $2n - 2 = p_1 + p_2$, then $2n = p_1 + p_2 + 2$ and $2n + 1 = p_1 + p_2 + 3$.]

7. A conjecture of Lagrange (1775) asserts that every odd integer greater than 5 can be written as a sum $p_1 + 2p_2$, where p_1, p_2 are both primes. Confirm this for all odd integers through 75.

8. Given a positive integer n, it can be shown that there exists an even integer a which is representable as the sum of two odd primes in n different ways. Confirm that the integers 60, 78, and 84 can be written as the sum of two primes in six, seven, and eight ways, respectively.

9. (a) For $n > 3$, show that the integers $n, n + 2, n + 4$ cannot all be prime.
   (b) Three integers $p, p + 2, p + 6$ which are all prime are called a *prime-triplet*. Find five sets of prime-triplets.

10. Establish that the sequence

$$(n + 1)! - 2, (n + 1)! - 3, \ldots, (n + 1)! - (n + 1)$$

produces n consecutive composite integers for $n > 1$.

11. Find the smallest positive integer n for which the function $f(n) = n^2 + n + 17$ is composite. Do the same for the functions $g(n) = n^2 + 21n + 1$ and $h(n) = 3n^2 + 3n + 23$.

12. Let p_n denote the n'th prime number. For $n \geq 3$, prove that $p_{n+3}^2 < p_n p_{n+1} p_{n+2}$. [*Hint:* Note that $p_{n+3}^2 < 4p_{n+2}^2 < 8p_{n+1} p_{n+2}$.]

13. Apply the same method of proof as in Theorem 3–6 to show that there are infinitely many primes of the form $6n + 5$.

14. Find a prime divisor of the integer $N = 4(3 \cdot 7 \cdot 11) - 1$ of the form $4n + 3$. Do the same for $N = 4(3 \cdot 7 \cdot 11 \cdot 15) - 1$.

15. Another unanswered question is whether there exist an infinite number of sets of five consecutive odd integers of which four are primes. Find five such sets of integers.

16. Let the sequence of primes, with 1 adjoined, be denoted by $p_0 = 1$, $p_1 = 2$, $p_2 = 3$, $p_3 = 5, \ldots$. For each $n \geq 1$, it is known that there exists a suitable choice of coefficients $\epsilon_k = \pm 1$ such that

$$p_{2n} = p_{2n-1} + \sum_{k=0}^{2n-2} \epsilon_k p_k, \qquad p_{2n+1} = 2p_{2n} + \sum_{k=0}^{2n-1} \epsilon_k p_k.$$

To illustrate:

$13 = 1 + 2 - 3 - 5 + 7 + 11$ and
$17 = 1 + 2 - 3 - 5 + 7 - 11 + 2 \cdot 13$.

Determine similar representations for the primes 23, 29, 31, and 37.

17. In 1848 de Polignac claimed that every odd integer is the sum of a prime and a power of 2. For example, $55 = 47 + 2^3 = 23 + 2^5$. Show that the integers 509 and 877 discredit this claim.

18. (a) If p is a prime and $p \nmid b$, prove that in the arithmetic progression

$$a, a + b, a + 2b, a + 3b, \ldots$$

every pth term is divisible by p. [*Hint:* Since $\gcd(p, b) = 1$, there exists integers r and s satisfying $pr + bs = 1$. Put $n_k = kp - as$ for $k = 1, 2, \ldots$ and show that $p \mid (a + n_k b)$.]

  (b) From part (a), conclude that if b is an odd integer, then every other term in the indicated progression is even.

19. In 1950, it was proven that any integer $n > 9$ can be written as a sum of distinct odd primes. Express the integers 25, 69, 81, and 125 in this fashion.

20. If p and $p^2 + 8$ are both prime numbers, prove that $p^3 + 4$ is also prime.

21. (a) For any integer $k > 0$, establish that the arithmetic progression

$$a + b, a + 2b, a + 3b, \ldots,$$

where $\gcd(a, b) = 1$, contains k consecutive terms which are composite. [*Hint:* Put $n = (a + b)(a + 2b) \cdots (a + kb)$ and consider the k terms

$$a + (n + 1)b, a + (n + 2)b, \ldots, a + (n + k)b.]$$

  (b) Find five consecutive composite terms in the arithmetic progression

$$6, 11, 16, 21, 26, 31, 36, \ldots.$$

22. Show that 13 is the largest prime that can divide two successive integers of the form $n^2 + 3$.

23. (a) The arithmetic mean of the twin primes 5 and 7 is the triangular number 6. Are there any other twin primes with a triangular mean?

(b) The arithmetic mean of the twin primes 3 and 5 is the perfect square 4. Are there any other twin primes with a square mean?

24. Determine all twin primes p and $q = p + 2$ for which $pq - 2$ is also prime.

25. Let p_n denote the nth prime. For $n > 3$, show that

$$p_n < p_1 + p_2 + \cdots + p_{n-1}.$$

[*Hint:* Use induction and Bertrand's Conjecture.]

26. Verify the following:

(a) There exist infinitely many primes ending in 33, such as 233, 433, 733, 1033, ... [*Hint:* Apply Dirichlet's Theorem.]

(b) There exist infinitely many primes which do not belong to any pair of twin primes. [*Hint:* Consider the arithmetic progression $21k + 5$ for $k = 1, 2, \ldots$.]

(c) There exists a prime ending in as many consecutive 1's as desired. [*Hint:* To obtain a prime ending in n consecutive 1's, consider the arithmetic progression $10^n k + R_n$ for $k = 1, 2, \ldots$.]

(d) There exist infinitely many primes which contain the block of digits 123456789. [*Hint:* Consider the arithmetic progression $(10^{11})k + 1234567891$ for $k = 1, 2, \ldots$.]

27. Prove that for every $n \geq 2$ there exists a prime p with $p \leq n < 2p$. [*Hint:* In the case $n = 2k + 1$, then by Bertrand's Conjecture there exists a prime p such that $k < p < 2k$.]

28. (a) If $n > 1$, show that $n!$ is never a perfect square.

(b) Find the values of $n \geq 1$ for which

$$n! + (n + 1)! + (n + 2)!$$

is a perfect square. [*Hint:* Note that $n! + (n + 1)! + (n + 2)! = n!(n + 2)^2$.]

4

The Theory of Congruences

"Gauss once said 'Mathematics is the queen of the sciences and number-theory the queen of mathematics.'
If this be true we may add that the Disquisitiones *is the Magna Charta of number-theory."*
M. CANTOR

4.1 CARL FRIEDRICH GAUSS

Another approach to divisibility questions is through the arithmetic of remainders, or the *theory of congruences* as it is now commonly known. The concept, and the notation that makes it such a powerful tool, was first introduced by the German mathematician Carl Friedrich Gauss (1777–1855) in his *Disquisitiones Arithmeticae;* this monumental work, which appeared in 1801 when Gauss was 24 years old, laid the foundations of modern number theory. Legend has it that a large part of the *Disquisitiones Arithmeticae* had been submitted as a memoir to the French Academy the previous year and had been rejected in a manner which, even if the work had been as worthless as the referees believed, would have been inexcusable. (In an attempt to lay this defamatory tale to rest, the officers of the Academy made an exhaustive search of their permanent records in 1935 and concluded that the *Disquisitiones* was never submitted, much less rejected.) "It is really astonishing," said Kronecker, "to think that a single man of such young years was able to bring to light such a wealth of results, and above all to present such a profound and well-organized treatment of an entirely new discipline."

Gauss was one of those remarkable infant prodigies whose natural aptitude for mathematics soon becomes apparent. As a child of three, according to a well-authenticated story, he corrected an error in his father's payroll calculations. His arithmetical powers so overwhelmed his schoolmasters that, by the time Gauss was 10 years old, they admitted that there was nothing more they could teach the boy. It is said that in his first arithmetic class Gauss astonished his teacher by instantly solving what was intended to be a "busy work" problem: Find the sum of all the numbers from 1 to 100. The young Gauss later confessed to having recognized the pattern

$$1 + 100 = 101, 2 + 99 = 101, 3 + 98 = 101, \ldots, 50 + 51 = 101.$$

Carl Friedrich Gauss

(1777–1855)

(From A Concise History of Mathematics *by Dirk Struik, 1967, Dover publications, Inc., N.Y.)*

Since there are 50 pairs of numbers, each of which adds up to 101, the sum of all the numbers must be $50 \cdot 101 = 5050$. This technique provides another way of deriving the formula

$$1 + 2 + 3 + \cdots + n = \frac{n(n + 1)}{2}$$

for the sum of the first n positive integers. One need only display the consecutive integers 1 through n in two rows as follows:

$$
\begin{array}{ccccc}
1 & 2 & 3 & \cdots n - 1 & n \\
n & n - 1 & n - 2 & \cdots \ 2 & 1
\end{array}
$$

Addition of the vertical columns produces n terms, each of which is equal to $n + 1$; when these terms are added, we get the value $n(n + 1)$. Because the same sum is obtained on adding the two rows horizontally, what occurs is the formula $n(n + 1) = 2(1 + 2 + 3 + \cdots + n)$.

Gauss went on to a succession of triumphs, each new discovery following on the heels of a previous one. The problem of constructing regular polygons with only "Euclidean tools," that is to say, with ruler and compass alone, had long been laid aside in the belief that the ancients had exhausted all the possible constructions. In 1796, Gauss showed that the 17-sided regular polygon is so constructible, the first advance in this area since Euclid's time. Gauss' doctoral thesis of 1799 provided a rigorous proof of the Fundamental Theorem of Algebra, which had been stated first by Girard in 1629 and then proved imperfectly by d'Alembert (1746) and later by Euler (1749). The theorem (it asserts that an algebraic equation of degree n has exactly n complex roots) was always a favorite with Gauss, and he gave, in all, four distinct demonstrations of it. The publication of *Disquisitiones Arithmeticae* in 1801 at once placed Gauss in the front rank of mathematicians.

The most extraordinary achievement of Gauss was more in the realm of theoretical astronomy than of mathematics. On the opening night of the 19th century, January 1, 1801, the Italian astronomer Piazzi discovered the first of the so-called minor planets (planetoids or asteroids), later called Ceres. But after the course of this newly found body, visible only by telescope, passed the sun, neither Piazzi nor any other astronomer could locate it again. Piazzi's observations extended over a period of 41 days, during which the orbit swept out an angle of only nine degrees. From the scanty data available, Gauss was able to calculate Ceres' orbit with amazing accuracy and the elusive planet was rediscovered at the end of the year in almost exactly the positions he had forecast. This success brought Gauss world-wide fame, and led to his appointment as director of Göttingen Observatory.

By the middle of the 19th century, mathematics had grown into an enormous and unwieldy structure, divided into a large number of fields in which only the specialist knew his way. Gauss was the last complete mathematician, and it is no exaggeration to say that he was in some degree connected with nearly every aspect of the subject. His contemporaries regarded him as Princeps Mathematicorum (Prince of Mathematicians), on a par with Archimedes and Isaac Newton. This is revealed in a small incident: On being asked who was the greatest mathematician in Germany, Laplace answered, "Why, Pfaff." When the questioner indicated that he would have thought Gauss was, Laplace replied, "Pfaff is by far the greatest in Germany, but Gauss is the greatest in all Europe."

Although Gauss adorned every branch of mathematics, he always held number theory in high esteem and affection. He insisted that, "Mathematics is the Queen of the Sciences, and the theory of numbers is the Queen of Mathematics."

4.2 BASIC PROPERTIES OF CONGRUENCE

In the first chapter of *Disquisitiones Arithmeticae,* Gauss introduces the concept of congruence and the notation which makes it such a powerful technique (he explains that he was induced to adopt the symbol $\equiv$ because of the close analogy with algebraic equality). According to Gauss, "If a number n measures the difference between two numbers a and $b,$ then a and b are said to be congruent with respect to $n;$ if not, incongruent." Putting this into the form of a definition, we have

DEFINITION 4-1.

Let n be a fixed positive integer. Two integers a and b are said to be *congruent modulo n,* symbolized by

$$a \equiv b \ (\mathrm{mod}\ n)$$

if n divides the difference $a - b;$ that is, provided that $a - b = kn$ for some integer k.

To fix the idea, consider $n = 7$. It is routine to check that

$$3 \equiv 24 \;(\text{mod } 7), \; -31 \equiv 11 \;(\text{mod } 7), \; -15 \equiv -64 \;(\text{mod } 7),$$

since $3 - 24 = (-3)7$, $-31 - 11 = (-6)7$, and $-15 - (-64) = 7 \cdot 7$. When $n \nmid (a - b)$, then we say that a is *incongruent to b modulo n* and in this case we write $a \not\equiv b \;(\text{mod } n)$. For example: $25 \not\equiv 12 \;(\text{mod } 7)$, since 7 fails to divide $25 - 12 = 13$.

It is to be noted that any two integers are congruent modulo 1, whereas two integers are congruent modulo 2 when they are both even or both odd. Inasmuch as congruence modulo 1 is not particularly interesting, the usual practice is to assume that $n > 1$.

Given an integer a, let q and r be its quotient and remainder upon division by n, so that

$$a = qn + r, \qquad\qquad 0 \le r < n.$$

Then, by definition of congruence, $a \equiv r \;(\text{mod } n)$. Since there are n choices for r, we see that every integer is congruent modulo n to exactly one of the values 0, 1, 2, ..., $n - 1$; in particular, $a \equiv 0 \;(\text{mod } n)$ if and only if $n \mid a$. The set of n integers 0, 1, 2, ..., $n - 1$ is called the set of *least positive residues modulo n*.

In general, a collection of n integers $a_1, a_2, \ldots, a_n$ is said to form a *complete set of residues* (or a *complete system of residues*) *modulo n* if every integer is congruent modulo n to one and only one of the a_k. To put it another way, $a_1, a_2, \ldots, a_n$ are congruent modulo n to 0, 1, 2, ..., $n - 1$, taken in some order. For instance,

$$-12, -4, 11, 13, 22, 82, 91$$

constitute a complete set of residues modulo 7; here, we have

$$-12 \equiv 2, -4 \equiv 3, 11 \equiv 4, 13 \equiv 6, 22 \equiv 1, 82 \equiv 5, 91 \equiv 0,$$

all modulo 7. An observation of some importance is that any n integers form a complete set of residues modulo n if and only if no two of the integers are congruent modulo n. We shall need this fact later on.

Our first theorem provides a useful characterization of congruence modulo n in terms of remainders upon division by n.

THEOREM 4-1.

For arbitrary integers a and b, $a \equiv b \;(\text{mod } n)$ if and only if a and b leave the same nonnegative remainder when divided by n.

Proof: First, take $a \equiv b \;(\text{mod } n)$, so that $a = b + kn$ for some integer k. Upon division by n, b leaves a certain remainder r; that is, $b = qn + r$, where $0 \le r < n$. Therefore,

$$a = b + kn = (qn + r) + kn = (q + k)n + r,$$

which indicates that a has the same remainder as b.

On the other hand, suppose we can write $a = q_1 n + r$ and $b = q_2 n + r$, with the same remainder r $(0 \le r < n)$. Then

$$a - b = (q_1 n + r) - (q_2 n + r) = (q_1 - q_2)n,$$

whence $n \mid a - b$. In the language of congruences, we have $a \equiv b \;(\text{mod } n)$.

Example 4-1

Since the integers -56 and -11 can be expressed in the form

$$-56 = (-7)9 + 7, \quad -11 = (-2)9 + 7$$

with the same remainder 7, Theorem 4-1 tells us that $-56 \equiv -11 \pmod 9$. Going in the other direction, the congruence $-31 \equiv 11 \pmod 7$ implies that -31 and 11 have the same remainder when divided by 7; this is clear from the relations

$$-31 = (-5)7 + 4, \quad 11 = 1 \cdot 7 + 4.$$

Congruence may be viewed as a generalized form of equality, in the sense that its behavior with respect to addition and multiplication is reminiscent of ordinary equality. Some of the elementary properties of equality that carry over to congruences appear in the next theorem.

THEOREM 4-2.

Let $n > 0$ be fixed and a, b, c, d be arbitrary integers. Then the following properties hold:

(1) $a \equiv a \pmod n$.
(2) *If $a \equiv b \pmod n$, then $b \equiv a \pmod n$.*
(3) *If $a \equiv b \pmod n$ and $b \equiv c \pmod n$, then $a \equiv c \pmod n$.*
(4) *If $a \equiv b \pmod n$ and $c \equiv d \pmod n$, then $a + c \equiv b + d \pmod n$ and $ac \equiv bd \pmod n$.*
(5) *If $a \equiv b \pmod n$, then $a + c \equiv b + c \pmod n$ and $ac \equiv bc \pmod n$.*
(6) *If $a \equiv b \pmod n$, then $a^k \equiv b^k \pmod n$ for any positive integer k.*

Proof: For any integer a, we have $a - a = 0 \cdot n$, so that $a \equiv a \pmod n$. Now if $a \equiv b \pmod n$, then $a - b = kn$ for some integer k. Hence, $b - a = -(kn) = (-k)n$ and, since $-k$ is an integer, this yields property (2).

Property (3) is slightly less obvious: Suppose that $a \equiv b \pmod n$ and also $b \equiv c \pmod n$. Then there exist integers h and k satisfying $a - b = hn$ and $b - c = kn$. It follows that

$$a - c = (a - b) + (b - c) = hn + kn = (h + k)n,$$

which is $a \equiv c \pmod n$ in congruence notation.

In the same vein, if $a \equiv b \pmod n$ and $c \equiv d \pmod n$, then we are assured that $a - b = k_1 n$ and $c - d = k_2 n$ for some choice of k_1 and k_2. Adding these equations, one gets

$$(a + c) - (b + d) = (a - b) + (c - d)$$
$$= k_1 n + k_2 n = (k_1 + k_2)n$$

or, as a congruence statement, $a + c \equiv b + d \pmod n$. As regards the second assertion of property (4), note that

$$ac = (b + k_1 n)(d + k_2 n) = bd + (bk_2 + dk_1 + k_1 k_2 n)n.$$

Since $bk_2 + dk_1 + k_1k_2n$ is an integer, this says that $ac - bd$ is divisible by n, whence $ac \equiv bd \pmod{n}$.

The proof of property (5) is covered by (4) and the fact that $c \equiv c \pmod{n}$. Finally, we obtain property (6) by making an induction argument. The statement certainly holds for $k = 1$, and we will assume it is true for some fixed k. From (4), we know that $a \equiv b \pmod{n}$ and $a^k \equiv b^k \pmod{n}$ together imply that $aa^k \equiv bb^k \pmod{n}$, or equivalently, $a^{k+1} \equiv b^{k+1} \pmod{n}$. This is the form the statement should take for $k + 1$, so the induction step is complete.

Before going further, we should illustrate that congruences can be a great help in carrying out certain types of computations.

Example 4-2

Let us endeavor to show that 41 divides $2^{20} - 1$. We begin by noting that $2^5 \equiv -9 \pmod{41}$, whence $(2^5)^4 \equiv (-9)^4 \pmod{41}$ by Theorem 4-2(6); in other words, $2^{20} \equiv 81 \cdot 81 \pmod{41}$. But $81 \equiv -1 \pmod{41}$ and so $81 \cdot 81 \equiv 1 \pmod{41}$. Using parts (2) and (5) of Theorem 4-2, we finally arrive at

$$2^{20} - 1 \equiv 81 \cdot 81 - 1 \equiv 1 - 1 \equiv 0 \pmod{41}.$$

Thus $41 \mid 2^{20} - 1$, as desired.

Example 4-3

For another example in the same spirit, suppose that we are asked to find the remainder obtained upon dividing the sum

$$1! + 2! + 3! + 4! + \cdots + 99! + 100!$$

by 12. Without the aid of congruences this would be an awesome calculation. The observation that starts us off is that $4! \equiv 24 \equiv 0 \pmod{12}$; thus, for $k \geq 4$,

$$k! \equiv 4! \cdot 5 \cdot 6 \cdots k \equiv 0 \cdot 5 \cdot 6 \cdots k \equiv 0 \pmod{12}.$$

One finds in this way that

$$1! + 2! + 3! + 4! + \cdots + 100!$$
$$\equiv 1! + 2! + 3! + 0 + \cdots + 0 \equiv 9 \pmod{12}.$$

Accordingly, the sum in question leaves a remainder of 9 when divided by 12.

In the last theorem, it was seen that if $a \equiv b \pmod{n}$, then $ca \equiv cb \pmod{n}$ for any integer c. The converse, however, fails to hold. For an example perhaps as simple as any, note that $2 \cdot 4 \equiv 2 \cdot 1 \pmod{6}$, while $4 \not\equiv 1 \pmod{6}$. In brief: one cannot unrestrictedly cancel a common factor in the arithmetic of congruences.

With suitable precautions, cancellation can be allowed; one step in this direction, and an important one, is provided by the following theorem.

THEOREM 4-3.
> *If ca ≡ cb* (mod *n*), *then a ≡ b* (mod *n/d*), *where d* = gcd(*c* , *n*).

> *Proof:* By hypothesis, we can write

$$c(a - b) = ca - cb = kn$$

> for some integer *k*. Knowing that gcd(*c* , *n*) = *d*, there exist relatively prime integers *r* and *s* satisfying *c* = *dr*, *n* = *ds*. When these values are substituted in the displayed equation and the common factor *d* cancelled, the net result is

$$r(a - b) = ks.$$

> Hence, *s* | *r*(*a* − *b*) and gcd(*r* , *s*) = 1. Euclid's Lemma yields *s* | *a* − *b*, which may be recast as *a* ≡ *b* (mod *s*); in other words, *a* ≡ *b* (mod *n/d*).

> Theorem 4-3 gets its maximum force when the requirement that gcd(*c* , *n*) = 1 is added, for then the cancellation may be accomplished without a change in modulus.

COROLLARY 1.
> *If ca ≡ cb* (mod *n*) *and* gcd(*c* , *n*) = 1, *then a ≡ b* (mod *n*).

> We take the moment to record a special case of Corollary 1 which we shall have frequent occasion to use, namely,

COROLLARY 2.
> *If ca ≡ cb* (mod *p*) *and p ∤ c, where p is a prime number, then a ≡ b* (mod *p*).

> *Proof:* The conditions *p ∤ c* and *p* a prime imply that gcd(*c* , *p*) = 1.

Example 4-4

> Consider the congruence 33 ≡ 15 (mod 9) or, if one prefers, 3 · 11 ≡ 3 · 5 (mod 9). Since gcd(3 , 9) = 3, Theorem 4-3 leads to the conclusion that 11 ≡ 5 (mod 3). A further illustration is given by the congruence −35 ≡ 45 (mod 8), which is the same as 5 · (−7) ≡ 5 · 9 (mod 8). The integers 5 and 8 being relatively prime, we may cancel the Factor 5 to obtain a correct congruence −7 ≡ 9 (mod 8).

Let us call attention to the fact that, in Theorem 4-3, it is unnecessary to stipulate that *c* ≢ 0 (mod *n*). Indeed, were *c* ≡ 0 (mod *n*), then gcd(*c* , *n*) = *n* and the conclusion of the theorem would state that *a* ≡ *b* (mod 1); but, as we remarked earlier, this holds trivially for all integers *a* and *b*.

There is another curious situation that can arise with congruences: the product of two integers, neither of which is congruent to zero, may turn out to be congruent to zero. For instance, 4 · 3 ≡ 0 (mod 12), but 4 ≢ 0 (mod 12) and 3 ≢ 0 (mod 12). It is a simple matter to show that if *ab* ≡ 0 (mod *n*) and gcd(*a* , *n*) = 1, then *b* ≡ 0 (mod *n*); for, Corollary 1 above permits us legitimately to cancel the factor *a* from both sides of the congruence *ab* ≡ *a* · 0 (mod *n*). A variation on this is that when *ab* ≡ 0 (mod *p*), with *p* a prime, then either *a* ≡ 0 (mod *p*) or *b* ≡ 0 (mod *p*).

PROBLEMS 4.2

1. Prove each of the following assertions:
    (a) If $a \equiv b \pmod{n}$ and $m \mid n$, then $a \equiv b \pmod{m}$.
    (b) If $a \equiv b \pmod{n}$ and $c > 0$, then $ca \equiv cb \pmod{cn}$.
    (c) If $a \equiv b \pmod{n}$ and the integers a, b, n are all divisible by $d > 0$, then $a/d \equiv b/d \pmod{n/d}$.

2. Give an example to show that $a^2 \equiv b^2 \pmod{n}$ need not imply that $a \equiv b \pmod{n}$.

3. If $a \equiv b \pmod{n}$, prove that $\gcd(a, n) = \gcd(b, n)$.

4. (a) Find the remainders when 2^{50} and 41^{65} are divided by 7.
    (b) What is the remainder when the sum

    $$1^5 + 2^5 + 3^5 + \cdots + 99^5 + 100^5$$

    is divided by 4?

5. Prove that the integer $53^{103} + 103^{53}$ is divisible by 39, and that $111^{333} + 333^{111}$ is divisible by 7.

6. For $n \geq 1$, use congruence theory to establish each of the following divisibility statements:
    (a) $7 \mid 5^{2n} + 3 \cdot 2^{5n-2}$;
    (b) $13 \mid 3^{n+2} + 4^{2n+1}$;
    (c) $27 \mid 2^{5n+1} + 5^{n+2}$;
    (d) $43 \mid 6^{n+2} + 7^{2n+1}$.

7. For $n \geq 1$, show that

    $$(-13)^{n+1} \equiv (-13)^n + (-13)^{n-1} \pmod{181}.$$

    [*Hint:* Notice that $(-13)^2 \equiv -13 + 1 \pmod{181}$; use induction on n.]

8. Prove the assertions below:
    (a) If a is an odd integer, then $a^2 \equiv 1 \pmod{8}$.
    (b) For any integer a, $a^3 \equiv 0$, 1, or 6 $\pmod{7}$.
    (c) For any integer a, $a^4 \equiv 0$ or 1 $\pmod{5}$.
    (d) If the integer a is not divisible by 2 or 3, then $a^2 \equiv 1 \pmod{24}$.

9. If p is a prime satisfying $n < p < 2n$, show that

    $$\binom{2n}{n} \equiv 0 \pmod{p}.$$

10. If $a_1, a_2, \ldots, a_n$ is a complete set of residues modulo n and $\gcd(a, n) = 1$, prove that $aa_1, aa_2, \ldots, aa_n$ is also a complete set of residues modulo n. [*Hint:* It suffices to show that the numbers in question are incongruent modulo n.]

11. Verify that $0, 1, 2, 2^2, 2^3, \ldots, 2^9$ form a complete set of residues modulo 11, but $0, 1^2, 2^2, 3^2, \ldots, 10^2$ do not.

12. Prove the following statements:
    (a) If $\gcd(a, n) = 1$, then the integers

    $$c, c + a, c + 2a, c + 3a, \ldots, c + (n-1)a$$

    form a complete set of residues modulo n for any c.

(b)   Any n consecutive integers form a complete set of residues modulo n. [*Hint:* Use part (a).]

(c)   The product of any set of n consecutive integers is divisible by n.

13.   Verify that if $a \equiv b \pmod{n_1}$ and $a \equiv b \pmod{n_2}$, then $a \equiv b \pmod{n}$, where the integer $n = \operatorname{lcm}(n_1 , n_2)$. Hence, whenever n_1 and n_2 are relatively prime, $a \equiv b \pmod{n_1 n_2}$.

14.   Give an example to show that $a^k \equiv b^k \pmod{n}$ and $k \equiv j \pmod{n}$ need not imply that $a^j \equiv b^j \pmod{n}$.

15.   Establish that if a is an odd integer, then

$$a^{2^n} \equiv 1 \pmod{2^{n+2}}$$

for any $n \geq 1$. [*Hint:* Proceed by induction on n.]

16.   Use the theory of congruences to verify that

$$89 \,|\, 2^{44} - 1 \quad \text{and} \quad 97 \,|\, 2^{48} - 1.$$

17.   Prove that whenever $ab \equiv cd \pmod{n}$ and $b \equiv d \pmod{n}$, with $\gcd(b , n) = 1$, then $a \equiv c \pmod{n}$.

18.   If $a \equiv b \pmod{n_1}$ and $a \equiv c \pmod{n_2}$, prove that $b \equiv c \pmod{n}$, where the integer $n = \gcd(n_1 , n_2)$.

4.3 SPECIAL DIVISIBILITY TESTS

One of the more interesting applications of congruence theory involves finding special criteria under which a given integer is divisible by another integer. At their heart, these divisibility tests depend on the notational system used to assign "names" to integers and, more particularly, to the fact that 10 is taken as the base for our number system. Let us, therefore, start by showing that, given an integer $b > 1$, any positive integer N can be written uniquely in terms of powers of b as

$$N = a_m b^m + a_{m-1} b^{m-1} + \cdots + a_2 b^2 + a_1 b + a_0,$$

where the coefficients a_k can take on the b different values $0, 1, 2, \ldots, b - 1$. For, the Division Algorithm yields integers q_1 and a_0 satisfying

$$N = q_1 b + a_0, \qquad\qquad 0 \leq a_0 < b.$$

If $q_1 \geq b$, we can divide once more, obtaining

$$q_1 = q_2 b + a_1, \qquad\qquad 0 \leq a_1 < b.$$

Now substitute for q_1 in the earlier equation to get

$$N = (q_2 b + a_1)b + a_0 = q_2 b^2 + a_1 b + a_0.$$

As long as $q_2 \geq b$, we can continue in the same fashion. Going one more step: $q_2 = q_3 b + a_2$, where $0 \leq a_2 < b$, hence

$$N = q_3 b^3 + a_2 b^2 + a_1 b + a_0.$$

Since $N > q_1 > q_2 > \cdots \geq 0$ is a strictly decreasing sequence of integers, this process must eventually terminate; say, at the $(m - 1)$th stage, where

$$q_{m-1} = q_m b + a_{m-1}, \qquad\qquad 0 \leq a_{m-1} < b$$

and $0 \leq q_m < b$. Setting $a_m = q_m$, we reach the representation

$$N = a_m b^m + a_{m-1} b^{m-1} + \cdots + a_1 b + a_0$$

which was our aim.

To show uniqueness, let us suppose that N has two distinct representations; say,

$$N = a_m b^m + \cdots + a_1 b + a_0 = c_m b^m + \cdots + c_1 b + c_0,$$

with $0 \leq a_i < b$ for each i and $0 \leq c_j < b$ for each j (we can use the same m by simply adding terms with coefficients $a_i = 0$ or $c_j = 0$ if necessary). Subtracting the second representation from the first gives the equation

$$0 = d_m b^m + \cdots + d_1 b + d_0,$$

where $d_i = a_i - c_i$ for $i = 0, 1, \ldots, m$. Because the two representations for N are assumed different, we must have $d_i \neq 0$ for some value of i. Take k to be the smallest subscript for which $d_k \neq 0$. Then

$$0 = d_m b^m + \cdots + d_{k+1} b^{k+1} + d_k b^k$$

and so, after dividing by b^k,

$$d_k = -b(d_m b^{m-k-1} + \cdots + d_{k+1}).$$

This tells us that $b \mid d_k$. Now the inequalities $0 \leq a_k < b$ and $0 \leq c_k < b$ lead us to $-b < a_k - c_k < b$, or $|d_k| < b$. The only way of reconciling the conditions $b \mid d_k$ and $|d_k| < b$ is to have $d_k = 0$, which is impossible. From this contradiction, we conclude that the representation of N is unique.

The essential feature in all of this is that the integer N is completely determined by the ordered array $a_m, a_{m-1}, \ldots, a_1, a_0$ of coefficients, with the plus signs and the powers of b being superfluous. Thus, the number

$$N = a_m b^m + a_{m-1} b^{m-1} + \cdots + a_2 b^2 + a_1 b + a_0$$

may be replaced by the simpler symbol

$$N = (a_m a_{m-1} \cdots a_2 a_1 a_0)_b$$

(the right-hand side is not to be interpreted as a product, but only as an abbreviation for N). We call this the *base b place-value notation for N*.

Small values of b give rise to lengthy representation of numbers, but have the advantage of requiring fewer choices for coefficients. The simplest case occurs when the base $b = 2$, and the resulting system of enumeration is called the *binary number system* (from the Latin *binarius*, two). The fact that when a number is written in the binary system only the integers 0 and 1 can appear as coefficients means: every positive integer is expressible in exactly one way as a sum of distinct powers of 2. For example, the integer 105 can be written as

$$105 = 1 \cdot 2^6 + 1 \cdot 2^5 + 0 \cdot 2^4 + 1 \cdot 2^3 + 0 \cdot 2^2 + 0 \cdot 2 + 1$$
$$= 2^6 + 2^5 + 2^3 + 1$$

or, in abbreviated form,

$$105 = (1101001)_2.$$

In the other direction, $(1001111)_2$ translates into

$$1 \cdot 2^6 + 0 \cdot 2^5 + 0 \cdot 2^4 + 1 \cdot 2^3 + 1 \cdot 2^2 + 1 \cdot 2 + 1 = 79.$$

The binary system is most convenient for use in modern electronic computing machines, since binary numbers are represented by strings of zeros and ones; 0 and 1 can be expressed in the machine by a switch (or a similar electronic device) being either on or off.

We ordinarily record numbers in the *decimal system* of notation, where $b = 10$, omitting the 10-subscript which specifies the base. For instance, the symbol 1492 stands for the more awkward expression

$$1 \cdot 10^3 + 4 \cdot 10^2 + 9 \cdot 10 + 2.$$

The integers 1, 4, 9, and 2 are called the *digits* of the given number, 1 being the thousands digit, 4 the hundreds digit, 9 the tens digit, and 2 the units digit. In technical language we refer to the representation of the positive integers as sums of powers of 10, with coefficients at most 9, as their *decimal representation* (from the Latin *decem, ten*).

We are about ready to derive criteria for determining whether an integer is divisible by 9 or 11, without performing the actual division. For this, we need a result having to do with congruences involving polynomials with integral coefficients.

THEOREM 4-4.

Let $P(x) = \sum_{k=0}^{m} c_k x^k$ be a polynomial function of x with integral coefficients c_k. If $a \equiv b \pmod{n}$, then $P(a) \equiv P(b) \pmod{n}$.

Proof: Since $a \equiv b \pmod{n}$, part (6) of Theorem 4-2 can be applied to give $a^k \equiv b^k \pmod{n}$ for $k = 0, 1, \ldots, m$. Therefore

$$c_k a^k \equiv c_k b^k \pmod{n}$$

for all such k. Adding these $m + 1$ congruences, we conclude that

$$\sum_{k=0}^{m} c_k a^k \equiv \sum_{k=0}^{m} c_k b^k \pmod{n}$$

or, in different notation, $P(a) \equiv P(b) \pmod{n}$.

If $P(x)$ is a polynomial with integral coefficients, one says that a is a solution of the congruence $P(x) \equiv 0 \pmod{n}$ if $P(a) \equiv 0 \pmod{n}$.

COROLLARY.

If a is a solution of $P(x) \equiv 0 \pmod{n}$ and $a \equiv b \pmod{n}$, then b is also a solution.

Proof: From the last theorem, it is known that $P(a) \equiv P(b) \pmod{n}$. Hence, if a is a solution of $P(x) \equiv 0 \pmod{n}$, then $P(b) \equiv P(a) \equiv 0 \pmod{n}$, making b a solution.

One divisibility test that we have in mind is this: A positive integer is divisible by 9 if and only if the sum of the digits in its decimal representation is divisible by 9.

THEOREM 4-5.

Let $N = a_m 10^m + a_{m-1} 10^{m-1} + \cdots + a_1 10 + a_0$ be the decimal expansion of the positive integer N, $0 \le a_k < 10$, and let $S = a_0 + a_1 + \cdots + a_m$. Then $9 \mid N$ if and only if $9 \mid S$.

Proof: Consider $P(x) = \sum_{k=0}^{m} a_k x^k$, a polynomial with integral coefficients. The key observation is that $10 \equiv 1 \pmod 9$, whence by Theorem 4-4, $P(10) \equiv P(1) \pmod 9$. But $P(10) = N$ and $P(1) = a_0 + a_1 + \cdots + a_m = S$, so that $N \equiv S \pmod 9$. It follows that $N \equiv 0 \pmod 9$ if and only if $S \equiv 0 \pmod 9$, which is what we wanted to prove.

Theorem 4-4 also serves as the basis for a well-known test for divisibility by 11; to wit, an integer is divisible by 11 if and only if the alternating sum of its digits is divisible by 11. Stated more precisely:

THEOREM 4-6.

Let $N = a_m 10^m + a_{m-1} 10^{m-1} + \cdots + a_1 10 + a_0$ be the decimal representation of the positive integer N, $0 \le a_k < 10$, and let $T = a_0 - a_1 + a_2 - \cdots + (-1)^m a_m$. Then $11 \mid N$ if and only if $11 \mid T$.

Proof: As in the proof of Theorem 4-5, put $P(x) = \sum_{k=0}^{m} a_k x^k$. Since $10 \equiv -1 \pmod{11}$, we get $P(10) \equiv P(-1) \pmod{11}$. But $P(10) = N$, whereas $P(-1) = a_0 - a_1 + a_2 - \cdots + (-1)^m a_m = T$, so that $N \equiv T \pmod{11}$. The implication is that both N and T are divisible by 11 or neither is divisible by 11.

Example 4-5

To see an illustration of the last two results, take the integer $N = 1,571,724$. Since the sum

$$1 + 5 + 7 + 1 + 7 + 2 + 4 = 27$$

is divisible by 9, Theorem 4–5 guarantees that 9 divides N. It can also be divided by 11; for, the alternating sum

$$4 - 2 + 7 - 1 + 7 - 5 + 1 = 11$$

is divisible by 11.

PROBLEMS 4.3

1. Prove the following statements:
   (a) For any integer a, the units digit of a^2 is 0, 1, 4, 5, 6, or 9.
   (b) Any one of the integers 0, 1, 2, 3, 4, 5, 6, 7, 8, 9 can occur as the units digit of a^3.
   (c) For any integer a, the units digit of a^4 is 0, 1, 5, or 6.
   (d) The units digit of a triangular number is 0, 1, 3, 5, 6, or 8.

2. Find the last two digits of the number 9^{9^9}. [*Hint:* $9^9 \equiv 9 \pmod{10}$, hence $9^{9^9} = 9^{9 + 10k}$; now use the fact that $9^9 \equiv 89 \pmod{100}$.]

3. Without performing the divisions, determine whether the integers 176,521,221 and 149,235,678 are divisible by 9 or 11.

4. (a) Obtain the following generalization of Theorem 4-5: If the integer N is represented in the base b by

   $$N = a_m b^m + \cdots + a_2 b^2 + a_1 b + a_0, \quad 0 \le a_k \le b - 1$$

   then $b - 1 \mid N$ if and only if $b - 1 \mid (a_m + \cdots + a_2 + a_1 + a_0)$.
   (b) Give criteria for the divisibility of N by 3 and 8 which depend on the digits of N when written in the base 9.
   (c) Is the integer $(447836)_9$ divisible by 3 and 8?

5. Working modulo 9 or 11, find the missing digits in the calculations below:
   (a) $51840 \cdot 273581 = 1418243x040$;
   (b) $2x99561 = [3(523 + x)]^2$;
   (c) $2784x = x \cdot 5569$;
   (d) $512 \cdot 1x53125 = 1000000000$.

6. Establish the following divisibility criteria:
   (a) An integer is divisible by 2 if and only if its units digit is 0, 2, 4, 6, or 8.
   (b) An integer is divisible by 3 if and only if the sum of its digits is divisible by 3.
   (c) An integer is divisible by 4 if and only if the number formed by its tens and units digits is divisible by 4. [*Hint:* $10^k \equiv 0 \pmod 4$ for $k \ge 2$.]
   (d) An integer is divisible by 5 if and only if its units digit is 0 or 5.

7. For any integer a, show that $a^2 - a + 7$ ends in one of the digits 3, 7, or 9.

8. Find the remainder when 4444^{4444} is divided by 9. [*Hint:* Observe that $2^3 \equiv -1 \pmod 9$.]

9. Prove that no integer whose digits add up to 15 can be a square or a cube. [*Hint:* For any a, $a^3 \equiv 0$, 1, or 8 $\pmod 9$.]

10. Assuming that 495 divides $273x49y5$, obtain the digits x and y.

11. Determine the last three digits of the number 7^{999}.
    [*Hint:* $7^{4n} \equiv (1 + 400)^n \equiv 1 + 400n \pmod{1000}$.]

12. If t_n denotes the nth triangular number, show that $t_{n + 2k} \equiv t_n \pmod k$; hence, t_n and $t_{n + 20}$ must have the same last digit.

13. For any $n > 1$, prove that there exists a prime with at least n of its digits equal to 0. [*Hint:* Consider the arithmetic progression $10^{n + 1} k + 1$ for $k = 1, 2, \ldots$.]

14. Find the values of $n \ge 1$ for which $1! + 2! + 3! + \cdots + n!$ is a perfect square. [*Hint:* Problem 1(a).]

15. Show that 2^n divides an integer N if and only if 2^n divides the number made up of the last n digits of N. [*Hint:* $10^k = 2^k 5^k \equiv 0 \pmod{2^n}$ for $k \ge n$.]

16. Let $N = a_m 10^m + \cdots + a_2 10^2 + a_1 10 + a_0$, where $0 \le a_k \le 9$, be the decimal expansion of a positive integer N.
    (a) Prove that 7, 11, and 13 all divide N if and only if 7, 11, and 13 divide the integer

    $$M = (100a_2 + 10a_1 + a_0) - (100a_5 + 10a_4 + a_3)$$
    $$+ (100a_8 + 10a_7 + a_6) - \cdots.$$

    [Hint: If n is even, then $10^{3n} \equiv 1$, $10^{3n+1} \equiv 10$, $10^{3n+2} \equiv 100 \pmod{1001}$; if n is odd, then $10^{3n} \equiv -1$, $10^{3n+1} \equiv -10$, $10^{3n+2} \equiv -100 \pmod{1001}$.]

    (b) Prove that 6 divides N if and only if 6 divides the integer

    $$M = a_0 + 4a_1 + 4a_2 + \cdots + 4a_m.$$

17. Without performing the divisions, determine whether the integer 1,010,908,899 is divisible by 7, 11, and 13.

18. (a) Given an integer N, let M be the integer formed by reversing the order of the digits of N (for example, if $N = 6923$, then $M = 3296$). Verify that $N - M$ is divisible by 9.
    (b) A *palindrome* is a number that reads the same backwards as forwards (for instance, 373 and 521125 are palindromes). Prove that any palindrome with an even number of digits is divisible by 11.

19. Given a repunit R_n, show that
    (a) $9 \mid R_n$ if and only if $9 \mid n$.
    (b) $11 \mid R_n$ if and only if n is even.

20. Factor the repunit $R_6 = 111111$ into a product of primes. [Hint: Problem 16.]

21. Explain why the following curious calculations hold:

    $$1 \cdot 9 + 2 = 11$$
    $$12 \cdot 9 + 3 = 111$$
    $$123 \cdot 9 + 4 = 1111$$
    $$1234 \cdot 9 + 5 = 11111$$
    $$12345 \cdot 9 + 6 = 111111$$
    $$123456 \cdot 9 + 7 = 1111111$$
    $$1234567 \cdot 9 + 8 = 11111111$$
    $$12345678 \cdot 9 + 9 = 111111111$$
    $$123456789 \cdot 9 + 10 = 1111111111.$$

    [Hint: Show that

    $$(10^{n-1} + 2 \cdot 10^{n-2} + 3 \cdot 10^{n-3} + \cdots + n)(10 - 1)$$
    $$+ (n + 1) = (10^{n+1} - 1)/9.]$$

22. An old and somewhat illegible invoice shows that 72 canned hams were purchased for $x 67.9y$. Find the missing digits.

23. If 792 divides the integer $13xy\,45z$, find the digits x, y, and z. [Hint: By Problem 15, $8 \mid 45z$.]

24. For any prime $p > 3$ prove that 13 divides $10^{2p} - 10^p + 1$. [Hint: By Problem 16(a), $10^6 \equiv 1 \pmod{13}$.]

4.4 LINEAR CONGRUENCES

This is a convenient place in our development of number theory at which to investigate the theory of linear congruences: An equation of the form $ax \equiv b \pmod{n}$ is called a *linear congruence,* and by a solution of such an equation we mean an integer x_0 for which $ax_0 \equiv b \pmod{n}$. By definition, $ax_0 \equiv b \pmod{n}$ if and only if $n \mid ax_0 - b$ or, what amounts to the same thing, if and only if $ax_0 - b = ny_0$ for some integer y_0. Thus, the problem of finding all integers which will satisfy the linear congruence $ax \equiv b \pmod{n}$ is identical with that of obtaining all solutions of the linear Diophantine equation $ax - ny = b$. This allows us to bring the results of Chapter 2 into play.

It is convenient to treat two solutions of $ax \equiv b \pmod{n}$ which are congruent modulo n as being "equal" even though they are not equal in the usual sense. For instance, $x = 3$ and $x = -9$ both satisfy the congruence $3x \equiv 9 \pmod{12}$; because $3 \equiv -9 \pmod{12}$, they are not counted as different solutions. In short: When we refer to the number of solutions of $ax \equiv b \pmod{n}$, we mean the number of incongruent integers satisfying this congruence.

With these remarks in mind, the principal result is easy to state.

THEOREM 4-7.

The linear congruence $ax \equiv b \pmod{n}$ has a solution if and only if $d \mid b$, where $d = \gcd(a, n)$. If $d \mid b$, then it has d mutually incongruent solutions modulo n.

Proof: We have already observed that the given congruence is equivalent to the linear Diophantine equation $ax - ny = b$. From Theorem 2-9, it is known that the latter equation can be solved if and only if $d \mid b$; moreover, if it is solvable and x_0, y_0 is one specific solution, then any other solution has the form

$$x = x_0 + \frac{n}{d}t, \quad y = y_0 + \frac{a}{d}t$$

for some choice of t.

Among the various integers satisfying the first of these formulas, consider those that occur when t takes on the successive values $t = 0, 1, 2, \ldots, d - 1$:

$$x_0, \; x_0 + \frac{n}{d}, \; x_0 + \frac{2n}{d}, \; \ldots, \; x_0 + \frac{(d-1)n}{d}.$$

We claim that these integers are incongruent modulo n, while all other such integers x are congruent to some one of them. If it happened that

$$x_0 + \frac{n}{d}t_1 \equiv x_0 + \frac{n}{d}t_2 \pmod{n},$$

where $0 \leq t_1 < t_2 \leq d - 1$, then one would have

$$\frac{n}{d}t_1 \equiv \frac{n}{d}t_2 \pmod{n}.$$

Now $\gcd(n/d, n) = n/d$ and so, by Theorem 4-3, the factor n/d could be cancelled to arrive at the congruence

$$t_1 \equiv t_2 \ (\text{mod } d).$$

which is to say that $d \mid t_2 - t_1$. But this is impossible, in view of the inequality $0 < t_2 - t_1 < d$.

It remains to argue that any other solution $x_0 + (n/d)t$ is congruent modulo n to one of the d integers listed above. The Division Algorithm permits us to write t as $t = qd + r$, where $0 \le r \le d - 1$. Hence

$$x_0 + \frac{n}{d}t = x_0 + \frac{n}{d}(qd + r)$$

$$= x_0 + nq + \frac{n}{d}r$$

$$\equiv x_0 + \frac{n}{d}r \ (\text{mod } n),$$

with $x_0 + (n/d)r$ being one of our d selected solutions. This ends the proof.

The argument that we gave in Theorem 4-7 brings out a point worth stating explicitly: If x_0 is any solution of $ax \equiv b \ (\text{mod } n)$, then the $d = \gcd(a, n)$ incongruent solutions are given by

$$x_0, x_0 + n/d, x_0 + 2(n/d), \ldots, x_0 + (d - 1)(n/d).$$

For the reader's convenience, let us also record the form Theorem 4-7 takes in the special case in which a and n are assumed to be relatively prime.

COROLLARY.

If $\gcd(a, n) = 1$, then the linear congruence $ax \equiv b \ (\text{mod } n)$ has a unique solution modulo n.

We now pause to look at two concrete examples.

Example 4-6

Consider the linear congruence $18x \equiv 30 \ (\text{mod } 42)$. Since $\gcd(18, 42) = 6$ and 6 surely divides 30, Theorem 4-7 guarantees the existence of exactly six solutions, which are incongruent modulo 42. By inspection, one solution is found to be $x = 4$. Our analysis tells us that the six solutions are as follows:

$$x \equiv 4 + (42/6)t \equiv 4 + 7t \ (\text{mod } 42), \qquad t = 0, 1, \ldots, 5$$

or, plainly enumerated,

$$x \equiv 4, 11, 18, 25, 32, 39 \ (\text{mod } 42).$$

Example 4-7

Let us solve the linear congruence $9x \equiv 21 \pmod{30}$. At the outset, since $\gcd(9, 30) = 3$ and $3 \mid 21$, we know that there must be three incongruent solutions.

One way to find these solutions is to divide the given congruence through by 3, thereby replacing it by the equivalent congruence $3x \equiv 7 \pmod{10}$. The relative primeness of 3 and 10 implies that the latter congruence admits a unique solution modulo 10. Although it is not the most efficient method, we could test the integers $0, 1, 2, \ldots, 9$ in turn until the solution is obtained. A better way is this: multiply both sides of the congruence $3x \equiv 7 \pmod{10}$ by 7 to get

$$21x \equiv 49 \pmod{10},$$

which reduces to $x \equiv 9 \pmod{10}$. (This simplification is no accident, for the multiples $0 \cdot 3, 1 \cdot 3, 2 \cdot 3, \ldots, 9 \cdot 3$ form a complete set of residues modulo 10; hence, one of them is necessarily congruent to 1 modulo 10.) But the original congruence was given modulo 30, so that its incongruent solutions are sought among the integers $0, 1, 2, \ldots, 29$. Taking $t = 0, 1, 2$, in the formula

$$x = 9 + 10t,$$

one gets 9, 19, 29, whence

$$x \equiv 9 \pmod{30}, \ x \equiv 19 \pmod{30}, \ x \equiv 29 \pmod{30}$$

are the required three solutions of $9x \equiv 21 \pmod{30}$.

A different approach to the problem would be to use the method that is suggested in the proof of Theorem 4-7. Since the congruence $9x \equiv 21 \pmod{30}$ is equivalent to the linear Diophantine equation

$$9x - 30y = 21,$$

we begin by expressing $3 = \gcd(9, 30)$ as a linear combination of 9 and 30. It is found, either by inspection or by using the Euclidean Algorithm, that $3 = 9(-3) + 30 \cdot 1$, so that

$$21 = 7 \cdot 3 = 9(-21) - 30(-7).$$

Thus, $x = -21, y = -7$ satisfy the Diophantine equation and, in consequence, all solutions of the congruence in question are to be found from the formula

$$x = -21 + \frac{30}{3}t = -21 + 10t.$$

The integers $x = -21 + 10t$, where $t = 0, 1, 2$ are incongruent modulo 30 (but all are congruent modulo 10); thus, we end up with the incongruent solutions

$$x \equiv -21 \pmod{30}, \ x \equiv -11 \pmod{30}, \ x \equiv -1 \pmod{30}$$

or, if one prefers positive numbers, $x \equiv 9, 19, 29 \pmod{30}$.

Having considered a single linear congruence, it is natural to turn to the problem of solving a system

$$a_1 x \equiv b_1 \ (\text{mod } m_1),\ a_2 x \equiv b_2 \ (\text{mod } m_2),\ \ldots,\ a_r x \equiv b_r \ (\text{mod } m_r)$$

of simultaneous linear congruences. We shall assume that the moduli m_k are relatively prime in pairs. Evidently, the system will admit no solution unless each individual congruence is solvable; that is, unless $d_k \mid b_k$ for each k, where $d_k = \gcd(a_k, m_k)$. When these conditions are satisfied, the factor d_k can be cancelled in the kth congruence to produce a new system having the same set of solutions as the original one,

$$a'_1 x \equiv b'_1 \ (\text{mod } n_1),\ a'_2 x \equiv b'_2 \ (\text{mod } n_2),\ \ldots,\ a'_r x \equiv b'_r \ (\text{mod } n_r),$$

where $n_k = m_k/d_k$ and $\gcd(n_i, n_j) = 1$ for $i \neq j$; also, $\gcd(a'_i, n_i) = 1$. The solutions of the individual congruences assume the form

$$x \equiv c_1 \ (\text{mod } n_1),\ x \equiv c_2 \ (\text{mod } n_2),\ \ldots,\ x \equiv c_r \ (\text{mod } n_r).$$

Thus, the problem is reduced to one of finding a simultaneous solution of a system of congruences of this simpler type.

The kind of problem that can be solved by simultaneous congruences has a long history, appearing in the Chinese literature as early as the first century A.D. Sun-Tsu asked: Find a number which leaves the remainders 2, 3, 2 when divided by 3, 5, 7, respectively. (Such mathematical puzzles are by no means confined to a single cultural sphere; indeed, the same problem occurs in the *Introductio Arithmeticae* of the Greek mathematician Nicomachus, circa 100 A.D.) In honor of their early contributions, the rule for obtaining a solution usually goes by the name of the Chinese Remainder Theorem.

THEOREM 4-8 (Chinese Remainder Theorem).
   Let $n_1, n_2, \ldots, n_r$ be positive integers such that $\gcd(n_i, n_j) = 1$ for $i \neq j$. Then the system of linear congruences

$$x \equiv a_1 \ (\text{mod } n_1),$$

$$x \equiv a_2 \ (\text{mod } n_2),$$

$$\vdots$$

$$x \equiv a_r \ (\text{mod } n_r)$$

has a simultaneous solution, which is unique modulo the integer $n_1 n_2 \ldots n_r$.

   Proof: We start by forming the product $n = n_1 n_2 \cdots n_r$. For each $k = 1, 2, \ldots, r$, let

$$N_k = n/n_k = n_1 \cdots n_{k-1} n_{k+1} \cdots n_r;$$

in other words, N_k is the product of all the integers n_i with the factor n_k omitted. By hypothesis, the n_i are relatively prime in pairs, so that $\gcd(N_k, n_k) = 1$.

According to the theory of a single linear congruence, it is therefore possible to solve the congruence $N_k x \equiv 1 \pmod{n_k}$; call the unique solution x_k. Our aim is to prove that the integer

$$\bar{x} = a_1 N_1 x_1 + a_2 N_2 x_2 + \cdots + a_r N_r x_r$$

is a simultaneous solution of the given system.

First, it is to be observed that $N_i \equiv 0 \pmod{n_k}$ for $i \neq k$, since $n_k \mid N_i$ in this case. The result is that

$$\bar{x} = a_1 N_1 x_1 + \cdots + a_r N_r x_r \equiv a_k N_k x_k \pmod{n_k}.$$

But the integer x_k was chosen to satisfy the congruence $N_k x \equiv 1 \pmod{n_k}$, which forces

$$\bar{x} \equiv a_k \cdot 1 \equiv a_k \pmod{n_k}.$$

This shows that a solution to the given system of congruences exists.

As for the uniqueness assertion, suppose that x' is any other integer which satisfies these congruences. Then

$$\bar{x} \equiv a_k \equiv x' \pmod{n_k}, \qquad\qquad k = 1, 2, \ldots, r$$

and so $n_k \mid \bar{x} - x'$ for each value of k. Because $\gcd(n_i, n_j) = 1$, Corollary 2 to Theorem 2-4 supplies us with the crucial point that $n_1 n_2 \cdots n_r \mid \bar{x} - x'$; hence $\bar{x} \equiv x' \pmod{n}$. With this, the Chinese Remainder Theorem is proven.

Example 4-8

The problem posed by Sun-Tsu corresponds to the system of three congruences

$$x \equiv 2 \pmod{3},$$
$$x \equiv 3 \pmod{5},$$
$$x \equiv 2 \pmod{7}.$$

In the notation of Theorem 4-8, we have $n = 3 \cdot 5 \cdot 7 = 105$ and

$$N_1 = n/3 = 35, \; N_2 = n/5 = 21, \; N_3 = n/7 = 15.$$

Now the linear congruences

$$35x \equiv 1 \pmod{3}, \; 21x \equiv 1 \pmod{5}, \; 15x \equiv 1 \pmod{7}$$

are satisfied by $x_1 = 2$, $x_2 = 1$, $x_3 = 1$, respectively. Thus, a solution of the system is given by

$$x = 2 \cdot 35 \cdot 2 + 3 \cdot 21 \cdot 1 + 2 \cdot 15 \cdot 1 = 233.$$

Modulo 105, we get the unique solution $x = 233 \equiv 23 \pmod{105}$.

Example 4-9

For a second illustration, let us solve the linear congruence

$$17x \equiv 9 \pmod{276}.$$

Since $276 = 3 \cdot 4 \cdot 23$, this is equivalent to finding a solution of the system of congruences

$$17x \equiv 9 \pmod 3 \qquad \text{or} \qquad x \equiv 0 \pmod 3$$
$$17x \equiv 9 \pmod 4 \qquad\qquad\qquad x \equiv 1 \pmod 4$$
$$17x \equiv 9 \pmod{23} \qquad\qquad 17x \equiv 9 \pmod{23}.$$

Note that if $x \equiv 0 \pmod 3$, then $x = 3k$ for any integer k. We substitute into the second congruence of the system and obtain

$$3k \equiv 1 \pmod 4.$$

Multiplication of both sides of this congruence by 3 gives us

$$k \equiv 9k \equiv 3 \pmod 4,$$

so that $k = 3 + 4j$, where j is an integer. Then

$$x = 3(3 + 4j) = 9 + 12j.$$

For x to satisfy the last congruence, we must have

$$17(9 + 12j) \equiv 9 \pmod{23}$$

or $204j \equiv -144 \pmod{23}$, which reduces to $3j \equiv 6 \pmod{23}$; that is, $j \equiv 2 \pmod{23}$. This yields $j = 2 + 23t$, t an integer, whence

$$x = 9 + 12(2 + 23t) = 33 + 276t.$$

All in all, $x \equiv 33 \pmod{276}$ provides a solution to the system of congruences and, in turn, a solution to $17x \equiv 9 \pmod{276}$.

PROBLEMS 4.4

1. Solve the following linear congruences:
   (a) $25x \equiv 15 \pmod{29}$.
   (b) $5x \equiv 2 \pmod{26}$.
   (c) $6x \equiv 15 \pmod{21}$.
   (d) $36x \equiv 8 \pmod{102}$.
   (e) $34x \equiv 60 \pmod{98}$.
   (f) $140x \equiv 133 \pmod{301}$. [*Hint:* gcd(140 , 301) = 7.]

2. Using congruences, solve the Diophantine equations below:
   (a) $4x + 51y = 9$. [*Hint:* $4x \equiv 9 \pmod{51}$ gives $x = 15 + 51t$, while $51y \equiv 9 \pmod 4$ gives $y = 3 + 4s$. Find the relation between s and t.]
   (b) $12x + 25y = 331$.
   (c) $5x - 53y = 17$.

3. Find all solutions of the linear congruence $3x - 7y \equiv 11 \pmod{13}$.

4. Solve each of the following sets of simultaneous congruences:
   (a) $x \equiv 1 \pmod 3$, $x \equiv 2 \pmod 5$, $x \equiv 3 \pmod 7$
   (b) $x \equiv 5 \pmod{11}$, $x \equiv 14 \pmod{29}$, $x \equiv 15 \pmod{31}$
   (c) $x \equiv 5 \pmod 6$, $x \equiv 4 \pmod{11}$, $x \equiv 3 \pmod{17}$
   (d) $2x \equiv 1 \pmod 5$, $3x \equiv 9 \pmod 6$, $4x \equiv 1 \pmod 7$, $5x \equiv 9 \pmod{11}$

5. Solve the linear congruence $17x \equiv 3 \pmod{2 \cdot 3 \cdot 5 \cdot 7}$ by solving the system

   $$17x \equiv 3 \pmod 2, \qquad 17x \equiv 3 \pmod 3,$$
   $$17x \equiv 3 \pmod 5, \qquad 17x \equiv 3 \pmod 7.$$

6. Find the smallest integer $a > 2$ such that

   $$2 \mid a, \; 3 \mid a + 1, \; 4 \mid a + 2, \; 5 \mid a + 3, \; 6 \mid a + 4.$$

7. (a) Obtain three consecutive integers each having a square factor. [*Hint:* Find an integer a such that $2^2 \mid a$, $3^2 \mid a + 1$, $5^2 \mid a + 2$.]
   (b) Obtain three consecutive integers, the first of which is divisible by a square, the second by a cube, and the third by a fourth power.

8. (Brahmagupta, 7th century A.D.) When eggs in a basket are removed 2, 3, 4, 5, 6 at a time there remain, respectively, 1, 2, 3, 4, 5 eggs. When they are taken out 7 at a time, none are left over. Find the smallest number of eggs that could have been contained in the basket.

9. The basket-of-eggs problem is often phrased in the following form: One egg remains when the eggs are removed from the basket 2, 3, 4, 5, or 6 at a time; but, no eggs remain if they are removed 7 at a time. Find the smallest number of eggs that could have been in the basket.

10. (Ancient Chinese Problem.) A band of 17 pirates stole a sack of gold coins. When they tried to divide the fortune into equal portions, 3 coins remained. In the ensuing brawl over who should get the extra coins, one pirate was killed. The wealth was redistributed, but this time an equal division left 10 coins. Again an argument developed in which another pirate was killed. But now the total fortune was evenly distributed among the survivors. What was the least number of coins that could have been stolen?

11. Prove that the congruences

    $$x \equiv a \pmod n \text{ and } x \equiv b \pmod m$$

    admit a simultaneous solution if and only if $\gcd(n, m) \mid a - b$; if a solution exists, confirm that it is unique modulo $\text{lcm}(n, m)$.

12. Use Problem 11 to show that the system

    $$x \equiv 5 \pmod 6 \text{ and } x \equiv 7 \pmod{15}$$

    does not possess a solution.

13. If $x \equiv a \pmod n$, prove that either $x \equiv a \pmod{2n}$ or $x \equiv a + n \pmod{2n}$.

14. A certain integer between 1 and 1200 leaves the remainders 1, 2, 6 when divided by 9, 11, 13 respectively. What is the integer?

15. (a) Find an integer having the remainders 1, 2, 5, 5 when divided by 2, 3, 6, 12, respectively. (Yih-hing, died 717.)

   (b) Find an integer having the remainders 2, 3, 4, 5 when divided by 3, 4, 5, 6, respectively. (Bhaskara, born 1114.)

   (c) Find an integer having the remainders 3, 11, 15 when divided by 10, 13, 17, respectively. (Regiomontanus, 1436–1473.)

16. Let t_n denote the nth triangular number. For which values of n does t_n divide

$$t_1^2 + t_2^2 + \cdots + t_n^2?$$

[*Hint:* Since $t_1^2 + t_2^2 + \cdots + t_n^2 = t_n(3n^3 + 12n^2 + 13n + 2)/30$, it suffices to determine those n satisfying $3n^3 + 12n^2 + 13n + 2 \equiv 0 \pmod{2 \cdot 3 \cdot 5}$.]

17. Find the solutions of the system of congruences

$$3x + 4y \equiv 5 \pmod{13}$$
$$2x + 5y \equiv 7 \pmod{13}.$$

18. Obtain the two incongruent solutions modulo 210 of the system

$$2x \equiv 3 \pmod 5$$
$$4x \equiv 2 \pmod 6$$
$$3x \equiv 2 \pmod 7.$$

5

Fermat's Theorem

*"And perhaps posterity will thank me for having shown it that the ancients
did not know everything."*

P. de FERMAT

5.1 PIERRE DE FERMAT

What the ancient world had known was largely forgotten during the intellectual torpor
of the Dark Ages, and it was only after the twelfth century that Western Europe again
became conscious of mathematics. The revival of classical scholarship was stimulated
by Latin translations from the Greek and, more especially, from the Arabic. The Lat-
inization of Arabic versions of Euclid's great treatise, the *Elements,* first appeared in
1120. The translation was not a faithful rendering of the *Elements,* having suffered
successive, inaccurate translations from the Greek—first into Arabic, then into Cas-
tilian, and finally into Latin—done by copyists not versed in the content of the work.
Nevertheless this much-used copy, with its accumulation of errors, served as the foundation of all editions known in Europe until 1505, when the Greek text was recovered.

With the fall of Constantinople to the Turks in 1453, the Byzantine scholars who
had served as the major custodians of mathematics brought the ancient masterpieces
of Greek learning to the West. It is reported that a copy of what survived of Diophantus' *Arithmetica* was found in the Vatican library around 1462 by Johannes Müller
(better known as Regiomontanus from the Latin name of his native town, Königsberg).
Presumably, it had been brought to Rome by the refugees from Byzantium. Regiomontanus observed that "In these books the very flower of the whole of arithmetic lies
hid," and tried to interest others in translating it. Notwithstanding the attention that
was called to the work, it remained practically a closed book until 1572 when the first
translation and printed edition was brought out by the German professor Wilhelm
Holzmann, who wrote under the Grecian form of his name, Xylander. The *Arithmetica*
became fully accessible to European mathematicians when Claude Bachet—
borrowing liberally from Xylander—published (1621) the original Greek text, along
with a Latin translation containing notes and comments. The Bachet edition probably
has the distinction of being the work that first directed the attention of Fermat to the
problems of number theory.

Pierre de Fermat

(1601–1665)

(By courtesy of Columbia University, David Eugene Smith Collection.)

Few if any periods were so fruitful for mathematics as the seventeenth century; Northern Europe alone produced as many men of outstanding ability as had appeared during the preceding millennium. At a time when such names as Desargues, Descartes, Pascal, Wallis, Bernoulli, Leibniz, and Newton were becoming famous, a certain French civil servant, Pierre de Fermat (1601–1665), stood as an equal among these brilliant scholars. Fermat, the "Prince of Amateurs," was the last great mathematician to pursue the subject as a sideline to a nonscientific career. By profession a lawyer and magistrate attached to the provincial parliament at Toulouse, he sought refuge from controversy in the abstraction of mathematics. Fermat evidently had no particular mathematical training and he evidenced no interest in its study until he was past 30; to him, it was merely a hobby to be cultivated in leisure time. Yet no practitioner of his day made greater discoveries or contributed more to the advancement of the discipline: one of the inventors of analytic geometry (the actual term was coined in the early 19th century), he laid the technical foundations of differential and integral calculus, and with Pascal established the conceptual guidelines of the theory of probability. Fermat's real love in mathematics was undoubtedly number theory, which he rescued from the realm of superstition and occultism where it had long been imprisoned. His contributions here overshadow all else; it may well be said that the revival of interest in the abstract side of number theory began with Fermat.

Fermat preferred the pleasure which he derived from mathematical research itself to any reputation that it might bring him; indeed, he published only one major manuscript during his lifetime and that just five years before his death using the concealing initials M.P.E.A.S. Adamantly refusing to put his work in finished form, he thwarted several efforts by others to make the results available in print under his name. In partial compensation for his lack of interest in publication, Fermat carried on a voluminous correspondence with contemporary mathematicians. Most of what little we know about his investigations is found in the letters to friends with whom he exchanged problems and to whom he reported his successes. They did their best to publicize Fermat's talents by passing these letters from hand to hand or by making copies, which were dispatched over the Continent.

As his parliamentary duties demanded an ever greater portion of his time, Fermat was given to inserting notes in the margin of whatever book he happened to be using. Fermat's personal copy of the Bachet edition of Diophantus held in its margin many of his famous theorems in number theory. These were discovered five years after Fermat's death by his son Samuel, who brought out a new edition of the *Arithmetica* incorporating his father's celebrated marginalia. Since there was little space available, Fermat's habit had been to jot down some result and omit all steps leading to the conclusion. Posterity has wished many times that the margins of the *Arithmetica* had been wider or that Fermat had been a little less secretive about his methods.

5.2 FERMAT'S FACTORIZATION METHOD

In a fragment of a letter, written in all probability to Father Marin Mersenne in 1643, Fermat described a technique of his for factoring large numbers. This represented the first real improvement over the classical method of attempting to find a factor of n by dividing by all primes not exceeding $\sqrt{n}$. Fermat's factorization scheme has at its heart the observation that the search for factors of an odd integer n (since powers of 2 are easily recognizable and may be removed at the outset, there is no loss in assuming that n is odd) is equivalent to obtaining integral solutions x and y of the equation

$$n = x^2 - y^2.$$

If n is the difference of two squares, then it is apparent that n can be factored as

$$n = x^2 - y^2 = (x + y)(x - y).$$

Conversely, when n has the factorization $n = ab$, with $a \geq b \geq 1$, then we may write

$$n = \left(\frac{a + b}{2} \right)^2 - \left(\frac{a - b}{2} \right)^2.$$

Moreover, because n is taken to be an odd integer, a and b are themselves odd; hence $(a + b)/2$ and $(a - b)/2$ will be nonnegative integers.

One begins the search for possible x and y satisfying the equation $n = x^2 - y^2$, or what is the same thing, the equation

$$x^2 - n = y^2$$

by first determining the smallest integer k for which $k^2 \geq n$. Now look successively at the numbers

$$k^2 - n, \ (k + 1)^2 - n, \ (k + 2)^2 - n, \ (k + 3)^2 - n, \ \ldots$$

until a value of $m \geq \sqrt{n}$ is found making $m^2 - n$ a square. The process cannot go on indefinitely, since we eventually arrive at

$$\left(\frac{n + 1}{2} \right)^2 - n = \left(\frac{n - 1}{2} \right)^2,$$

the representation of n corresponding to the trivial factorization $n = n \cdot 1$. If this point is reached without a square difference having been discovered earlier, then n has no factors other than n and 1, in which case it is a prime.

Fermat used the procedure just described to factor

$$2027651281 = 44021 \cdot 46061$$

in only 11 steps, as compared to making 4850 divisions by the odd primes up to 44021. This was probably a favorable case devised on purpose to show the chief virtue of his method: it does not require one to know all the primes less than $\sqrt{n}$ in order to find factors of n.

Example 5-1

To illustrate the application of Fermat's method, let us factor the integer $n = 119143$. From a table of squares, we find that $345^2 < 119143 < 346^2$; thus it suffices to consider values of $k^2 - 119143$ for k in the range $346 \leq k < (119143 + 1)/2 = 59572$. The calculations begin as follows:

$$346^2 - 119143 = 119716 - 119143 = 573,$$
$$347^2 - 119143 = 120409 - 119143 = 1266,$$
$$348^2 - 119143 = 121104 - 119143 = 1961,$$
$$349^2 - 119143 = 121801 - 119143 = 2658,$$
$$350^2 - 119143 = 122500 - 119143 = 3357,$$
$$351^2 - 119143 = 123201 - 119143 = 4058,$$
$$352^2 - 119143 = 123904 - 119143 = 4761 = 69^2.$$

This last line exhibits the factorization

$$119143 = 352^2 - 69^2 = (352 + 69)(352 - 69) = 421 \cdot 283,$$

the two factors themselves being prime. In only seven trials, we have obtained the prime factorization of the number 119143. Of course, one does not always fare so luckily; it may take many steps before a difference turns out to be a square.

Fermat's method is most effective when the two factors of n are of nearly the same magnitude, for in this case a suitable square will appear quickly. To illustrate, let us suppose that $n = 23449$ is to be factored. The smallest square exceeding n is 154^2, so that the sequence $k^2 - n$ starts with

$$154^2 - 23449 = 23716 - 23449 = 267,$$
$$155^2 - 23449 = 24025 - 23449 = 576 = 24^2.$$

Hence, factors of 23449 are

$$23449 = (155 + 24)(155 - 24) = 179 \cdot 131.$$

When examining the differences $k^2 - n$ as possible squares, many values can be immediately excluded by inspection of the final digits. We know, for instance, that a square must end in one of the six digits 0, 1, 4, 5, 6, 9 (Problem 1a, Section 4.3). This allows us to exclude all values in the above example, save for 1266, 1961, and 4761.

By calculating the squares of the integers from 0 to 99 modulo 100, one sees further that, for a square, the last two digits are limited to the following twenty-two possibilities:

00	21	41	64	89
01	24	44	69	96
04	25	49	76	
09	29	56	81	
16	36	61	84	

The integer 1266 can be eliminated from consideration in this way. Since 61 is among the last two digits allowable in a square, it is only necessary to look at the numbers 1961 and 4761; the former is not a square, but $4761 = 69^2$.

There is a generalization of Fermat's factorization method that has been used with some success. Here, we look for distinct integers x and y such that $x^2 - y^2$ is a multiple of n rather than n itself; that is,

$$x^2 \equiv y^2 \pmod{n}.$$

Having obtained such integers, then $d = \gcd(x - y, n)$ (or $d = \gcd(x + y, n)$) can be calculated by means of the Euclidean Algorithm. Clearly d is a divisor of n, but is it a nontrivial divisor? In other words, do we have $1 < d < n$?

In practice, n is usually the product of two primes p and q, with $p < q$, so that d is equal to 1, p, q, or pq. Now the congruence $x^2 \equiv y^2 \pmod{n}$ translates into $pq \mid (x - y)(x + y)$. Euclid's Lemma tells us that p and q must divide one of the factors. If it happened that $p \mid x - y$ and $q \mid x - y$, then $pq \mid x - y$, or expressed as a congruence $x \equiv y \pmod{n}$. Also, $p \mid x + y$ and $q \mid x + y$ yields $x \equiv -y \pmod{n}$. By seeking integers x and y satisfying $x^2 \equiv y^2 \pmod{n}$, where $x \not\equiv \pm y \pmod{n}$, these two situations are ruled out. The result of all this is that d is either p or q, giving us a nontrivial divisor of n.

Example 5-2

Suppose we wish to factor the positive integer $n = 2189$ and happen to notice that $579^2 \equiv 18^2 \pmod{2189}$. Then we compute

$$\gcd(579 - 18, 2189) = \gcd(561, 2189) = 11$$

using the Euclidean Algorithm:

$$2189 = 3 \cdot 561 + 506$$
$$561 = 1 \cdot 506 + 55$$
$$506 = 9 \cdot 55 + 11$$
$$55 = 5 \cdot 11.$$

This leads to the prime divisor 11 of 2189. The other factor, namely 199, can be obtained by observing that

$$\gcd(579 + 18, 2189) = \gcd(597, 2189) = 199.$$

The reader might wonder how we ever arrived at a number, such as 579, whose square modulo 2189 also turns out to be a perfect square. In looking for squares close to multiples of 2189, it was observed that

$$81^2 - 3 \cdot 2189 = 6 \text{ and } 155^2 - 11 \cdot 2189 = 54,$$

which translates into

$$81^2 \equiv 2 \cdot 3 \pmod{2189} \text{ and } 155^2 \equiv 2 \cdot 3^3 \pmod{2189}.$$

When these congruences are multiplied, they produce

$$(81 \cdot 155)^2 \equiv (2 \cdot 3^2)^2 \pmod{2189}.$$

Because the product $81 \cdot 155 = 12555 \equiv -579 \pmod{2189}$, we ended up with the congruence $579^2 \equiv 18^2 \pmod{2189}$.

The basis of our approach is to find several x_i which have the property that each x_i^2 is, modulo n, the product of small prime powers; and such that their product's square is congruent to a perfect square.

When n has more than two prime factors, our factorization algorithm may still be applied; but there is no guarantee that a particular solution of $x^2 \equiv y^2 \pmod{n}$, with $x \not\equiv \pm y \pmod{n}$, will result in a nontrivial divisor of n. Of course the more solutions of this congruence that are available, the better the chance of finding the desired factors of n.

PROBLEMS 5.2

1. Use Fermat's method to factor
    (a) 2279;
    (b) 10541;
    (c) 340663. [*Hint:* The smallest square just exceeding 340663 is 584^2.]

2. Prove that a perfect square must end in one of the following pairs of digits: 00, 01, 04, 09, 16, 21, 24, 25, 29, 36, 41, 44, 49, 56, 61, 64, 69, 76, 81, 84, 89, 96. [*Hint:* Since $x^2 \equiv (50 + x)^2 \pmod{100}$ and $x^2 \equiv (50 - x)^2 \pmod{100}$, it suffices to examine the final digits of x^2 for the 26 values $x = 0, 1, 2, \ldots, 25$.]

3. Factor the number $2^{11} - 1$ by Fermat's factorization method.

4. In 1647, Mersenne noted that when a number can be written as a sum of two relatively prime squares in two distinct ways, it is composite and can be factored as follows: if $n = a^2 + b^2 = c^2 + d^2$, then

    $$n = (ac + bd)(ac - bd)/(a + d)(a - d).$$

    Use this result to factor the numbers

    $$493 = 18^2 + 13^2 = 22^2 + 3^2,$$
    and $\qquad\qquad 38025 = 168^2 + 99^2 = 156^2 + 117^2.$

5. Employ the generalized Fermat method to factor the numbers below:
    (a) 2911 [*Hint:* $138^2 \equiv 67^2 \pmod{2911}$.]
    (b) 4573 [*Hint:* $177^2 \equiv 92^2 \pmod{4573}$.]
    (c) 6923 [*Hint:* $208^2 \equiv 93^2 \pmod{6923}$.]

6. Factor 13561 with the help of the congruences

    $$233^2 \equiv 3^2 \cdot 5 \pmod{13561} \text{ and } 1281^2 \equiv 2^4 \cdot 5 \pmod{13561}.$$

7.   (a)   Factor the number 4537 by searching for x such that

$$x^2 - k \cdot 4537$$

is the product of small prime powers.

(b)   Use the procedure indicated in part (a) to factor 14429. [*Hint:* $120^2 - 14429 = -29$ and $3003^2 - 625 \cdot 14429 = -116$.]

5.3 THE LITTLE THEOREM

The most significant of Fermat's correspondents in number theory was Bernhard Frénicle de Bessy (1605–1675), an official at the French mint who was renowned for his gift of manipulating large numbers. (Frénicle's facility in numerical calculation is revealed by the following incident: On hearing that Fermat had proposed the problem of finding cubes which when increased by their proper divisors become squares, as is the case with $7^3 + (1 + 7 + 7^2) = 20^2$, he immediately gave four different solutions; and supplied six more the next day.) Though in no way Fermat's equal as a mathematician, Frénicle alone among his contemporaries could challenge him in number theory and his challenges had the distinction of coaxing out of Fermat some of his carefully guarded secrets. One of the most striking is the theorem which states: If p is a prime and a is any integer not divisible by p, then p divides $a^{p-1} - 1$. Fermat communicated the result in a letter to Frénicle dated October 18, 1640, along with the comment, "I would send you the demonstration, if I did not fear its being too long." This theorem has since become known as "Fermat's Little Theorem," or just "Fermat's Theorem," to distinguish it from Fermat's "Great" or "Last Theorem," which is the subject of Chapter 11. Almost 100 years were to elapse before Euler published the first proof of the Little Theorem in 1736. Leibniz, however, seems not to have received his share of recognition; for he left an identical argument in an unpublished manuscript sometime before 1683.

We now proceed to a proof of Fermat's Theorem.

THEOREM 5-1 (Fermat's Theorem).

If p is a prime and $p \nmid a$, then $a^{p-1} \equiv 1 \pmod{p}$.

Proof: We begin by considering the first $p - 1$ positive multiples of a; that is, the integers

$$a, 2a, 3a, \ldots, (p-1)a.$$

None of these numbers is congruent modulo p to any other, nor is any congruent to zero. Indeed, if it happened that

$$ra \equiv sa \pmod{p}, \qquad 1 \leq r < s \leq p - 1$$

then a could be cancelled to give $r \equiv s \pmod{p}$, which is impossible. Therefore, the above set of integers must be congruent modulo p to $1, 2, 3, \ldots, p - 1$, taken in some order. Multiplying all these congruences together, we find that

$$a \cdot 2a \cdot 3a \cdots (p-1)a \equiv 1 \cdot 2 \cdot 3 \cdots (p-1)\pmod{p},$$

whence

$$a^{p-1}(p-1)! \equiv (p-1)! \pmod{p}.$$

Once $(p - 1)!$ is cancelled from both sides of the preceding congruence (this is possible since $p \nmid (p - 1)!$), our line of reasoning culminates in the statement $a^{p-1} \equiv 1 \pmod{p}$, which is Fermat's Theorem.

This result can be stated in a slightly more general way in which the requirement that $p \nmid a$ is dropped.

COROLLARY.

If p is a prime, then $a^p \equiv a \pmod{p}$ for any integer a.

Proof: When $p \mid a$, the statement obviously holds; for, in this setting, $a^p \equiv 0 \equiv a \pmod{p}$. If $p \nmid a$, then in accordance with Fermat's Theorem, we have $a^{p-1} \equiv 1 \pmod{p}$. When this congruence is multiplied by a, the conclusion $a^p \equiv a \pmod{p}$ follows.

There is a different proof of the fact that $a^p \equiv a \pmod{p}$, involving induction on a. If $a = 1$, the assertion is that $1^p \equiv 1 \pmod{p}$, which is clearly true, as is the case $a = 0$. Assuming that the result holds for a, we must confirm its validity for $a + 1$. In light of the Binomial Theorem,

$$(a + 1)^p = a^p + \binom{p}{1}a^{p-1} + \cdots + \binom{p}{k}a^{p-k} + \cdots + \binom{p}{p-1}a + 1,$$

where the coefficient $\binom{p}{k}$ is given by

$$\binom{p}{k} = \frac{p!}{k!(p-k)!} = \frac{p(p-1)\cdots(p-k+1)}{1 \cdot 2 \cdot 3 \cdots k}.$$

Our argument hinges on the observation that $\binom{p}{k} \equiv 0 \pmod{p}$ for $1 \le k \le p - 1$. To see this, note that

$$k!\binom{p}{k} = p(p-1) \cdots (p-k+1) \equiv 0 \pmod{p},$$

by virtue of which $p \mid k!$ or $p \mid \binom{p}{k}$. But $p \mid k!$ implies that $p \mid j$ for some j satisfying $1 \le j \le k \le p - 1$, an absurdity. Therefore, $p \mid \binom{p}{k}$ or, converting to a congruence statement,

$$\binom{p}{k} \equiv 0 \pmod{p}.$$

The point we wish to make is that

$$(a + 1)^p \equiv a^p + 1 \equiv a + 1 \pmod{p},$$

where the right-most congruence uses our inductive assumption. Thus, the desired conclusion holds for $a + 1$ and, in consequence, for all $a \ge 0$. If a is a negative integer, there is no problem: since $a \equiv r \pmod{p}$ for some r, where $0 \le r \le p - 1$, we get $a^p \equiv r^p \equiv r \equiv a \pmod{p}$.

Fermat's Theorem has many applications and is central to much of what is done in number theory. On one hand, it can be a labor-saving device in certain calculations. If asked to verify that $5^{38} \equiv 4 \pmod{11}$, for instance, we would take the congruence $5^{10} \equiv 1 \pmod{11}$ as our starting point. Knowing this,

$$5^{38} = 5^{10 \cdot 3 + 8} = (5^{10})^3 (5^2)^4$$
$$\equiv 1^3 \cdot 3^4 \equiv 81 \equiv 4 \pmod{11},$$

as desired.

Another use of Fermat's Theorem is as a tool in testing the primality of a given integer n. For, if it could be shown that the congruence

$$a^n \equiv a \pmod{n}$$

fails to hold for some choice of a, then n is necessarily composite. As an example of this approach, let us look at $n = 117$. The computation is kept under control by selecting a small integer for a; say, $a = 2$. Since 2^{117} may be written as

$$2^{117} = 2^{7 \cdot 16 + 5} = (2^7)^{16} 2^5$$

and $2^7 = 128 \equiv 11 \pmod{117}$, we have

$$2^{117} \equiv 11^{16} \cdot 2^5 \equiv (121)^8 2^5 \equiv 4^8 \cdot 2^5 \equiv 2^{21} \pmod{117}.$$

But $2^{21} = (2^7)^3$, which leads to

$$2^{21} \equiv 11^3 \equiv 121 \cdot 11 \equiv 4 \cdot 11 \equiv 44 \pmod{117}.$$

Combining these congruences, we finally obtain

$$2^{117} \equiv 44 \not\equiv 2 \pmod{117},$$

so that 117 must be composite; actually, $117 = 13 \cdot 9$.

It might be worthwhile to give an example illustrating the failure of the converse of Fermat's Theorem to hold; in other words, to show that if $a^{n-1} \equiv 1 \pmod{n}$ for some integer a, then n need not be prime. As a prelude we require a technical lemma:

LEMMA.

If p and q are distinct primes such that $a^p \equiv a \pmod{q}$ and $a^q \equiv a \pmod{p}$, then $a^{pq} \equiv a \pmod{pq}$.

Proof: It is known from the last corollary that $(a^q)^p \equiv a^q \pmod{p}$, while $a^q \equiv a \pmod{p}$ holds by hypothesis. Combining these congruences, we obtain $a^{pq} \equiv a \pmod{p}$ or, in different terms, $p \mid a^{pq} - a$. In an entirely similar manner, $q \mid a^{pq} - a$. The corollary to Theorem 2-4 now yields $pq \mid a^{pq} - a$, which can be recast as $a^{pq} \equiv a \pmod{pq}$.

Our contention is that $2^{340} \equiv 1 \pmod{341}$ where $341 = 11 \cdot 31$. In working towards this end, notice that $2^{10} = 1024 = 31 \cdot 33 + 1$. Thus,

$$2^{11} = 2 \cdot 2^{10} \equiv 2 \cdot 1 \equiv 2 \pmod{31}$$

and

$$2^{31} = 2 (2^{10})^3 \equiv 2 \cdot 1^3 \equiv 2 \pmod{11}.$$

Exploiting the lemma,

$$2^{11 \cdot 31} \equiv 2 \;(\mathrm{mod}\; 11 \cdot 31)$$

or $2^{341} \equiv 2 \;(\mathrm{mod}\; 341)$. After cancelling a factor of 2, we pass to

$$2^{340} \equiv 1 \;(\mathrm{mod}\; 341),$$

so that the converse to Fermat's Theorem is false.

The historical interest in numbers of the form $2^n - 2$ resides in the claim made by the Chinese mathematicians over 25 centuries ago that n is prime if and only if $n \mid 2^n - 2$ (in point of fact, this criterion is reliable for all integers $n \leq 340$). Needless to say, our example, where $341 \mid 2^{341} - 2$ although $341 = 11 \cdot 31$, lays the conjecture to rest; this was discovered in the year 1819. The situation in which $n \mid 2^n - 2$ occurs often enough to merit a name though: call a composite integer n *pseudoprime* whenever $n \mid 2^n - 2$. It can be shown that there are infinitely many pseudoprimes, the smallest four being 341, 561, 645, and 1105.

The following theorem allows us to construct an increasing sequence of pseudoprimes.

THEOREM 5-2.

If n is an odd pseudoprime, then $M_n = 2^n - 1$ is a larger one.

Proof: Since n is a composite number, we can write $n = rs$, with $1 < r \leq s < n$. Then, according to Problem 21, Section 2.2, $2^r - 1 \mid 2^n - 1$, or equivalently $2^r - 1 \mid M_n$, making M_n composite. By our hypotheses, $2^n \equiv 2 \;(\mathrm{mod}\; n)$, and hence $2^n - 2 = kn$ for some integer k. It follows that

$$2^{M_n - 1} = 2^{2^n - 2} = 2^{kn}.$$

This yields

$$\begin{aligned}
2^{M_n - 1} - 1 &= 2^{kn} - 1 \\
&= (2^n - 1)(2^{n(k-1)} + 2^{n(k-2)} + \cdots + 2^n + 1) \\
&= M_n(2^{n(k-1)} + 2^{n(k-2)} + \cdots + 2^n + 1) \\
&\equiv 0 \;(\mathrm{mod}\; M_n).
\end{aligned}$$

We see immediately that $2^{M_n} - 2 \equiv 0 \;(\mathrm{mod}\; M_n)$, in light of which M_n is a pseudoprime.

In analogy with Dirichlet's Theorem, it has been shown (1963) that any arithmetic progression $an + b$ $(n = 1, 2, \ldots)$ with $\gcd(a, b) = 1$ contains infinitely many pseudoprimes. These "false primes" are much rarer than actual primes; for instance, there are only 245 pseudoprimes smaller than one million, in comparison with 78492 primes. The first example of an even pseudoprime, namely the number

$$161038 = 2 \cdot 73 \cdot 1103,$$

was found in 1950.

There exist composite numbers, n, with the property that $a^n \equiv a \;(\mathrm{mod}\; n)$ for all integers a. The least such n is 561. These exceptional numbers are called *absolute pseudoprimes* or *Carmichael numbers,* for R. C. Carmichael, who was the first (1909) to notice their existence. To see that $561 = 3 \cdot 11 \cdot 17$ is an absolute pseudoprime, notice that $\gcd(a, 561) = 1$ gives $\gcd(a, 3) = \gcd(a, 11) = \gcd(a, 17) = 1$. An application of Fermat's Theorem leads to the congruences

$$a^2 \equiv 1 \ (\text{mod } 3), \ a^{10} \equiv 1 \ (\text{mod } 11), \ a^{16} \equiv 1 \ (\text{mod } 17),$$

and, in turn, to

$$a^{560} \equiv (a^2)^{280} \equiv 1 \ (\text{mod } 3),$$

$$a^{560} \equiv (a^{10})^{56} \equiv 1 \ (\text{mod } 11),$$

$$a^{560} \equiv (a^{16})^{35} \equiv 1 \ (\text{mod } 17).$$

These give rise to the single congruence $a^{560} \equiv 1 \ (\text{mod } 561)$, where $\gcd(a, 561) = 1$. But then $a^{561} \equiv a \ (\text{mod } 561)$ for all a, showing 561 to be an absolute pseudoprime.

We next present a theorem which furnishes a means for producing absolute pseudoprimes.

THEOREM 5-3.

Let n be a composite square-free integer; say, $n = p_1 p_2 \cdots p_r$, where the p_i are distinct primes. If $p_i - 1 \mid n - 1$ for $i = 1, 2, \cdots, r$, then n is an absolute pseudoprime.

Proof: Suppose that a is an integer satisfying $\gcd(a, n) = 1$, so that $\gcd(a, p_i) = 1$ for each i. Then Fermat's Theorem yields $p_i \mid a^{p_i - 1} - 1$. From the hypothesis $p_i - 1 \mid n - 1$, we have $p_i \mid a^{n-1} - 1$, and therefore $p_i \mid a^n - a$ for all a and $i = 1, 2, \cdots, r$. As a result of the corollary to Theorem 2-4, we end up with $n \mid a^n - a$, which makes n an absolute pseudoprime.

Examples of integers which satisfy the conditions of Theorem 5-3 are

$$1729 = 7 \cdot 13 \cdot 19, \ 6601 = 7 \cdot 23 \cdot 41, \text{ and } 10585 = 5 \cdot 29 \cdot 73.$$

It was proven in 1992 that infinitely many absolute pseudoprimes exist; but they are fairly rare. There are just 43 of them less than one million and 105,212 less than 10^{15}.

PROBLEMS 5.3

1. Use Fermat's Theorem to verify that 17 divides $11^{204} + 1$.

2. (a) If $\gcd(a, 35) = 1$, show that $a^{12} \equiv 1 \ (\text{mod } 35)$. [*Hint:* From Fermat's Theorem $a^6 \equiv 1 \ (\text{mod } 7)$ and $a^4 \equiv 1 \ (\text{mod } 5)$.]
    (b) If $\gcd(a, 42) = 1$, show that $168 = 3 \cdot 7 \cdot 8$ divides $a^6 - 1$.
    (c) If $\gcd(a, 133) = \gcd(b, 133) = 1$, show that $133 \mid a^{18} - b^{18}$.

3. From Fermat's Theorem deduce that, for any integer $n \geq 0$, $13 \mid 11^{12n+6} + 1$.

4. Derive each of the following congruences:
    (a) $a^{21} \equiv a \ (\text{mod } 15)$ for all a. [*Hint:* By Fermat's Theorem, $a^5 \equiv a \ (\text{mod } 5)$.]
    (b) $a^7 \equiv a \ (\text{mod } 42)$ for all a.
    (c) $a^{13} \equiv a \ (\text{mod } 3 \cdot 7 \cdot 13)$ for all a.
    (d) $a^9 \equiv a \ (\text{mod } 30)$ for all a.

5. If $\gcd(a, 30) = 1$, show that 60 divides $a^4 + 59$.

6. (a) Find the units digit of 3^{100} by the use of Fermat's Theorem.
    (b) For any integer a, verify that a^5 and a have the same units digit.

7. If $7 \nmid a$, prove that either $a^3 + 1$ or $a^3 - 1$ is divisible by 7. [*Hint:* Apply Fermat's Theorem.]

8. The three most recent appearances of Halley's comet were in the years 1835, 1910, and 1986; the next occurrence will be in 2061. Prove that

$$1835^{1910} + 1986^{2061} \equiv 0 \pmod 7.$$

9. (a) Let p be a prime and $\gcd(a, p) = 1$. Use Fermat's Theorem to verify that $x \equiv a^{p-2}b \pmod p$ is a solution of the linear congruence $ax \equiv b \pmod p$.
 (b) By applying part (a), solve the congruences $2x \equiv 1 \pmod{31}$, $6x \equiv 5 \pmod{11}$, and $3x \equiv 17 \pmod{29}$.

10. Assuming that a and b are integers not divisible by the prime p, establish the following:
 (a) If $a^p \equiv b^p \pmod p$, then $a \equiv b \pmod p$.
 (b) If $a^p \equiv b^p \pmod p$, then $a^p \equiv b^p \pmod{p^2}$. [*Hint:* By (a), $a = b + pk$ for some k, so that $a^p - b^p = (b + pk)^p - b^p$; now show that p^2 divides the latter expression.]

11. Employ Fermat's Theorem to prove that, if p is an odd prime, then
 (a) $1^{p-1} + 2^{p-1} + 3^{p-1} + \cdots + (p-1)^{p-1} \equiv -1 \pmod p$.
 (b) $1^p + 2^p + 3^p + \cdots + (p-1)^p \equiv 0 \pmod p$.
 [*Hint:* Recall the identity $1 + 2 + 3 + \cdots + (p-1) = p(p-1)/2$.]

12. Prove that if p is an odd prime and k is an integer satisfying $1 \le k \le p - 1$, then the binomial coefficient

$$\binom{p-1}{k} \equiv (-1)^k \pmod p.$$

13. Assume that p and q are distinct odd primes such that $p - 1 \mid q - 1$. If $\gcd(a, pq) = 1$, show that $a^{q-1} \equiv 1 \pmod{pq}$.

14. If p and q are distinct primes, prove that

$$p^{q-1} + q^{p-1} \equiv 1 \pmod{pq}.$$

15. Establish the statements below:
 (a) If the number $M_p = 2^p - 1$ is composite, where p is a prime, then M_p is a pseudoprime.
 (b) Every composite number $F_n = 2^{2^n} + 1$ is a pseudoprime ($n = 0, 1, 2 \ldots$). [*Hint:* By Problem 21, Section 2.2, $2^{n+1} \mid 2^{2^n}$ implies that $2^{2^{n+1}} - 1 \mid 2^{F_n - 1} - 1$; but $F_n \mid 2^{2^{n+1}} - 1$.]

16. Confirm that the following integers are absolute pseudoprimes:
 (a) $1105 = 5 \cdot 13 \cdot 17$,
 (b) $2821 = 7 \cdot 13 \cdot 31$,
 (c) $2465 = 5 \cdot 17 \cdot 29$.

17. Show that the pseudoprime 341 is not an absolute pseudoprime by showing that $11^{341} \not\equiv 11 \pmod{341}$. [*Hint:* $31 \nmid 11^{341} - 11$.]

18. (a) When $n = 2p$, where p is an odd prime, prove that $a^{n-1} \equiv a \pmod n$ for any integer a.
 (b) For $n = 195 = 3 \cdot 5 \cdot 13$, verify that $a^{n-2} \equiv a \pmod n$ for any integer a.

19. Prove that any integer of the form

$$n = (6k + 1)(12k + 1)(18k + 1)$$

is an absolute pseudoprime if all three factors are prime; hence, $1729 = 7 \cdot 13 \cdot 19$ is an absolute pseudoprime.

20. Show that $561 \mid 2^{561} - 2$ and $561 \mid 3^{561} - 3$. It is an unanswered question whether there exist infinitely many composite numbers n with the property that $n \mid 2^n - 2$ and $n \mid 3^n - 3$.

21. Establish the congruence

$$2222^{5555} + 5555^{2222} \equiv 0 \pmod 7.$$

[*Hint:* First evaluate 1111 modulo 7.]

5.4 WILSON'S THEOREM

We now turn to another milestone in the development of number theory. In his *Meditationes Algebraicae* of 1770, the English mathematician Edward Waring (1741–1793) announced several new theorems. Foremost among these is an interesting property of primes reported to him by one of his former students, a certain John Wilson. The property is the following: if p is a prime number, then p divides $(p - 1)! + 1$. Wilson appears to have guessed this on the basis of numerical computations; at any rate, neither he nor Waring knew how to prove it. Confessing his inability to supply a demonstration, Waring added, "Theorems of this kind will be very hard to prove, because of the absence of a notation to express prime numbers." (Reading the passage, Gauss uttered his telling comment on "notationes versus notiones," implying that in questions of this nature it was the notion that really mattered, not the notation.) Despite Waring's pessimistic forecast, Lagrange soon afterwards (1771) gave a proof of what in the literature is called "Wilson's Theorem" and observed that the converse also holds. It would be perhaps more just to name the theorem after Leibniz, for there is evidence that he was aware of the result almost a century earlier, but published nothing upon the subject.

Now to a proof of Wilson's Theorem.

THEOREM 5-4 (Wilson).

If p is a prime, then

$$(p - 1)! \equiv -1 \pmod p.$$

Proof: Dismissing the cases $p = 2$ and $p = 3$ as being evident, let us take $p > 3$. Suppose that a is any one of the $p - 1$ positive integers

$$1, 2, 3, \ldots, p - 1$$

and consider the linear congruence $ax \equiv 1 \pmod p$. Then $\gcd(a, p) = 1$. By Theorem 4-7, this congruence admits a unique solution modulo p; hence, there is a unique integer a', with $1 \le a' \le p - 1$, satisfying $aa' \equiv 1 \pmod p$.

Since p is prime, $a = a'$ if and only if $a = 1$ or $a = p - 1$. Indeed, the congruence $a^2 \equiv 1 \pmod p$ is equivalent to $(a - 1) \cdot (a + 1) \equiv 0 \pmod p$. Therefore, either $a - 1 \equiv 0 \pmod p$, in which case $a = 1$, or else $a + 1 \equiv 0 \pmod p$, in which case $a = p - 1$.

If we omit the numbers 1 and $p - 1$, the effect is to group the remaining integers $2, 3, \ldots, p - 2$ into pairs a, a', where $a \neq a'$, such that their product $aa' \equiv 1 \pmod{p}$. When these $(p - 3)/2$ congruences are multiplied together and the factors rearranged, we get

$$2 \cdot 3 \cdots (p - 2) \equiv 1 \pmod{p}$$

or rather

$$(p - 2)! \equiv 1 \pmod{p}.$$

Now multiply by $p - 1$ to obtain the congruence

$$(p - 1)! \equiv p - 1 \equiv -1 \pmod{p},$$

as was to be proved.

Example 5-3

A concrete example should help to clarify the proof of Wilson's Theorem. Specifically, let us take $p = 13$. It is possible to divide the integers $2, 3, \ldots, 11$ into $(p - 3)/2 = 5$ pairs each of whose products is congruent to 1 modulo 13. To write these congruences out explicitly:

$$2 \cdot 7 \equiv 1 \pmod{13},$$
$$3 \cdot 9 \equiv 1 \pmod{13},$$
$$4 \cdot 10 \equiv 1 \pmod{13},$$
$$5 \cdot 8 \equiv 1 \pmod{13},$$
$$6 \cdot 11 \equiv 1 \pmod{13}.$$

Multiplying the above congruences gives the result

$$11! = (2 \cdot 7)(3 \cdot 9)(4 \cdot 10)(5 \cdot 8)(6 \cdot 11) \equiv 1 \pmod{13}$$

and so

$$12! \equiv 12 \equiv -1 \pmod{13}.$$

Thus, $(p - 1)! \equiv -1 \pmod{p}$, with $p = 13$.

The converse of Wilson's Theorem is also true: If $(n - 1)! \equiv -1 \pmod{n}$, then n must be prime. For, if n is not a prime, then n has a divisor d with $1 < d < n$. Furthermore, since $d \leq n - 1$, d occurs as one of the factors in $(n - 1)!$, whence $d \mid (n - 1)!$. Now we are assuming that $n \mid (n - 1)! + 1$, and so $d \mid (n - 1)! + 1$ too. The conclusion is that $d \mid 1$, which is nonsense.

Taken together, Wilson's Theorem and its converse provide a necessary and sufficient condition for determining primality; namely, an integer $n > 1$ is prime if and only if $(n - 1)! \equiv -1 \pmod{n}$. Unfortunately, this test is of more theoretical than practical interest since as n increases, $(n - 1)!$ rapidly becomes unmanageable in size.

We would like to close this chapter with an application of Wilson's Theorem to the study of quadratic congruences. [It is understood that *quadratic congruence* means a congruence of the form $ax^2 + bx + c \equiv 0 \pmod{n}$, with $a \not\equiv 0 \pmod{n}$.] This is the content of

THEOREM 5-5.

The quadratic congruence $x^2 + 1 \equiv 0 \pmod{p}$, where p is an odd prime, has a solution if and only if $p \equiv 1 \pmod 4$.

Proof: Let a be any solution of $x^2 + 1 \equiv 0 \pmod{p}$, so that $a^2 \equiv -1 \pmod{p}$. Since $p \nmid a$, the outcome of applying Fermat's Theorem is:

$$1 \equiv a^{p-1} \equiv (a^2)^{(p-1)/2} \equiv (-1)^{(p-1)/2} \pmod{p}.$$

The possibility that $p = 4k + 3$ for some k does not arise. If it did, we would have

$$(-1)^{(p-1)/2} = (-1)^{2k+1} = -1;$$

hence $1 \equiv -1 \pmod{p}$. The net result of this is that $p \mid 2$, which is patently false. Therefore, p must be of the form $4k + 1$.

Now for the opposite direction. In the product

$$(p - 1)! = 1 \cdot 2 \cdots \frac{p-1}{2} \cdot \frac{p+1}{2} \cdots (p - 2)(p - 1),$$

we have the congruences

$$p - 1 \equiv -1 \pmod{p},$$
$$p - 2 \equiv -2 \pmod{p},$$
$$\vdots$$
$$\frac{p+1}{2} \equiv -\frac{p-1}{2} \pmod{p}.$$

Rearranging the factors produces

$$(p - 1)! \equiv 1 \cdot (-1) \cdot 2 \cdot (-2) \cdots \frac{p-1}{2} \cdot \left(-\frac{p-1}{2}\right) \pmod{p}$$
$$\equiv (-1)^{(p-1)/2} \left(1 \cdot 2 \cdots \frac{p-1}{2}\right)^2 \pmod{p},$$

since there are $(p - 1)/2$ minus signs involved. It is at this point that Wilson's Theorem can be brought to bear; for, $(p - 1)! \equiv -1 \pmod{p}$, whence

$$-1 \equiv (-1)^{(p-1)/2}\left[\left(\frac{p-1}{2}\right)!\right]^2 \pmod{p}.$$

If we assume that p is of the form $4k + 1$, then $(-1)^{(p-1)/2} = 1$, leaving us with the congruence

$$-1 \equiv \left[\left(\frac{p-1}{2}\right)!\right]^2 \pmod{p}.$$

The conclusion: $[(p-1)/2]!$ satisfies the quadratic congruence $x^2 + 1 \equiv 0 \pmod{p}$.

Let us take a look at an actual example; say, the case $p = 13$, which is a prime of the form $4k + 1$. Here, we have $(p-1)/2 = 6$ and it is easy to see that

$$6! = 720 \equiv 5 \pmod{13},$$

while

$$5^2 + 1 = 26 \equiv 0 \pmod{13}.$$

Thus the assertion that $\left[\left(\frac{p-1}{2}\right)!\right]^2 + 1 \equiv 0 \pmod{p}$ is correct for $p = 13$.

Wilson's Theorem implies that there exists an infinitude of composite numbers of the form $n! + 1$. On the other hand, it is an open question whether $n! + 1$ is prime for infinitely many values of n. The only values of n in the range $1 \le n \le 100$ for which $n! + 1$ is known to be a prime number are $n = 1, 2, 3, 11, 27, 37, 41, 73$, and 77. Currently the largest prime of the form $n! + 1$ is $1477! + 1$, discovered in 1984.

PROBLEMS 5.4

1. (a) Find the remainder when 15! is divided by 17.
   (b) Find the remainder when 2(26!) is divided by 29.

2. Determine whether 17 is a prime by deciding whether or not $16! \equiv -1 \pmod{17}$.

3. Arrange the integers $2, 3, 4, \ldots, 21$ in pairs a and b which satisfy $ab \equiv 1 \pmod{23}$.

4. Show that $18! \equiv -1 \pmod{437}$.

5. (a) Prove that an integer $n > 1$ is prime if and only if $(n-2)! \equiv 1 \pmod{n}$.
   (b) If n is a composite integer, show that $(n-1)! \equiv 0 \pmod{n}$, except when $n = 4$.

6. Given a prime number p, establish the congruence

   $$(p-1)! \equiv p - 1 \pmod{1 + 2 + 3 + \cdots + (p-1)}.$$

7. If p is a prime, prove that

   $$p \mid a^p + (p-1)!a \quad \text{and} \quad p \mid (p-1)!a^p + a$$

   for any integer a. [*Hint:* By Wilson's Theorem, $a^p + (p-1)!a \equiv a^p - a \pmod{p}$.]

8. Find two odd primes $p \le 13$ for which the congruence $(p-1)! \equiv -1 \pmod{p^2}$ holds.

9. Using Wilson's Theorem, prove that

$$1^2 \cdot 3^2 \cdot 5^2 \cdots (p-2)^2 \equiv (-1)^{(p+1)/2} \pmod{p}$$

for any odd prime p. [*Hint:* Since $k \equiv -(p-k) \pmod{p}$, it follows that $2 \cdot 4 \cdot 6 \cdots (p-1) \equiv (-1)^{(p-1)/2} 1 \cdot 3 \cdot 5 \cdots (p-2) \pmod{p}$.]

10.   (a)   For a prime p of the form $4k+3$, prove that either

$$\left(\frac{p-1}{2}\right)! \equiv 1 \pmod{p} \quad \text{or} \quad \left(\frac{p-1}{2}\right)! \equiv -1 \pmod{p};$$

hence, $[(p-1)/2]!$ satisfies the quadratic congruence $x^2 \equiv 1 \pmod{p}$.

(b)   Use part (a) to show that if $p = 4k+3$ is prime, then the product of all the even integers less than p is congruent modulo p to either 1 or -1. [*Hint:* Fermat's Theorem implies that $2^{(p-1)/2} \equiv \pm 1 \pmod{p}$.]

11.   Apply Theorem 5-5 to find two solutions to the quadratic congruences $x^2 \equiv -1 \pmod{29}$ and $x^2 \equiv -1 \pmod{37}$.

12.   Show that if $p = 4k+3$ is prime and $a^2 + b^2 \equiv 0 \pmod{p}$, then $a \equiv b \equiv 0 \pmod{p}$. [*Hint:* If $a \not\equiv 0 \pmod{p}$, then there exists an integer c such that $ac \equiv 1 \pmod{p}$; use this fact to contradict Theorem 5-5.]

13.   Supply any missing details in the following proof of the irrationality of $\sqrt{2}$: Suppose $\sqrt{2} = a/b$, with $\gcd(a,b) = 1$. Then $a^2 = 2b^2$, so that $a^2 + b^2 = 3b^2$. But $3 \mid (a^2 + b^2)$ implies that $3 \mid a$ and $3 \mid b$, a contradiction.

14.   Prove that the odd prime divisors of the integer $n^2 + 1$ are of the form $4k+1$. [*Hint:* Theorem 5-5.]

15.   Verify that $4(29!) + 5!$ is divisible by 31.

16.   For a prime p and $0 \le k \le p - 1$, show that $k!(p-k-1)! \equiv (-1)^{k+1} \pmod{p}$.

17.   If p and q are distinct primes, prove that

$$pq \mid a^{pq} - a^p - a^q + a$$

for any integer a.

18.   Prove that if p and $p+2$ are a pair of twin primes, then

$$4((p-1)! + 1) + p \equiv 0 \pmod{p(p+2)}.$$

6

Number-Theoretic Functions

"Mathematicians are like Frenchmen: whatever you say to them they translate into their own language and forthwith it is something entirely different."
GOETHE

6.1 THE FUNCTIONS τ AND σ

Certain functions are found to be of special importance in connection with the study of the divisors of an integer. Any function whose domain of definition is the set of positive integers is said to be a *number-theoretic (or arithmetic) function*. While the value of a number-theoretic function is not required to be a positive integer or, for that matter, even an integer, most of the number-theoretic functions that we shall encounter are integer-valued. Among the easiest to handle, as well as the most natural, are the functions τ and σ.

DEFINITION 6-1.
Given a positive integer n, let $\tau(n)$ denote the number of positive divisors of n and $\sigma(n)$ denote the sum of these divisors.

For an example of these notions, consider $n = 12$. Since 12 has the positive divisors 1, 2, 3, 4, 6, 12, we find that

$$\tau(12) = 6 \quad \text{and} \quad \sigma(12) = 1 + 2 + 3 + 4 + 6 + 12 = 28.$$

For the first few integers,

$$\tau(1) = 1, \tau(2) = 2, \tau(3) = 2, \tau(4) = 3, \tau(5) = 2, \tau(6) = 4, \ldots$$

and

$$\sigma(1) = 1, \sigma(2) = 3, \sigma(3) = 4, \sigma(4) = 7, \sigma(5) = 6, \sigma(6) = 12, \ldots.$$

It is not difficult to see that $\tau(n) = 2$ if and only if n is a prime number; also, $\sigma(n) = n + 1$ if and only if n is a prime.

Before studying the functions τ and σ in more detail, we wish to introduce a notation that will clarify a number of situations later on. It is customary to interpret the symbol

$$\sum_{d \mid n} f(d)$$

to mean, "Sum the values $f(d)$ as d runs over all the positive divisors of the positive integer n." For instance, we have

$$\sum_{d \mid 20} f(d) = f(1) + f(2) + f(4) + f(5) + f(10) + f(20).$$

With this understanding, τ and σ may be expressed in the form

$$\tau(n) = \sum_{d \mid n} 1, \quad \sigma(n) = \sum_{d \mid n} d.$$

The notation $\sum_{d \mid n} 1$, in particular, says that we are to add together as many 1's as there are positive divisors of n. To illustrate: the integer 10 has the four positive divisors 1, 2, 5, 10, whence

$$\tau(10) = \sum_{d \mid 10} 1 = 1 + 1 + 1 + 1 = 4,$$

while

$$\sigma(10) = \sum_{d \mid 10} d = 1 + 2 + 5 + 10 = 18.$$

Our first theorem makes it easy to obtain the positive divisors of a positive integer n once its prime factorization is known.

THEOREM 6-1.

If $n = p_1^{k_1} p_2^{k_2} \cdots p_r^{k_r}$ is the prime factorization of $n > 1$, then the positive divisors of n are precisely those integers d of the form

$$d = p_1^{a_1} p_2^{a_2} \cdots p_r^{a_r},$$

where $0 \le a_i \le k_i$ $(i = 1, 2, \ldots, r)$.

Proof: Note that the divisor $d = 1$ is obtained when $a_1 = a_2 = \cdots = a_r = 0$, and n itself occurs when $a_1 = k_1, a_2 = k_2, \ldots, a_r = k_r$. Suppose that d divides n nontrivially; say $n = dd'$, where $d > 1$, $d' > 1$. Express both d and d' as products of (not necessarily distinct) primes:

$$d = q_1 q_2 \cdots q_s, \quad d' = t_1 t_2 \cdots t_u,$$

with q_i, t_j prime. Then

$$p_1^{k_1} p_2^{k_2} \cdots p_r^{k_r} = q_1 \cdots q_s t_1 \cdots t_u$$

are two prime factorizations of the positive integer n. By the uniqueness of the prime factorization, each prime q_i must be one of the p_j. Collecting the equal primes into a single integral power, we get

$$d = q_1 q_2 \cdots q_s = p_1^{a_1} p_2^{a_2} \cdots p_r^{a_r},$$

where the possibility that $a_i = 0$ is allowed.

Conversely, every number $d = p_1^{a_1}p_2^{a_2} \cdots p_r^{a_r}$ $(0 \le a_i \le k_i)$ turns out to be a divisor of n. For we can write

$$n = p_1^{k_1}p_2^{k_2} \cdots p_r^{k_r}$$
$$= (p_1^{a_1}p^{a_2} \cdots p_r^{a_r})(p_1^{k_1 - a_1}p_2^{k_2 - a_2} \cdots p_r^{k_r - a_r})$$
$$= dd',$$

with $d' = p_1^{k_1 - a_1}p_2^{k_2 - a_2} \cdots p_r^{k_r - a_r}$ and $k_i - a_i \ge 0$ for each i. Then $d' > 0$ and $d \mid n$.

We put this theorem to work at once.

THEOREM 6-2.

If $n = p_1^{k_1}p_2^{k_2} \cdots p_r^{k_r}$ is the prime factorization of $n > 1$, then

(a) $\tau(n) = (k_1 + 1)(k_2 + 1) \cdots (k_r + 1)$, *and*

(b) $\sigma(n) = \dfrac{p_1^{k_1 + 1} - 1}{p_1 - 1} \dfrac{p_2^{k_2 + 1} - 1}{p_2 - 1} \cdots \dfrac{p_r^{k_r + 1} - 1}{p_r - 1}.$

Proof: According to Theorem 6-1, the positive divisors of n are precisely those integers

$$d = p_1^{a_1}p_2^{a_2} \cdots p_r^{a_r},$$

where $0 \le a_i \le k_i$. There are $k_1 + 1$ choices for the exponent a_1; $k_2 + 1$ choices for $a_2, \ldots$; $k_r + 1$ choices for a_r; hence, there are

$$(k_1 + 1)(k_2 + 1) \cdots (k_r + 1)$$

possible divisors of n.

In order to evaluate $\sigma(n)$, consider the product

$$(1 + p_1 + p_1^2 + \cdots + p_1^{k_1})(1 + p_2 + p_2^2 + \cdots + p_2^{k_2})$$
$$\cdots (1 + p_r + p_r^2 + \cdots + p_r^{k_r}).$$

Each positive divisor of n appears once and only once as a term in the expansion of this product, so that

$$\sigma(n) = (1 + p_1 + p_1^2 + \cdots + p_1^{k_1}) \cdots (1 + p_r + p_r^2 + \cdots + p_r^{k_r}).$$

Applying the formula for the sum of a finite geometric series to the ith factor on the right-hand side, we get

$$1 + p_i + p_i^2 + \cdots + p_i^{k_i} = \frac{p_i^{k_i + 1} - 1}{p_i - 1}.$$

It follows that

$$\sigma(n) = \frac{p_1^{k_1 + 1} - 1}{p_1 - 1} \frac{p_2^{k_2 + 1} - 1}{p_2 - 1} \cdots \frac{p_r^{k_r + 1} - 1}{p_r - 1}.$$

Corresponding to the $\sum$ notation for sums, a notation for products may be defined using Π, the Greek capital letter "pi." The restriction delimiting the numbers over which the product is to be made is usually put under the Π-sign. Examples are

$$\prod_{1 \le d \le 5} f(d) = f(1)f(2)f(3)f(4)f(5),$$

$$\prod_{d \mid 9} f(d) = f(1)f(3)f(9),$$

$$\prod_{\substack{p \mid 30 \\ p \text{ prime}}} f(d) = f(2)f(3)f(5).$$

With this convention, the conclusion to Theorem 6-2 takes the compact form: if $n = p_1^{k_1}p_2^{k_2} \cdots p_r^{k_r}$ is the prime factorization of $n > 1$, then

$$\tau(n) = \prod_{1 \le i \le r} (k_i + 1)$$

and

$$\sigma(n) = \prod_{1 \le i \le r} \frac{p_i^{k_i + 1} - 1}{p_i - 1}.$$

Example 6-1

The number $180 = 2^2 \cdot 3^2 \cdot 5$ has

$$\tau(180) = (2 + 1)(2 + 1)(1 + 1) = 18$$

positive divisors. These are integers of the form

$$2^{a_1} \cdot 3^{a_2} \cdot 5^{a_3},$$

where $a_1 = 0, 1, 2$; $a_2 = 0, 1, 2$; $a_3 = 0, 1$. Specifically, we obtain

$$1, 2, 3, 4, 5, 6, 9, 10, 12, 15, 18, 20, 30, 36, 45, 60, 90, 180.$$

The sum of these integers is

$$\sigma(180) = \frac{2^3 - 1}{2 - 1} \frac{3^3 - 1}{3 - 1} \frac{5^2 - 1}{5 - 1} = \frac{7}{1} \frac{26}{2} \frac{24}{4} = 7 \cdot 13 \cdot 6 = 546.$$

One of the more interesting properties of the divisor function τ is that the product of the positive divisors of an integer $n > 1$ is equal to $n^{\tau(n)/2}$. It is not difficult to get at this fact: Let d denote an arbitrary positive divisor of n, so that $n = dd'$ for some d'. As d ranges over all $\tau(n)$ positive divisors of n, $\tau(n)$ such equations occur. Multiplying these together, we get

$$n^{\tau(n)} = \prod_{d \mid n} d \cdot \prod_{d' \mid n} d'.$$

But as d runs through the divisors of n, so does d'; hence, $\prod_{d \mid n} d = \prod_{d' \mid n} d'$. The situation is now this:

$$n^{\tau(n)} = \left(\prod_{d \mid n} d \right)^2$$

or equivalently,

$$n^{\tau(n)/2} = \prod_{d \mid n} d.$$

The reader might (or, at any rate, should) have one lingering doubt concerning this equation. For it is by no means obvious that the left-hand side is always an integer. If $\tau(n)$ is even, there is certainly no problem. When $\tau(n)$ is odd, n turns out to be a perfect square (Problem 7), say $n = m^2$; thus $n^{\tau(n)/2} = m^{\tau(n)}$, settling all suspicions.

For a numerical example, the product of the five divisors of 16 (namely, 1, 2, 4, 8, 16) is

$$\prod_{d \mid 16} d = 16^{\tau(16)/2} = 16^{5/2} = 4^5 = 1024.$$

Multiplicative functions arise naturally in the study of the prime factorization of an integer. Before presenting the definition, we observe that

$$\tau(2 \cdot 10) = \tau(20) = 6 \neq 2 \cdot 4 = \tau(2) \cdot \tau(10).$$

At the same time

$$\sigma(2 \cdot 10) = \sigma(20) = 42 \neq 3 \cdot 18 = \sigma(2) \cdot \sigma(10).$$

These calculations bring out the nasty fact that, in general, it need not be true that

$$\tau(mn) = \tau(m)\tau(n) \quad \text{and} \quad \sigma(mn) = \sigma(m)\sigma(n).$$

On the positive side of the ledger, equality always holds provided we stick to relatively prime m and n. This circumstance is what prompts

DEFINITION 6-2.
A number-theoretic function f is said to be *multiplicative* if

$$f(mn) = f(m)f(n)$$

whenever $\gcd(m, n) = 1$.

For simple illustrations of multiplicative functions, one need only consider the functions given by $f(n) = 1$ and $g(n) = n$ for all $n \geq 1$. It follows by induction that if f is multiplicative and $n_1, n_2, \ldots, n_r$ are positive integers which are pairwise relatively prime, then

$$f(n_1 n_2 \cdots n_r) = f(n_1)f(n_2) \cdots f(n_r).$$

Multiplicative functions have one big advantage for us: they are completely determined once their values at prime powers are known. Indeed, if $n > 1$ is a given positive integer, then we can write $n = p_1^{k_1} p_2^{k_2} \cdots p_r^{k_r}$ in canonical form; since the $p_i^{k_i}$ are relatively prime in pairs, the multiplicative property ensures that

$$f(n) = f(p_1^{k_1}) f(p_2^{k_2}) \cdots f(p_r^{k_r}).$$

If f is a multiplicative function which does not vanish identically, then there exists an integer n such that $f(n) \neq 0$. But

$$f(n) = f(n \cdot 1) = f(n)f(1).$$

Being nonzero, $f(n)$ may be cancelled from both sides of this equation to give $f(1) = 1$. The point to which we wish to call attention is that $f(1) = 1$ for any multiplicative function not identically zero.

We now establish that τ and σ have the multiplicative property.

THEOREM 6-3.

The functions τ and σ are both multiplicative functions.

Proof: Let m and n be relatively prime integers. Since the result is trivially true if either m or n is equal to 1, we may assume that $m > 1$ and $n > 1$. If

$$m = p_1^{k_1} p_2^{k_2} \cdots p_r^{k_r} \quad \text{and} \quad n = q_1^{j_1} q_2^{j_2} \cdots q_s^{j_s}$$

are the prime factorizations of m and n, then, since $\gcd(m, n) = 1$, no p_i can occur among the q_j. It follows that the prime factorization of the product mn is given by

$$mn = p_1^{k_1} \cdots p_r^{k_r} q_1^{j_1} \cdots q_s^{j_s}.$$

Appealing to Theorem 6-2, we obtain

$$\tau(mn) = [(k_1 + 1) \cdots (k_r + 1)][(j_1 + 1) \cdots (j_s + 1)]$$
$$= \tau(m)\tau(n).$$

In a similar fashion, Theorem 6–2 gives

$$\sigma(mn) = \left[\frac{p_1^{k_1 + 1} - 1}{p_1 - 1} \cdots \frac{p_r^{k_r + 1} - 1}{p_r - 1} \right] \left[\frac{q_1^{j_1 + 1} - 1}{q_1 - 1} \cdots \frac{q_s^{j_s + 1} - 1}{q_s - 1} \right]$$
$$= \sigma(m)\sigma(n).$$

Thus, τ and σ are multiplicative functions.

We continue our program by proving a general result on multiplicative functions. This requires a preparatory lemma.

LEMMA.

If $\gcd(m , n) = 1$, then the set of positive divisors of mn consists of all products $d_1 d_2$, where $d_1 \mid n$, $d_2 \mid m$ and $\gcd(d_1 , d_2) = 1$; furthermore, these products are all distinct.

Proof: It is harmless to assume that $m > 1$ and $n > 1$; let $m = p_1^{k_1} p_2^{k_2} \cdots p_r^{k_r}$ and $n = q_1^{j_1} q_2^{j_2} \cdots q_s^{j_s}$ be their respective prime factorizations. Inasmuch as the primes $p_1, \ldots, p_r, q_1, \ldots, q_s$ are all distinct, the prime factorization of mn is

$$mn = p_1^{k_1} \cdots p_r^{k_r} q_1^{j_1} \cdots q_s^{j_s}.$$

Hence, any positive divisor d of mn will be uniquely representable in the form

$$d = p_1^{a_1} \cdots p_r^{a_r} q_1^{b_1} \cdots q_s^{b_s}, \ 0 \le a_i \le k_i, 0 \le b_i \le j_i.$$

This allows us to write d as $d = d_1 d_2$, where $d_1 = p_1^{a_1} \cdots p_r^{a_r}$ divides m and $d_2 = q_1^{b_1} \cdots q_s^{b_s}$ divides n. Since no p_i is equal to any q_j, we surely must have $\gcd(d_1 , d_2) = 1$.

A keystone in much of our subsequent work is

THEOREM 6-4.

If f is a multiplicative function and F is defined by

$$F(n) = \sum_{d \mid n} f(d),$$

then F is also multiplicative.

Proof: Let m and n be relatively prime positive integers. Then

$$F(mn) = \sum_{d \mid mn} f(d) = \sum_{\substack{d_1 \mid m \\ d_2 \mid n}} f(d_1 d_2),$$

since every divisor d of mn can be uniquely written as a product of a divisor d_1 of m and a divisor d_2 of n, where $\gcd(d_1 , d_2) = 1$. By the definition of a multiplicative function,

$$f(d_1 d_2) = f(d_1) f(d_2).$$

It follows that

$$F(mn) = \sum_{\substack{d_1 \mid m \\ d_2 \mid n}} f(d_1) f(d_2)$$

$$= \left(\sum_{d_1 \mid m} f(d_1) \right) \left(\sum_{d_2 \mid n} f(d_2) \right) = F(m) F(n).$$

It might be helpful to take time out and run through the proof of Theorem 6-4 in a concrete case. Letting $m = 8$ and $n = 3$, we have

$$F(8 \cdot 3) = \sum_{d \mid 24} f(d)$$

$$= f(1) + f(2) + f(3) + f(4) + f(6) + f(8) + f(12) + f(24)$$

$$= f(1 \cdot 1) + f(2 \cdot 1) + f(1 \cdot 3) + f(4 \cdot 1) + f(2 \cdot 3)$$
$$+ f(8 \cdot 1) + f(4 \cdot 3) + f(8 \cdot 3)$$

$$= f(1)f(1) + f(2)f(1) + f(1)f(3) + f(4)f(1) + f(2)f(3)$$
$$+ f(8)f(1) + f(4)f(3) + f(8)f(3)$$

$$= [f(1) + f(2) + f(4) + f(8)][f(1) + f(3)]$$
$$= \sum_{d \mid 8} f(d) \cdot \sum_{d \mid 3} f(d) = F(8)F(3).$$

Theorem 6-4 provides a deceptively short way of drawing the conclusion that τ and σ are multiplicative.

COROLLARY.

The functions τ and σ are multiplicative functions.

Proof: We have mentioned before that the constant function $f(n) = 1$ is multiplicative, as is the identity function $f(n) = n$. Since τ and σ may be represented in the form

$$\tau(n) = \sum_{d \mid n} 1 \quad \text{and} \quad \sigma(n) = \sum_{d \mid n} d,$$

the stated result follows immediately from Theorem 6-4.

PROBLEMS 6.1

1. Let m and n be positive integers and $p_1, p_2, \ldots, p_r$ be the distinct primes which divide at least one of m or n. Then m and n may be written in the form

$$m = p_1^{k_1} p_2^{k_2} \cdots p_r^{k_r}, \qquad \text{with } k_i \geq 0 \text{ for } i = 1, 2, \ldots, r$$
$$n = p_1^{j_1} p_2^{j_2} \cdots p_r^{j_r}, \qquad \text{with } j_i \geq 0 \text{ for } i = 1, 2, \ldots, r$$

Prove that

$$\gcd(m, n) = p_1^{u_1} p_2^{u_2} \cdots p_r^{u_r}, \quad \text{lcm}(m, n) = p_1^{v_1} p_2^{v_2} \cdots p_r^{v_r},$$

where $u_i = \min \{k_i, j_i\}$, the smaller of k_i and j_i; and $v_i = \max \{k_i, j_i\}$, the larger of k_i and j_i.

2. Use the result of Problem 1 to calculate $\gcd(12378, 3054)$ and $\text{lcm}(12378, 3054)$.

3. Deduce from Problem 1 that $\gcd(m, n) \text{ lcm}(m, n) = mn$ for positive integers m and n.

4. In the notation of Problem 1, show that $\gcd(m, n) = 1$ if and only if $k_i j_i = 0$ for $i = 1$, $2, \ldots, r$.

5. (a) Verify that $\tau(n) = \tau(n + 1) = \tau(n + 2) = \tau(n + 3)$ holds for $n = 3655$ and 4503.
    (b) When $n = 14$, 206, and 957, show that $\sigma(n) = \sigma(n + 1)$.

6. For any integer $n \geq 1$, establish the inequality $\tau(n) \leq 2\sqrt{n}$. [*Hint:* If $d \mid n$, then one of d or n/d is less than or equal to $\sqrt{n}$.]

7. Prove that:
    (a) $\tau(n)$ is an odd integer if and only if n is a perfect square;
    (b) $\sigma(n)$ is an odd integer if and only if n is a perfect square or twice a perfect square.
    [*Hint:* If p is an odd prime, then $1 + p + p^2 + \cdots + p^k$ is odd only when k is even.]

8. Show that $\sum_{d \mid n} 1/d = \sigma(n)/n$ for every positive integer n.

9. If n is a square-free integer, prove that $\tau(n) = 2^r$, where r is the number of prime divisors of n.

10. Establish the assertions below:
    (a) If $n = p_1^{k_1} p_2^{k_2} \cdots p_r^{k_r}$ is the prime factorization of $n > 1$, then

    $$1 > \frac{n}{\sigma(n)} > \left(1 - \frac{1}{p_1}\right)\left(1 - \frac{1}{p_2}\right) \cdots \left(1 - \frac{1}{p_r}\right).$$

    (b) For any positive integer n,

    $$\sigma(n!)/n! \geq 1 + 1/2 + 1/3 + \cdots + 1/n.$$

    [*Hint:* See Problem 8.]

    (c) If $n > 1$ is a composite number, then $\sigma(n) > n + \sqrt{n}$. [*Hint:* Let $d \mid n$, where $1 < d < n$, so $1 < n/d < n$. If $d \leq \sqrt{n}$, then $n/d \geq \sqrt{n}$.]

11. Given a positive integer $k > 1$, show that there are infinitely many integers n for which $\tau(n) = k$, but at most finitely many n with $\sigma(n) = k$. [*Hint:* Utilize Problem 10(a).]

12. (a) Find the form of all positive integers n satisfying $\tau(n) = 10$. What is the smallest positive integer for which this is true?
    (b) Show that there are no positive integers n satisfying $\sigma(n) = 10$. [*Hint:* Note that for $n > 1$, $\sigma(n) > n$.]

13. Prove that there are infinitely many pairs of integers m and n with $\sigma(m^2) = \sigma(n^2)$. [*Hint:* Choose k such that $\gcd(k, 10) = 1$ and consider the integers $m = 5k$, $n = 4k$.]

14. For $k \geq 2$, show each of the following:
    (a) $n = 2^{k-1}$ satisfies the equation $\sigma(n) = 2n - 1$;
    (b) if $2^k - 1$ is prime, then $n = 2^{k-1}(2^k - 1)$ satisfies the equation $\sigma(n) = 2n$;
    (c) if $2^k - 3$ is prime, then $n = 2^{k-1}(2^k - 3)$ satisfies the equation $\sigma(n) = 2n + 2$.
    It is not known if there are any positive integers n for which $\sigma(n) = 2n + 1$.

15. If n and $n + 2$ are a pair of twin primes, establish that $\sigma(n + 2) = \sigma(n) + 2$; this also holds for $n = 434$ and 8575.

16. (a) For any integer $n > 1$, prove that there exist integers n_1 and n_2 with $\tau(n_1) + \tau(n_2) = n$.
    (b) Prove that Goldbach's Conjecture implies that for each even integer $2n$ there exist integers n_1 and n_2 with $\sigma(n_1) + \sigma(n_2) = 2n$.

17. For a fixed integer k, show that the function f defined by $f(n) = n^k$ is multiplicative.

18. Let f and g be multiplicative functions which are not identically zero and such that $f(p^k) = g(p^k)$ for each prime p and $k \geq 1$. Prove that $f = g$.

19. Prove that if f and g are multiplicative functions, then so is their product fg and quotient f/g (whenever the latter function is defined).

20. Let $\omega(n)$ denote the number of distinct prime divisors of $n > 1$, with $\omega(1) = 0$. For instance, $\omega(360) = \omega(2^3 \cdot 3^2 \cdot 5) = 3$.
     (a) Show that $2^{\omega(n)}$ is a multiplicative function.
     (b) For a positive integer n, establish the formula

$$\tau(n^2) = \sum_{d \mid n} 2^{\omega(d)}.$$

21. For any positive integer n, prove that $\sum_{d \mid n} \tau(d)^3 = (\sum_{d \mid n} \tau(d))^2$. [*Hint:* Both sides of the equation in question are multiplicative functions of n, so that it suffices to consider the case $n = p^k$, where p is a prime.]

22. Given $n \geq 1$, let $\sigma_s(n)$ denote the sum of the sth powers of the positive divisors of n; that is,

$$\sigma_s(n) = \sum_{d \mid n} d^s.$$

Verify the following:
     (a) $\sigma_0 = \tau$ and $\sigma_1 = \sigma$.
     (b) σ_s is a multiplicative function. [*Hint:* The function f, defined by $f(n) = n^s$, is multiplicative.]
     (c) If $n = p_1^{k_1} p_2^{k_2} \cdots p_r^{k_r}$ is the prime factorization of n, then

$$\sigma_s(n) = \left(\frac{p_1^{s(k_1 + 1)} - 1}{p_1^s - 1} \right) \left(\frac{p_2^{s(k_2 + 1)} - 1}{p_2^s - 1} \right) \cdots \left(\frac{p_r^{s(k_r + 1)} - 1}{p_r^s - 1} \right).$$

23. For any positive integer n, show that
     (a) $\sum_{d \mid n} \sigma(d) = \sum_{d \mid n} \frac{n}{d} \tau(d)$, and

     (b) $\sum_{d \mid n} \frac{n}{d} \sigma(d) = \sum_{d \mid n} d \tau(d)$.

[*Hint:* Since the functions

$$F(n) = \sum_{d \mid n} \sigma(d) \quad \text{and} \quad G(n) = \sum_{d \mid n} (n/d) \tau(d)$$

are both multiplicative, it suffices to prove that $F(p^k) = G(p^k)$ for any prime p.]

6.2 THE MÖBIUS INVERSION FORMULA

We introduce another naturally defined function on the positive integers, the Möbius μ-function.

DEFINITION 6-3.

For a positive integer n, define μ by the rules

$$\mu(n) = \begin{cases} 1 \text{ if } n = 1 \\ 0 \text{ if } p^2 \mid n \text{ for some prime } p \\ (-1)^r \text{ if } n = p_1 p_2 \cdots p_r, \text{ where the } p_i \text{ are distinct primes} \end{cases}$$

Put somewhat differently, Definition 6-3 states that $\mu(n) = 0$ if n is not a square-free integer, while $\mu(n) = (-1)^r$ if n is square-free with r prime factors. For example: $\mu(30) = \mu(2 \cdot 3 \cdot 5) = (-1)^3 = -1$. The first few values of μ are

$$\mu(1) = 1, \mu(2) = -1, \mu(3) = -1, \mu(4) = 0, \mu(5) = -1, \mu(6) = 1, \ldots .$$

If p is a prime number, it is clear that $\mu(p) = -1$; also, $\mu(p^k) = 0$ for $k \geq 2$.

As the reader may have guessed already, the Möbius μ-function is multiplicative. This is the content of

THEOREM 6-5.

The function μ is a multiplicative function.

Proof: We want to show that $\mu(mn) = \mu(m)\mu(n)$, whenever m and n are relatively prime. If either $p^2 \mid m$ or $p^2 \mid n$, p a prime, then $p^2 \mid mn$; hence, $\mu(mn) = 0 = \mu(m)\mu(n)$, and the formula holds trivially. We may therefore assume that both m and n are square-free integers. Say, $m = p_1 p_2 \cdots p_r$, $n = q_1 q_2 \cdots q_s$, the primes p_i and q_j being all distinct. Then

$$\mu(mn) = \mu(p_1 \cdots p_r q_1 \cdots q_s) = (-1)^{r+s}$$
$$= (-1)^r (-1)^s = \mu(m)\mu(n),$$

which completes the proof.

Let us see what happens if $\mu(d)$ is evaluated for all the positive divisors d of an integer n and the results are added. In case $n = 1$, the answer is easy; here,

$$\sum_{d \mid 1} \mu(d) = \mu(1) = 1.$$

Suppose that $n > 1$ and put

$$F(n) = \sum_{d \mid n} \mu(d).$$

To prepare the ground, we first calculate $F(n)$ for the power of a prime, say, $n = p^k$. The positive divisors of p^k are just the $k + 1$ integers $1, p, p^2, \ldots, p^k$, so that

$$F(p^k) = \sum_{d \mid p^k} \mu(d) = \mu(1) + \mu(p) + \mu(p^2) + \cdots + \mu(p^k)$$

$$= \mu(1) + \mu(p) = 1 + (-1) = 0.$$

Since μ is known to be a multiplicative function, an appeal to Theorem 6-4 is legitimate; this result guarantees that F is multiplicative too. Thus, if the canonical factorization of n is $n = p_1^{k_1} p_2^{k_2} \cdots p_r^{k_r}$, then $F(n)$ is the product of the values assigned to F for the prime powers in this representation:

$$F(n) = F(p_1^{k_1}) F(p_2^{k_2}) \cdots F(p_r^{k_r}) = 0.$$

We record this result as

THEOREM 6-6.

For each positive integer $n \geq 1$,

$$\sum_{d \mid n} \mu(d) = \begin{cases} 1 \text{ if } n = 1 \\ 0 \text{ if } n > 1 \end{cases}$$

where d runs through the positive divisors of n.

For an illustration of this last theorem, consider $n = 10$. The divisors of 10 are 1, 2, 5, 10 and the desired sum is

$$\sum_{d \mid 10} \mu(d) = \mu(1) + \mu(2) + \mu(5) + \mu(10)$$

$$= 1 + (-1) + (-1) + 1 = 0.$$

The full significance of Möbius' μ-function should become apparent with the next theorem.

THEOREM 6-7.

(Möbius Inversion Formula). Let F and f be two number-theoretic functions related by the formula

$$F(n) = \sum_{d \mid n} f(d).$$

Then

$$f(n) = \sum_{d \mid n} \mu(d) F(n/d) = \sum_{d \mid n} \mu(n/d) F(d).$$

Proof: The two sums mentioned in the conclusion of the theorem are seen to be the same upon replacing the dummy index d by $d' = n/d$; as d ranges over all positive divisors of n, so does d'.

Carrying out the required computation, we get

$$(1) \quad \sum_{d \mid n} \mu(d) F(n/d) = \sum_{d \mid n} \left(\mu(d) \sum_{c \mid (n/d)} f(c) \right) = \sum_{d \mid n} \left(\sum_{c \mid (n/d)} \mu(d) f(c) \right).$$

It is easily verified that $d \mid n$ and $c \mid (n/d)$ if and only if $c \mid n$ and $d \mid (n/c)$. Because of this, the last expression in equation (1) becomes

$$(2) \qquad \sum_{d \mid n} \left(\sum_{c \mid (n/d)} \mu(d)f(c) \right) = \sum_{c \mid n} \left(\sum_{d \mid (n/c)} f(c)\mu(d) \right)$$

$$= \sum_{c \mid n} \left(f(c) \sum_{d \mid (n/c)} \mu(d) \right).$$

In compliance with Theorem 6-6, the sum $\sum_{d \mid (n/c)} \mu(d)$ must vanish except when $n/c = 1$ (that is, when $n = c$), in which case it is equal to 1; the upshot is that the right-hand side of equation (2) simplifies to

$$\sum_{c \mid n} \left(f(c) \sum_{d \mid (n/c)} \mu(d) \right) = \sum_{c = n} f(c) \cdot 1 = f(n),$$

giving us the stated result.

Let us use $n = 10$ again to illustrate how the double sum in equation (2) is turned around. In this instance, we find that

$$\sum_{d \mid 10} \left(\sum_{c \mid (10/d)} \mu(d)f(c) \right) = \mu(1)[f(1) + f(2) + f(5) + f(10)]$$

$$+ \mu(2)[f(1) + f(5)] + \mu(5)[f(1) + f(2)]$$

$$+ \mu(10)f(1)$$

$$= f(1)[\mu(1) + \mu(2) + \mu(5) + \mu(10)]$$

$$+ f(2)[\mu(1) + \mu(5)] + f(5)[\mu(1) + \mu(2)]$$

$$+ f(10)\mu(1)$$

$$= \sum_{c \mid 10} \left(\sum_{d \mid (10/c)} f(c)\mu(d) \right).$$

To see how the Möbius inversion formula works in a particular case, we remind the reader that the functions τ and σ may both be described as "sum functions":

$$\tau(n) = \sum_{d \mid n} 1 \quad \text{and} \quad \sigma(n) = \sum_{d \mid n} d.$$

Theorem 6-7 tells us that these formulas may be inverted to give

$$1 = \sum_{d \mid n} \mu(n/d)\tau(d) \quad \text{and} \quad n = \sum_{d \mid n} \mu(n/d)\sigma(d),$$

which are valid for all $n \geq 1$.

Theorem 6-4 insures that if f is a multiplicative function, then so is $F(n) = \sum_{d \mid n} f(d)$. Turning the situation around, one might ask whether the multiplicative nature of F forces that of f. Surprisingly enough, this is exactly what happens.

Theorem 6-8.

If F is a multiplicative function and

$$F(n) = \sum_{d \mid n} f(d),$$

then f is also multiplicative.

Proof: Let m and n be relatively prime positive integers. We recall that any divisor d of mn can be uniquely written as $d = d_1 d_2$, where $d_1 \mid m$, $d_2 \mid n$, and gcd $(d_1, d_2) = 1$. Thus, using the inversion formula,

$$f(mn) = \sum_{d \mid mn} \mu(d) F\left(\frac{mn}{d}\right)$$

$$= \sum_{\substack{d_1 \mid m \\ d_2 \mid n}} \mu(d_1 d_2) F\left(\frac{mn}{d_1 d_2}\right)$$

$$= \sum_{\substack{d_1 \mid m \\ d_2 \mid n}} \mu(d_1)\mu(d_2) F\left(\frac{m}{d_1}\right) F\left(\frac{n}{d_2}\right)$$

$$= \sum_{d_1 \mid m} \mu(d_1) F\left(\frac{m}{d_1}\right) \sum_{d_2 \mid n} \mu(d_2) F\left(\frac{n}{d_2}\right) = f(m)f(n),$$

which is the assertion of the theorem. Needless to say, the multiplicative character of μ and of F is crucial to the above calculation.

For $n \geq 1$, we define the sum

$$M(n) = \sum_{k=1}^{n} \mu(k).$$

Then $M(n)$ is the difference between the number of square-free positive integers $k \leq n$ with an even number of prime factors and those with an odd number of prime factors. For example, $M(9) = 2 - 4 = -2$. In 1897, Franz Mertens (1840–1927) published a paper with a fifty-page table of values of $M(n)$ for $n = 1, 2, \cdots, 10,000$. On the basis of the tabular evidence, Mertens concluded that the inequality

$$|M(n)| < \sqrt{n}, \quad n > 1,$$

is "very probable." (In the example above, $|M(9)| = 2 < \sqrt{9}$.) This conclusion later became known as the Mertens Conjecture. A computer search carried out in 1963 verified the conjecture for all n up to 10 billion. But in 1984, Andrew Odlyzko and Herman te Riele showed that the Mertens Conjecture is false. Their proof, which involved the use of a computer, was indirect and produced no specific value of n for which $|M(n)| \geq \sqrt{n}$; all it demonstrated was that such a number n has to exist somewhere. Subsequently, it has been shown that there is a counterexample to the Mertens Conjecture for at least one $n \leq (3.21)10^{64}$.

PROBLEMS 6.2

1. (a) For each positive integer n, show that

$$\mu(n)\mu(n+1)\mu(n+2)\mu(n+3) = 0.$$

   (b) For any integer $n \geq 3$, show that $\sum_{k=1}^{n} \mu(k!) = 1$.

2. The *Mangoldt function* Λ is defined by

$$\Lambda(n) = \begin{cases} \log p, & \text{if } n = p^k, \text{ where } p \text{ is a prime and } k \geq 1 \\ 0, & \text{otherwise} \end{cases}$$

   Prove that $\Lambda(n) = \sum_{d \mid n} \mu(n/d) \log d = -\sum_{d \mid n} \mu(d) \log d$. [*Hint:* First show that $\sum_{d \mid n} \Lambda(d) = \log n$ and then apply the Möbius Inversion Formula.]

3. Let $n = p_1^{k_1} p_2^{k_2} \cdots p_r^{k_r}$ be the prime factorization of the integer $n > 1$. If f is a multiplicative function, prove that

$$\sum_{d \mid n} \mu(d)f(d) = (1 - f(p_1))(1 - f(p_2)) \cdots (1 - f(p_r)).$$

   [*Hint:* By Theorem 6-4, the function F defined by $F(n) = \sum_{d \mid n} \mu(d)f(d)$ is multiplicative; hence, $F(n)$ is the product of the values $F(p_i^{k_i})$.]

4. If the integer $n > 1$ has the prime factorization $n = p_1^{k_1} p_2^{k_2} \cdots p_r^{k_r}$, use Problem 3 to establish the following:

   (a) $\displaystyle\sum_{d \mid n} \mu(d)\tau(d) = (-1)^r$;

   (b) $\displaystyle\sum_{d \mid n} \mu(d)\sigma(d) = (-1)^r p_1 p_2 \cdots p_r$;

   (c) $\displaystyle\sum_{d \mid n} \mu(d)/d = (1 - 1/p_1)(1 - 1/p_2) \cdots (1 - 1/p_r)$;

   (d) $\displaystyle\sum_{d \mid n} d\mu(d) = (1 - p_1)(1 - p_2) \cdots (1 - p_r)$.

5. Let $S(n)$ denote the number of square-free divisors of n. Establish that

$$S(n) = \sum_{d \mid n} |\mu(d)| = 2^{\omega(n)},$$

   where $\omega(n)$ is the number of distinct prime divisors of n. [*Hint:* S is a multiplicative function.]

6. Find formulas for $\sum_{d \mid n} \mu^2(d)/\tau(d)$ and $\sum_{d \mid n} \mu^2(d)/\sigma(d)$ in terms of the prime factorization of n.

7. The *Liouville λ-function* is defined by $\lambda(1) = 1$ and $\lambda(n) = (-1)^{k_1 + k_2 + \cdots + k_r}$, if the prime factorization of $n > 1$ is $n = p_1^{k_1} p_2^{k_2} \cdots p_r^{k_r}$. For instance,

$$\lambda(360) = \lambda(2^3 \cdot 3^2 \cdot 5) = (-1)^{3+2+1} = (-1)^6 = 1.$$

   (a) Prove that λ is a multiplicative function.
   (b) Given a positive integer n, verify that

$$\sum_{d \mid n} \lambda(d) = \begin{cases} 1 & \text{if } n = m^2 \text{ for some integer } m \\ 0 & \text{otherwise} \end{cases}$$

8. For an integer $n \geq 1$, verify the formulas below:

  (a) $\sum_{d \mid n} \mu(d)\lambda(d) = 2^{\omega(n)}$;

  (b) $\sum_{d \mid n} \lambda(n/d)\, 2^{\omega(d)} = 1$.

6.3 THE GREATEST INTEGER FUNCTION

The greatest integer or "bracket" function [ ] is especially suitable for treating divisibility problems. While not strictly a number-theoretic function, its study has a natural place in this chapter.

DEFINITION 6-4.

  For an arbitrary real number x, we denote by $[x]$ the largest integer less than or equal to x; that is, $[x]$ is the unique integer satisfying $x - 1 < [x] \leq x$.

By way of illustration, [ ] assumes the particular values

$$[-3/2] = -2, [\sqrt{2}] = 1, [1/3] = 0, [\pi] = 3, [-\pi] = -4.$$

The important observation to be made here is that the equality $[x] = x$ holds if and only if x is an integer. Definition 6-4 also makes plain that any real number x can be written as

$$x = [x] + \theta$$

for a suitable choice of θ, with $0 \leq \theta < 1$.

  We now plan to investigate the question of how many times a particular prime p appears in $n!$. For instance, if $p = 3$ and $n = 9$, then

$$9! = 1 \cdot 2 \cdot 3 \cdot 4 \cdot 5 \cdot 6 \cdot 7 \cdot 8 \cdot 9$$
$$= 2^7 \cdot 3^4 \cdot 5 \cdot 7,$$

so that the exact power of 3 which divides $9!$ is 4. It is desirable to have a formula that will give this count, without the necessity of always writing $n!$ in canonical form. This is accomplished by

THEOREM 6-9.

  If n is a positive integer and p a prime, then the exponent of the highest power of p that divides $n!$ is

$$\sum_{k=1}^{\infty} [n/p^k]$$

where the series is finite, since $[n/p^k] = 0$ for $p^k > n$.

Proof: Among the first n positive integers, those which are divisible by p are p, $2p, \ldots, tp$, where t is the largest integer such that $tp \leq n$; in other words, t is the largest integer less than or equal to n/p (which is to say $t = [n/p]$). Thus, there are exactly $[n/p]$ multiples of p occurring in the product that defines $n!$, namely,

(1) $\qquad\qquad\qquad\qquad p, 2p, \ldots, [n/p]p.$

The exponent of p in the prime factorization of $n!$ is obtained by adding to the number of integers in (1), the number of integers among $1, 2, \ldots, n$ which are divisible by p^2, and then the number divisible by p^3, and so on. Reasoning as in the first paragraph, the integers between 1 and n that are divisible by p^2 are

(2) $\qquad\qquad\qquad\qquad p^2, 2p^2, \ldots, [n/p^2]p^2,$

which are $[n/p^2]$ in number. Of these, $[n/p^3]$ are again divisible by p:

(3) $\qquad\qquad\qquad\qquad p^3, 2p^3, \ldots, [n/p^3]p^3.$

After a finite number of repetitions of this process, we are led to conclude that the total number of times p divides $n!$ is

$$\sum_{k=1}^{\infty} [n/p^k].$$

This result can be cast as the following equation, which usually appears under the name of Legendre's formula:

$$n! = \prod_{p \leq n} p^{\sum_{k=1}^{\infty} [n/p^k]}.$$

Example 6-2

We would like to find the number of zeroes with which the decimal representation of 50! terminates. In determining the number of times 10 enters into the product 50!, it is enough to find the exponents of 2 and 5 in the prime factorization of 50!, and then to select the smaller figure.

By direct calculation we see that

$$[50/2] + [50/2^2] + [50/2^3] + [50/2^4] + [50/2^5]$$
$$= 25 + 12 + 6 + 3 + 1 = 47.$$

Theorem 6-9 tells us that 2^{47} divides 50!, but 2^{48} does not. Similarly,

$$[50/5] + [50/5^2] = 10 + 2 = 12$$

and so the highest power of 5 dividing 50! is 12. This means that 50! ends with 12 zeroes.

We cannot resist using Theorem 6-9 to prove the following fact.

THEOREM 6-10.

If n and r are positive integers with $1 \leq r < n$, then the binomial coefficient

$$\binom{n}{r} = \frac{n!}{r!(n-r)!}$$

is also an integer.

Proof: The argument rests on the observation that if a and b are arbitrary real numbers, then $[a + b] \geq [a] + [b]$. In particular, for each prime factor p of $r!\,(n-r)!$,

$$[n/p^k] \geq [r/p^k] + [(n-r)/p^k], \qquad k = 1, 2, \ldots$$

Adding these inequalities together, we obtain

$$(1) \qquad \sum_{k \geq 1} [n/p^k] \geq \sum_{k \geq 1} [r/p^k] + \sum_{k \geq 1} [(n-r)/p^k].$$

The left-hand side of equation (1) gives the exponent of the highest power of the prime p that divides $n!$, whereas the right-hand side equals the highest power of this prime contained in $r!(n-r)!$. Hence, p will appear in the numerator of $n!/r!(n-r)!$ at least as many times as it occurs in the denominator. Since this holds true for every prime divisor of the denominator, $r!(n-r)!$ must divide $n!$, making $n!/r!(n-r)!$ an integer.

COROLLARY.

For a positive integer r, the product of any r consecutive positive integers is divisible by r!.

Proof: The product of r consecutive positive integers, the largest of which is n, is

$$n(n-1)(n-2) \cdots (n-r+1).$$

Now we have

$$n(n-1) \cdots (n-r+1) = \left(\frac{n!}{r!(n-r)!} \right) r!.$$

Since $n!/r!(n-r)!$ is an integer, it follows that $r!$ must divide the product $n(n-1) \cdots (n-r+1)$, as asserted.

We pick up a few loose threads. Having introduced the greatest integer function, let us see what it has to do with the study of number-theoretic functions. Their relationship is brought out by

THEOREM 6-11.

Let f and F be number-theoretic functions such that

$$F(n) = \sum_{d \mid n} f(d).$$

Then, for any positive integer N,

$$\sum_{n=1}^{N} F(n) = \sum_{k=1}^{N} f(k)[N/k].$$

Proof: We begin by noting that

(1) $$\sum_{n=1}^{N} F(n) = \sum_{n=1}^{N} \sum_{d \mid n} f(d).$$

The strategy is to collect terms with equal values of $f(d)$ in this double sum. For a fixed positive integer $k \leq N$, the term $f(k)$ appears in $\sum_{d \mid n} f(d)$ if and only if k is a divisor of n. (Since each integer has itself as a divisor, the right-hand side of equation (1) includes $f(k)$, at least once.) Now, in order to calculate the number of sums $\sum_{d \mid n} f(d)$ in which $f(k)$ occurs as a term, it is sufficient to find the number of integers among $1, 2, \ldots, N$ which are divisible by k. There are exactly $[N/k]$ of them:

$$k, 2k, 3k, \ldots, [N/k]k.$$

Thus, for each k such that $1 \leq k \leq N$, $f(k)$ is a term of the sum $\sum_{d \mid n} f(d)$ for $[N/k]$ different positive integers less than or equal to N. Knowing this, we may rewrite the double sum in equation (1) as

$$\sum_{n=1}^{N} \sum_{d \mid n} f(d) = \sum_{k=1}^{N} f(k)[N/k]$$

and our task is complete.

As an immediate application of Theorem 6-11, we deduce

COROLLARY 1.

If N is a positive integer, then

$$\sum_{n=1}^{N} \tau(n) = \sum_{n=1}^{N} [N/n].$$

Proof: Noting that $\tau(n) = \sum_{d \mid n} 1$, we may write τ for F and take f to be the constant function $f(n) = 1$ for all n.

In the same way, the relation $\sigma(n) = \sum_{d \mid n} d$ yields

COROLLARY 2.

If N is a positive integer, then

$$\sum_{n=1}^{N} \sigma(n) = \sum_{n=1}^{N} n[N/n].$$

These last two corollaries are perhaps clarified with an example.

Example 6-3

Consider the case $N = 6$. The definition of τ tells us that

$$\sum_{n=1}^{6} \tau(n) = 14.$$

From Corollary 1,

$$\sum_{n=1}^{6} [6/n] = [6] + [3] + [2] + [3/2] + [6/5] + [1]$$

$$= 6 + 3 + 2 + 1 + 1 + 1 = 14,$$

as it should. In the present case, we also have

$$\sum_{n=1}^{6} \sigma(n) = 33,$$

while a simple calculation leads to

$$\sum_{n=1}^{6} n[6/n] = 1[6] + 2[3] + 3[2] + 4[3/2] + 5[6/5] + 6[1]$$

$$= 1 \cdot 6 + 2 \cdot 3 + 3 \cdot 2 + 4 \cdot 1 + 5 \cdot 1 + 6 \cdot 1 = 33.$$

PROBLEMS 6.3

1. Given integers a and $b > 0$, show that there exists a unique integer r with $0 \le r < b$ satisfying $a = [a/b]b + r$.

2. Let x and y be real numbers. Prove that the greatest integer function satisfies the following properties:
   (a) $[x + n] = [x] + n$ for any integer n.
   (b) $[x] + [-x] = 0$ or -1, according as x is an integer or not. [*Hint:* Write $x = [x] + \theta$, with $0 \le \theta < 1$, so that $-x = -[x] - 1 + (1 - \theta)$.]
   (c) $[x] + [y] \le [x + y]$ and, when x and y are positive, $[x][y] \le [xy]$.
   (d) $[x/n] = [[x]/n]$ for any positive integer n. [*Hint:* Let $x/n = [x/n] + \theta$, where $0 \le \theta < 1$; then $[x] = n[x/n] + [n\theta]$.]
   (e) $[nm/k] \ge n[m/k]$ for positive integers, n, m, k.
   (f) $[x] + [y] + [x + y] \le [2x] + [2y]$. [*Hint:* Let $x = [x] + \theta, 0 \le \theta < 1$, and $y = [y] + \theta', 0 \le \theta' < 1$. Consider cases in which neither, one, or both of θ and θ' are greater than or equal to $\frac{1}{2}$.]

3. Find the highest power of 5 dividing 1000! and the highest power of 7 dividing 2000!.

4. For an integer $n \ge 0$, show that $[n/2] - [-n/2] = n$.

5. (a) Verify that 1000! terminates in 249 zeros.
   (b) For what values of n does $n!$ terminate in 37 zeros?

6. If $n \geq 1$ and p is a prime, prove that

    (a) $(2n)!/(n!)^2$ is an even integer. [*Hint:* Use Theorem 6–10.]

    (b) The exponent of the highest power of p that divides $(2n)!/(n!)^2$ is

$$\sum_{k=1}^{\infty} ([2n/p^k] - 2[n/p^k]).$$

    (c) In the prime factorization of $(2n)!/(n!)^2$ the exponent of any prime p such that we have $n < p < 2n$ is equal to 1.

7. Let the positive integer n be written in terms of powers of the prime p so that we have $n = a_k p^k + \cdots + a_2 p^2 + a_1 p + a_0$, where $0 \leq a_i < p$. Show that the exponent of the highest power of p appearing in the prime factorization of $n!$ is

$$\frac{n - (a_k + \cdots + a_2 + a_1 + a_0)}{p - 1}.$$

8. (a) Using Problem 7, show that the exponent of highest power of p dividing $(p^k - 1)!$ is $[p^k - (p - 1)k - 1]/(p - 1)$. [*Hint:* Recall the identity

$$p^k - 1 = (p - 1)(p^{k-1} + \cdots + p^2 + p + 1).]$$

    (b) Determine the highest power of 3 dividing 80! and the highest power of 7 dividing 2400!. [*Hint:* $2400 = 7^4 - 1$.]

9. Find an integer $n \geq 1$ such that the highest power of 5 contained in $n!$ is 100. [*Hint:* Since the sum of coefficients of the powers of 5 needed to express n in the base 5 is at least 1, begin by considering the equation $(n - 1)/4 = 100$.]

10. Given a positive integer N, show that

    (a) $\displaystyle\sum_{n=1}^{N} \mu(n)[N/n] = 1$;

    (b) $\displaystyle\left| \sum_{n=1}^{N} \mu(n)/n \right| \leq 1$.

11. Illustrate Problem 10 in the case $N = 6$.

12. Verify that the formula

$$\sum_{n=1}^{N} \lambda(n)[N/n] = [\sqrt{N}]$$

holds for any positive integer N. [*Hint:* Apply Theorem 6-11 to the multiplicative function $F(n) = \sum_{d \mid n} \lambda(d)$, noting that there are $[\sqrt{n}]$ perfect squares not exceeding n.]

13. If N is a positive integer, establish that

    (a) $\displaystyle N = \sum_{n=1}^{2N} \tau(n) - \sum_{n=1}^{N} [2N/n]$;

    (b) $\displaystyle \tau(N) = \sum_{n=1}^{N} ([N/n] - [(N - 1)/n])$.

7

Euler's Generalization of Fermat's Theorem

*"Euler calculated without apparent effort, just as men breathe,
as eagles sustain themselves in the air."*
ARAGO

7.1 LEONHARD EULER

The importance of Fermat's work resides, not so much in any contribution to the mathematics of his own day, but rather in its animating effect on later generations of mathematicians. Perhaps the greatest disappointment of Fermat's career was his inability to interest others in his new number theory. A century was to pass before a first class mathematician, Leonhard Euler (1707–1783), either understood or appreciated its significance. Many of the theorems announced without proof by Fermat yielded to Euler's skill, and it is likely that the arguments devised by Euler were not substantially different from those which Fermat said he possessed.

The key figure in 18th century mathematics, Euler was the son of a Lutheran pastor who lived in the vicinity of Basel, Switzerland. His father earnestly wished him to enter the ministry and, at the age of 13, sent his son to the University of Basel to study theology. There he came into contact with Johann Bernoulli—then one of Europe's leading mathematicians—and he befriended Bernoulli's two sons, Nicolaus and Daniel. Within a short time, Euler broke off the theological studies that had been selected for him in order to address himself exclusively to mathematics. He received his master's degree in 1723 and in 1727, when he was only 19, won a prize from the Paris Academy of Sciences for a treatise on the most efficient arrangement of ship masts.

Where the 17th century had been an age of great amateur mathematicians, the 18th century was almost exclusively an era of professionals—university professors and members of scientific academies. Many of the reigning monarchs delighted in regarding themselves as patrons of learning, and the academies served as the intellectual crown jewels of the royal courts. While the motives of these rulers may not have been

Leonhard Euler

(1707–1783)

(From A Concise History of Mathematics *by Dirk Struik, 1967, Dover Publications, Inc., N.Y.)*

entirely philanthropic, the fact remains that the learned societies constituted important agencies for the promotion of science. They provided salaries for distinguished scholars, published journals of research papers on a regular basis, and offered monetary prizes for scientific discoveries. Euler was at different times associated with two of the newly formed academies, the Imperial Academy at St. Petersburg (from 1727 to 1741, and again, from 1766 to 1783) and the Royal Academy in Berlin (from 1741 to 1766). In 1725, Peter the Great had founded the Academy of St. Petersburg and attracted a number of leading mathematicians to Russia, including Nicolaus and Daniel Bernoulli. On their recommendation an appointment was secured for Euler. Because of his youth, he had recently been denied a professorship in physics at the University of Basel and was only too ready to accept the invitation of the Academy. In St. Petersburg, he soon came in contact with the versatile scholar Christian Goldbach (of the famous conjecture), a man who subsequently rose from professor of mathematics to Russian Minister of Foreign Affairs. Given his interests, it seems likely that Goldbach was the one who first drew Euler's attention to the work of Fermat on the theory of numbers.

Euler eventually sickened of the political repression then prevalent in Russia and accepted the call of Frederick the Great to become a member of the Berlin Academy. The story is told that, during a reception at Court, he was kindly received by the Queen Mother who inquired why so distinguished a scholar should be so timid and reticent; he replied, "Madame, it is because I have just come from a country where, when one speaks, one is hanged." Flattered by the warmth of the Russian feeling towards him, however, and unendurably offended by the contrasting coolness of Frederick and his court, Euler returned to St. Petersburg in 1766 to spend his remaining days. Within two or three years of his going back, Euler had the misfortune to become totally blind.

However, Euler would not permit blindness to retard his scientific work; aided by a phenomenal memory, his writings grew to such enormous proportions as to be virtually unmanageable. Without a doubt, Euler was the most prolific writer in the entire history of mathematics. He wrote or dictated over 700 books and papers in his lifetime, and left so much unpublished material that the St. Petersburg Academy did not finish printing all his manuscripts until 47 years after his death. The publication of Euler's collected works was begun by the Swiss Society of Natural Sciences in 1911 and it is estimated that more than 75 large volumes will ultimately be required for the completion of this monumental project. The best testament to the quality of these papers may be the fact that on twelve occasions they won the coveted biennial prize of the French Academy in Paris.

During his stay in Berlin, Euler acquired the habit of writing memoir after memoir, placing each when finished at the top of a pile of manuscripts. Whenever material was needed to fill the Academy's journal, the printers would help themselves to a few papers from the top of the stack. As the height of the pile increased more rapidly than the demands made upon it, memoirs at the bottom tended to remain in place a long time. This explains how it happened that various papers of Euler were published, while extensions and improvements of the material contained in them had previously appeared in print under his name. We might also add that the manner in which Euler made his work public contrasts sharply with the secrecy customary in Fermat's time.

7.2 EULER'S PHI-FUNCTION

The present chapter deals with that part of the theory arising out of the result known as Euler's Generalization of Fermat's Theorem. In a nutshell, Euler extended Fermat's Theorem, which concerns congruences with prime moduli, to arbitrary moduli. While doing so, he introduced an important number-theoretic function, described as follows:

DEFINITION 7-1.
   For $n \geq 1$, let $\phi(n)$ denote the number of positive integers not exceeding n that are relatively prime to n.

As an illustration of the definition, we find that $\phi(30) = 8$; for, among the positive integers that do not exceed 30, there are eight which are relatively prime to 30; specifically

$$1, 7, 11, 13, 17, 19, 23, 29.$$

Similarly, for the first few positive integers, the reader may check that

$$\phi(1) = 1, \phi(2) = 1, \phi(3) = 2, \phi(4) = 2, \phi(5) = 4, \phi(6) = 2, \phi(7) = 6, \ldots.$$

Notice that $\phi(1) = 1$, since $\gcd(1, 1) = 1$. While if $n > 1$, then $\gcd(n, n) = n \neq 1$, so that $\phi(n)$ can be characterized as the number of integers less than n and relatively prime to it. The function ϕ is usually called the *Euler phi-function* (sometimes, the *indicator* or *totient*) after its originator; the functional notation $\phi(n)$, however, is credited to Gauss.

If n is a prime number, then every integer less than n is relatively prime to it; whence, $\phi(n) = n - 1$. On the other hand, if $n > 1$ is composite, then n has a divisor d such that $1 < d < n$. It follows that there are at least two integers among $1, 2, 3, \ldots, n$ which are not relatively prime to n, namely, d and n itself. As a result, $\phi(n) \leq n - 2$. This proves: for $n > 1$,

$$\phi(n) = n - 1 \text{ if and only if } n \text{ is prime.}$$

The first item on the agenda is to derive a formula that will allow us to calculate the value of $\phi(n)$ directly from the prime-power factorization of n. A large step in this direction stems from

THEOREM 7-1.

If p is a prime and $k > 0$, then

$$\phi(p^k) = p^k - p^{k-1} = p^k(1 - 1/p).$$

Proof: Clearly, $\gcd(n, p^k) = 1$ if and only if $p \nmid n$. There are p^{k-1} integers between 1 and p^k which are divisible by p, namely

$$p, 2p, 3p, \ldots, (p^{k-1})p.$$

Thus, the set $\{1, 2, \ldots, p^k\}$ contains exactly $p^k - p^{k-1}$ integers which are relatively prime to p^k and so, by the definition of the phi-function, $\phi(p^k) = p^k - p^{k-1}$.

For an example, we have

$$\phi(9) = \phi(3^2) = 3^2 - 3 = 6;$$

the six integers less than and relatively prime to 9 are 1, 2, 4, 5, 7, 8. To give a second illustration, there are 8 integers which are less than 16 and relatively prime to it, that is, 1, 3, 5, 7, 9, 11, 13, 15. Theorem 7–1 yields the same count:

$$\phi(16) = \phi(2^4) = 2^4 - 2^3 = 16 - 8 = 8.$$

We now know how to evaluate the phi-function for prime powers and our aim is to obtain a formula for $\phi(n)$ based on the factorization of n as a product of primes. The missing link in the chain is obvious: show that ϕ is a multiplicative function. We pave the way with an easy lemma.

LEMMA.

Given integers a, b, c, $\gcd(a, bc) = 1$ if and only if $\gcd(a, b) = 1$ and $\gcd(a, c) = 1$.

Proof: Suppose first that $\gcd(a, bc) = 1$, and put $d = \gcd(a, b)$. Then $d \mid a$ and $d \mid b$, whence $d \mid a$ and $d \mid bc$. This implies that $\gcd(a, bc) \geq d$, which forces $d = 1$. Similar reasoning gives rise to the statement $\gcd(a, c) = 1$.

For the other direction, let $\gcd(a, b) = 1 = \gcd(a, c)$ and assume that $\gcd(a, bc) = d_1 > 1$. Then d_1 must have a prime divisor p. Since $d_1 \mid bc$, it follows that $p \mid bc$; in consequence, $p \mid b$ or $p \mid c$. If $p \mid b$, then (by virtue of the fact that $p \mid a$) $\gcd(a , b) \geq p$, a contradiction. In the same way, the condition $p \mid c$ leads to the equally false conclusion that $\gcd(a , c) \geq p$. Thus $d_1 = 1$ and the lemma is proven.

THEOREM 7-2.

The function ϕ is a multiplicative function.

Proof: It is required to show that $\phi(mn) = \phi(m)\phi(n)$, wherever m and n have no common factor. Since $\phi(1) = 1$, the result obviously holds if either m or n equals 1. Thus we may assume that $m > 1$ and $n > 1$. Arrange the integers from 1 to mn in m columns of n integers each, as follows:

1	2	$\cdots$	r	$\cdots$	m
$m + 1$	$m + 2$		$m + r$		$2m$
$2m + 1$	$2m + 2$		$2m + r$		$3m$
$\vdots$	$\vdots$		$\vdots$		$\vdots$
$(n - 1)m + 1$	$(n - 1)m + 2$		$(n - 1)m + r$		nm

We know that $\phi(mn)$ is equal to the number of entries in the above array which are relatively prime to mn; by virtue of the lemma, this is the same as the number of integers which are relatively prime to both m and n.

Before embarking on the details, it is worth commenting on the tactics to be adopted: Since $\gcd(qm + r , m) = \gcd(r , m)$, the numbers in the rth column are relatively prime to m if and only if r itself is relatively prime to m. Therefore, only $\phi(m)$ columns contain integers relatively prime to m, and every entry in the column will be relatively prime to m. The problem is one of showing that in each of these $\phi(m)$ columns there are exactly $\phi(n)$ integers which are relatively prime to n; for then there would be altogether $\phi(m)\phi(n)$ numbers in the table which are relatively prime to both m and n.

Now the entries in the rth column (where it is assumed that $\gcd(r , m) = 1$) are

$$r, m + r, 2m + r . . . , (n - 1)m + r.$$

There are n integers in this sequence and no two are congruent modulo n. Indeed, were

$$km + r \equiv jm + r \pmod{n}$$

with $0 \leq k < j < n$, it would follow that $km \equiv jm \pmod{n}$. Since $\gcd(m , n) = 1$, we could cancel m from both sides of this congruence to arrive at the contradiction that $k \equiv j \pmod{n}$. Thus, the numbers in the rth column are congruent modulo n to $0, 1, 2, . . . , n - 1$, in some order. But if $s \equiv t \pmod{n}$, then $\gcd(s , n) = 1$ if and only if $\gcd(t , n) = 1$. The implication is that the rth column contains as many integers which are relatively prime to n as does the set $\{0, 1, 2, . . . , n - 1\}$, namely, $\phi(n)$ integers. Therefore, the total number of entries in the array that are relatively prime to both m and n is $\phi(m)\phi(n)$. This completes the proof of the theorem.

With these preliminaries in hand, we can now prove

THEOREM 7-3.

If the integer $n > 1$ has the prime factorization $n = p_1^{k_1} p_2^{k_2} \cdots p_r^{k_r}$, then

$$\phi(n) = (p_1^{k_1} - p_1^{k_1 - 1})(p_2^{k_2} - p_2^{k_2 - 1}) \cdots (p_r^{k_r} - p_r^{k_r - 1})$$
$$= n(1 - 1/p_1)(1 - 1/p_2) \cdots (1 - 1/p_r).$$

Proof: We intend to use induction on r, the number of distinct prime factors of n. By Theorem 7-1, the result is true for $r = 1$. Suppose that it holds for $r = i$. Since

$$\gcd(p_1^{k_1} p_2^{k_2} \cdots p_i^{k_i}, p_{i+1}^{k_i + 1}) = 1,$$

the definition of multiplicative function gives

$$\phi((p_1^{k_1} \cdots p_i^{k_i}) p_{i+1}^{k_i + 1}) = \phi(p_1^{k_1} \cdots p_i^{k_i}) \phi(p_{i+1}^{k_i + 1})$$
$$= \phi(p_1^{k_1} \cdots p_i^{k_i}) (p_{i+1}^{k_i + 1} - p_{i+1}^{k_i + 1 - 1}).$$

Invoking the induction assumption, the first factor on the right-hand side becomes

$$\phi(p_1^{k_1} p_2^{k_2} \cdots p_i^{k_i}) = (p_1^{k_1} - p_1^{k_1 - 1})(p_2^{k_2} - p_2^{k_2 - 1}) \cdots (p_i^{k_i} - p_i^{k_i - 1})$$

and this serves to complete the induction step, as well as the proof.

Example 7-1

Let us calculate the value $\phi(360)$, for instance. The prime-power decomposition of 360 is $2^3 \cdot 3^2 \cdot 5$, and Theorem 7-3 tells us that

$$\phi(360) = 360(1 - \tfrac{1}{2})(1 - \tfrac{1}{3})(1 - \tfrac{1}{5})$$
$$= 360 \cdot \tfrac{1}{2} \cdot \tfrac{2}{3} \cdot \tfrac{4}{5} = 96.$$

The sharp-eyed reader will have noticed that, save for $\phi(1)$ and $\phi(2)$, the values of $\phi(n)$ in our examples are always even. This is no accident, as the next theorem shows.

THEOREM 7-4.

For $n > 2$, $\phi(n)$ is an even integer.

Proof: First, assume that n is a power of 2, let us say $n = 2^k$, with $k \geq 2$. By Theorem 7-3,

$$\phi(n) = \phi(2^k) = 2^k(1 - \tfrac{1}{2}) = 2^{k-1},$$

an even integer. If n does not happen to be a power of 2, then it is divisible by an odd prime p; we may therefore write n as $n = p^k m$, where $k \geq 1$ and $\gcd(p^k, m) = 1$. Exploiting the multiplicative nature of the phi-function, one gets

$$\phi(n) = \phi(p^k)\phi(m) = p^{k-1}(p - 1)\phi(m),$$

which is again even since $2 \mid p - 1$.

We can establish Euclid's Theorem on the infinitude of primes in the following new way: As before, assume that there are only a finite number of primes. Call them p_1, $p_2, \ldots, p_r$ and consider the integer $n = p_1 p_2 \cdots p_r$. We argue that if $1 < a \leq n$, then $\gcd(a, n) \neq 1$. For, the Fundamental Theorem of Arithmetic tells us that a has a prime divisor q. Since $p_1, p_2, \ldots, p_r$ are the only primes, q must be one of these p_i, whence $q \mid n$; in other words, $\gcd(a, n) \geq q$. The implication of all this is that $\phi(n) = 1$, which is clearly impossible by Theorem 7-4.

PROBLEMS 7.2

1. Calculate $\phi(1001)$, $\phi(5040)$, and $\phi(36,000)$.

2. Verify that the equality $\phi(n) = \phi(n + 1) = \phi(n + 2)$ holds when $n = 5186$.

3. Show that the integers $m = 3^k \cdot 568$ and $n = 3^k \cdot 638$, where $k \geq 0$, satisfy simultaneously

    $$\tau(m) = \tau(n), \quad \sigma(m) = \sigma(n), \quad \phi(m) = \phi(n).$$

4. Establish each of the assertions below:
    (a) If n is an odd integer, then $\phi(2n) = \phi(n)$.
    (b) If n is an even integer, then $\phi(2n) = 2\phi(n)$.
    (c) $\phi(3n) = 3\phi(n)$ if and only if $3 \mid n$.
    (d) $\phi(3n) = 2\phi(n)$ if and only if $3 \nmid n$.
    (e) $\phi(n) = n/2$ if and only if $n = 2^k$ for some $k \geq 1$. [*Hint:* Write $n = 2^k N$, where N is odd, and use the condition $\phi(n) = n/2$ to show that $N = 1$.]

5. Prove that the equation $\phi(n) = \phi(n + 2)$ is satisfied by $n = 2(2p - 1)$ whenever p and $2p - 1$ are both odd primes.

6. Show that there are infinitely many integers n for which $\phi(n)$ is a perfect square. [*Hint:* Consider the integers $n = 2^{2k+1}$ for $k = 1, 2, \ldots$.]

7. Verify the following:
    (a) For any positive integer n, $\frac{1}{2} \sqrt{n} \leq \phi(n) \leq n$. [*Hint:* Write $n = 2^{k_0} p_1^{k_1} \cdots p_r^{k_r}$, so $\phi(n) = 2^{k_0 - 1} p_1^{k_1 - 1} \cdots p_r^{k_r - 1}(p_1 - 1) \cdots (p_r - 1)$. Now use the inequalities $p - 1 > \sqrt{p}$ and $k - \frac{1}{2} \geq k/2$ to obtain $\phi(n) \geq 2^{k_0 - 1} p_1^{k_1/2} \cdots p_r^{k_r/2}$.]
    (b) If the integer $n > 1$ has r distinct prime factors, then $\phi(n) \geq n/2^r$.
    (c) If $n > 1$ is a composite number, then $\phi(n) \leq n - \sqrt{n}$. [*Hint:* Let p be the smallest prime divisor of n, so that $p \leq \sqrt{n}$. Then $\phi(n) \leq n(1 - 1/p)$.]

8. Prove that if the integer n has r distinct odd prime factors, then $2^r \mid \phi(n)$.

9. Prove that:
    (a) If n and $n + 2$ are a pair of twin primes, then $\phi(n + 2) = \phi(n) + 2$; this also holds for $n = 12, 14$, and 20.
    (b) If p and $2p + 1$ are both odd primes, then $n = 4p$ satisfies $\phi(n + 2) = \phi(n) + 2$.

10. If every prime that divides n also divides m, establish that $\phi(nm) = n\phi(m)$; in particular, $\phi(n^2) = n\phi(n)$ for every positive integer n.

11. (a) If $\phi(n) \mid n - 1$, prove that n is a square-free integer. [*Hint:* Assume that n has the prime factorization $n = p_1^{k_1} p_2^{k_2} \cdots p_r^{k_r}$, where $k_1 \geq 2$. Then $p_1 \mid \phi(n)$, whence $p_1 \mid n - 1$, which leads to a contradiction.]
    (b) Show that if $n = 2^k$ or $2^k 3^j$, with k and j positive integers, then $\phi(n) \mid n$.

12. If $n = p_1^{k_1} p_2^{k_2} \cdots p_r^{k_r}$, derive the inequalities
    (a) $\sigma(n)\phi(n) \geq n^2(1 - 1/p_1^2)(1 - 1/p_2^2) \cdots (1 - 1/p_r^2)$, and
    (b) $\tau(n)\phi(n) \geq n$. [*Hint:* Show that $\tau(n)\phi(n) \geq 2^r \cdot n(1/2)^r$.]

13. Assuming that $d \mid n$, prove that $\phi(d) \mid \phi(n)$. [*Hint:* Work with the prime factorizations of d and n.]

14. Obtain the following two generalizations of Theorem 7-2:
    (a) For positive integers m and n,

$$\phi(m)\phi(n) = \phi(mn)\phi(d)/d,$$

    where $d = \gcd(m , n)$.
    (b) For positive integers m and n,

$$\phi(m)\phi(n) = \phi(\gcd(m , n))\phi(\mathrm{lcm}(m , n)).$$

15. Prove that:
    (a) There are infinitely many integers n for which $\phi(n) = n/3$. [*Hint:* Consider $n = 2^k 3^j$, where k and j are positive integers.]
    (b) There are no integers n for which $\phi(n) = n/4$.

16. Show that Goldbach's Conjecture implies that for each even integer $2n$ there exist integers n_1 and n_2 with $\phi(n_1) + \phi(n_2) = 2n$.

17. Given a positive integer k, show that
    (a) there are at most a finite number of integers n for which $\phi(n) = k$;
    (b) if the equation $\phi(n) = k$ has a unique solution, say $n = n_0$, then $4 \mid n_0$. [*Hint:* See Problem 4(a) and 4(b).]
    A famous conjecture of Carmichael is that the number of solutions of $\phi(n) = k$ cannot be equal to one.

18. Find all solutions of $\phi(n) = 16$ and $\phi(n) = 24$. [*Hint:* If $n = p_1^{k_1} p_2^{k_2} \cdots p_r^{k_r}$ satisfies $\phi(n) = k$, then $n = [k/\Pi(p_i - 1)] \Pi p_i$. Thus the integers $d_i = p_i - 1$ can be determined by the conditions (1) $d_i \mid k$, (2) $d_i + 1$ is prime and (3) $k/\Pi d_i$ contains no prime factor not in Πp_i.]

19. (a) Prove that the equation $\phi(n) = 2p$, where p is a prime number and $2p + 1$ is composite, is not solvable.
    (b) Prove that there is no solution to the equation $\phi(n) = 14$, and that 14 is the smallest (positive) even integer with this property.

20. If p is a prime and $k \geq 2$, show that $\phi(\phi(p^k)) = p^{k-2}\phi((p - 1)^2)$.

7.3 EULER'S THEOREM

As remarked earlier, the first published proof of Fermat's Theorem (that $a^{p-1} \equiv 1$ (mod p) if $p \nmid a$) was given by Euler in 1736. Somewhat later, in 1760, he succeeded in generalizing Fermat's Theorem from the case of a prime p to an arbitrary integer n. This landmark result states: if $\gcd(a , n) = 1$, then $a^{\phi(n)} \equiv 1$ (mod n).

For example, putting $n = 30$ and $a = 11$, we have

$$11^{\phi(30)} \equiv 11^8 \equiv (11^2)^4 \equiv (121)^4 \equiv 1^4 \equiv 1 \pmod{30}.$$

As a prelude to launching our proof of Euler's Generalization of Fermat's Theorem, we require a preliminary lemma.

LEMMA.

> Let $n > 1$ and $\gcd(a, n) = 1$. If $a_1, a_2, \ldots, a_{\phi(n)}$ are the positive integers less than n and relatively prime to n, then
>
> $$aa_1, aa_2, \ldots, aa_{\phi(n)}$$
>
> are congruent modulo n to $a_1, a_2, \ldots, a_{\phi(n)}$ in some order.

Proof: Observe that no two of the integers $aa_1, aa_2, \ldots, aa_{\phi(n)}$ are congruent modulo n. For if $aa_i \equiv aa_j \pmod{n}$, with $1 \leq i < j \leq \phi(n)$, then the cancellation law yields $a_i \equiv a_j \pmod{n}$ and thus $a_i = a_j$, a contradiction. Furthermore, since $\gcd(a_i, n) = 1$ for all i and $\gcd(a, n) = 1$, the lemma on page 126 guarantees that each of the aa_i is relatively prime to n.

Fixing on a particular aa_i, there exists a unique integer b, where $0 \leq b < n$, for which $aa_i \equiv b \pmod{n}$. Because

$$\gcd(b, n) = \gcd(aa_i, n) = 1,$$

b must be one of the integers $a_1, a_2, \ldots, a_{\phi(n)}$. All told, this proves that the numbers $aa_1, aa_2, \ldots, aa_{\phi(n)}$ and the numbers $a_1, a_2, \ldots, a_{\phi(n)}$ are identical (modulo n) in a certain order.

THEOREM 7-5 (Euler).

> If n is a positive integer and $\gcd(a, n) = 1$, then $a^{\phi(n)} \equiv 1 \pmod{n}$.

Proof: There is no harm in taking $n > 1$. Let $a_1, a_2, \ldots, a_{\phi(n)}$ be the positive integers less than n which are relatively prime to n. Since $\gcd(a, n) = 1$, it follows from the lemma that $aa_1, aa_2, \ldots, aa_{\phi(n)}$ are congruent, not necessarily in order of appearance, to $a_1, a_2, \ldots, a_{\phi(n)}$. Then

$$aa_1 \equiv a_1' \pmod{n},$$

$$aa_2 \equiv a_2' \pmod{n},$$

$$\vdots \qquad \vdots$$

$$aa_{\phi(n)} \equiv a_{\phi(n)}' \pmod{n},$$

where $a_1', a_2', \ldots, a_{\phi(n)}'$ are the integers $a_1, a_2, \ldots, a_{\phi(n)}$ in some order. On taking the product of these $\phi(n)$ congruences, we get

$$(aa_1)(aa_2) \cdots (aa_{\phi(n)}) \equiv a_1' a_2' \cdots a_{\phi(n)}' \pmod{n}$$

$$\equiv a_1 a_2 \cdots a_{\phi(n)} \pmod{n}$$

and so

$$a^{\phi(n)}(a_1 a_2 \cdots a_{\phi(n)}) \equiv a_1 a_2 \cdots a_{\phi(n)} \pmod{n}.$$

Since $\gcd(a_i, n) = 1$ for each i, the lemma preceding Theorem 7-2 implies that $\gcd(a_1 a_2 \cdots a_{\phi(n)}, n) = 1$. Therefore we may divide both sides of the foregoing congruence by the common factor $a_1 a_2 \cdots a_{\phi(n)}$, leaving us with

$$a^{\phi(n)} \equiv 1 \pmod{n}.$$

This proof can best be illustrated by carrying it out with some specific numbers. Let $n = 9$, for instance. The positive integers less than and relatively prime to 9 are

$$1, 2, 4, 5, 7, 8.$$

These play the role of the integers $a_1, a_2, \ldots, a_{\phi(n)}$ in the proof of Theorem 7-5. If $a = -4$, then the integers aa_i are

$$-4, -8, -16, -20, -28, -32,$$

where, modulo 9,

$$-4 \equiv 5, \; -8 \equiv 1, \; -16 \equiv 2, \; -20 \equiv 7, \; -28 \equiv 8, \; -32 \equiv 4.$$

When the above congruences are all multiplied together, we obtain

$$(-4)(-8)(-16)(-20)(-28)(-32) \equiv 5 \cdot 1 \cdot 2 \cdot 7 \cdot 8 \cdot 4 \; (\text{mod } 9),$$

which becomes

$$(1 \cdot 2 \cdot 4 \cdot 5 \cdot 7 \cdot 8)(-4)^6 \equiv (1 \cdot 2 \cdot 4 \cdot 5 \cdot 7 \cdot 8) \; (\text{mod } 9).$$

Being relatively prime to 9, the six integers 1, 2, 4, 5, 7, 8 may be successively cancelled to give

$$(-4)^6 \equiv 1 \; (\text{mod } 9).$$

The validity of this last congruence is confirmed by the calculation

$$(-4)^6 \equiv 4^6 \equiv (64)^2 \equiv 1^2 \equiv 1 \; (\text{mod } 9).$$

Note that Theorem 7-5 does indeed generalize the one due to Fermat, which we proved earlier. For if p is a prime, then $\phi(p) = p - 1$; hence, whenever $\gcd(a, p) = 1$, we get

$$a^{p-1} \equiv a^{\phi(p)} \equiv 1 \; (\text{mod } p)$$

and so:

COROLLARY (Fermat).
  If p is a prime and $p \nmid a$, then

$$a^{p-1} \equiv 1 \; (\text{mod } p).$$

Example 7-2

Euler's Theorem is helpful in reducing large powers modulo n. To cite a typical example, let us find the last two digits in the decimal representation of 3^{256}; this is equivalent to obtaining the smallest nonnegative integer to which 3^{256} is congruent modulo 100. Since $\gcd(3, 100) = 1$ and

$$\phi(100) = \phi(2^2 \cdot 5^2) = 100(1 - \tfrac{1}{2})(1 - \tfrac{1}{5}) = 40,$$

Euler's Theorem yields

$$3^{40} \equiv 1 \; (\text{mod } 100).$$

By the Division Algorithm, $256 = 6 \cdot 40 + 16$; whence

$$3^{256} \equiv 3^{6 \cdot 40 + 16} \equiv (3^{40})^6 3^{16} \equiv 3^{16} \pmod{100}$$

and our problem reduces to one of evaluating 3^{16}, modulo 100. The calculations, with reasons omitted, are as follows:

$$3^{16} \equiv (81)^4 \equiv (-19)^4 \equiv (361)^2 \equiv 61^2 \equiv 21 \pmod{100}.$$

There is another path to Euler's Theorem, one which requires the use of Fermat's Theorem.

Second Proof of Euler's Theorem: To start, we argue by induction that if $p \nmid a$ (p a prime), then

$$(1) \qquad\qquad a^{\phi(p^k)} \equiv 1 \pmod{p^k}, \qquad\qquad k > 0.$$

When $k = 1$, this assertion reduces to the statement of Fermat's Theorem. Assuming the truth of equation (1) for a fixed value of k, we wish to show that it is true with k replaced by $k + 1$.

Since (1) is assumed to hold, we may write

$$a^{\phi(p^k)} = 1 + qp^k$$

for some integer q. Notice too that

$$\phi(p^{k+1}) = p^{k+1} - p^k = p(p^k - p^{k-1}) = p\phi(p^k).$$

Using these facts, along with the Binomial Theorem, we obtain

$$a^{\phi(p^{k+1})} = a^{p\phi(p^k)} = (a^{\phi(p^k)})^p$$

$$= (1 + qp^k)^p$$

$$= 1 + \binom{p}{1}(qp^k) + \binom{p}{2}(qp^k)^2 + \cdots$$

$$+ \binom{p}{p-1}(qp^k)^{p-1} + (qp^k)^p$$

$$\equiv 1 + \binom{p}{1}(qp^k) \pmod{p^{k+1}}.$$

But $p \mid \binom{p}{1}$ and so $p^{k+1} \mid \binom{p}{1}(qp^k)$. Thus, the last-written congruence becomes

$$a^{\phi(p^{k+1})} \equiv 1 \pmod{p^{k+1}},$$

completing the induction step.

Now let $\gcd(a, n) = 1$ and n have the prime-power factorization $n = p_1^{k_1} p_2^{k_2} \cdots p_r^{k_r}$. In view of what has already been proved, each of the congruences

$$(2) \qquad\qquad a^{\phi(p_i^{k_i})} \equiv 1 \pmod{p_i^{k_i}}, \qquad\qquad i = 1, 2, \ldots, r$$

holds. Noting that $\phi(n)$ is divisible by $\phi(p_i^{k_i})$, we may raise both sides of equation (2) to the power $\phi(n)/\phi(p_i^{k_i})$ and arrive at

$$a^{\phi(n)} \equiv 1 \pmod{p_i^{k_i}}, \qquad\qquad i = 1, 2, \ldots, r.$$

Inasmuch as the moduli are relatively prime, this leads us to the relation

$$a^{\phi(n)} \equiv 1 \pmod{p_1{}^{k_1} p_2{}^{k_2} \cdots p_r{}^{k_r}}$$

or $a^{\phi(n)} \equiv 1 \pmod{n}$.

The usefulness of Euler's Theorem in number theory would be hard to exaggerate. It leads, for instance, to a different proof of the Chinese Remainder Theorem. In other words, we seek to establish that if $\gcd(n_i, n_j) = 1$ for $i \neq j$, then the system of linear congruences

$$x \equiv a_i \pmod{n_i}, \qquad\qquad i = 1, 2, \ldots, r$$

admits a simultaneous solution. Let $n = n_1 n_2 \cdots n_r$ and put $N_i = n/n_i$ for $n = 1, 2, \ldots, r$. Then the integer

$$x = a_1 N_1{}^{\phi(n_1)} + a_2 N_2{}^{\phi(n_2)} + \cdots + a_r N_r{}^{\phi(n_r)}$$

fulfills our requirements. To see this, first note that $N_j \equiv 0 \pmod{n_i}$ whenever $i \neq j$; whence,

$$x \equiv a_i N_i{}^{\phi(n_i)} \pmod{n_i}.$$

But, since $\gcd(N_i, n_i) = 1$, we have

$$N_i{}^{\phi(n_i)} \equiv 1 \pmod{n_i}$$

and so $x \equiv a_i \pmod{n_i}$ for each i.

As a second application of Euler's Theorem, let us show that if n is an odd integer which is not a multiple of 5, then n divides an integer all of whose digits are equal to 1. (For example: $7 \mid 111111$.) Since $\gcd(n, 10) = 1$ and $\gcd(9, 10) = 1$, we have $\gcd(9n, 10) = 1$. Quoting Theorem 7-5 again,

$$10^{\phi(9n)} \equiv 1 \pmod{9n}.$$

This says that $10^{\phi(9n)} - 1 = 9nk$ for some integer k or, what amounts to the same thing,

$$kn = \frac{10^{\phi(9n)} - 1}{9}.$$

The right-hand side of the above expression is an integer whose digits are all equal to 1, each digit of the numerator being clearly equal to 9.

PROBLEMS 7.3

1. Use Euler's Theorem to establish the following:
   (a) For any integer a, $a^{37} \equiv a \pmod{1729}$. [*Hint:* $1729 = 7 \cdot 13 \cdot 19$.]
   (b) For any integer a, $a^{13} \equiv a \pmod{2730}$. [*Hint:* $2730 = 2 \cdot 3 \cdot 5 \cdot 7 \cdot 13$.]
   (c) For any odd integer a, $a^{33} \equiv a \pmod{4080}$. [*Hint:* $4080 = 15 \cdot 16 \cdot 17$.]

2. Use Euler's Theorem to confirm that, for any integer $n \geq 0$,

$$51 \mid 10^{32n + 9} - 7.$$

3. Prove that $2^{15} - 2^3$ divides $a^{15} - a^3$ for any integer a.
   [*Hint:* $2^{15} - 2^3 = 5 \cdot 7 \cdot 8 \cdot 9 \cdot 13$.]

4. Show that if $\gcd(a, n) = \gcd(a - 1, n) = 1$, then

$$1 + a + a^2 + \cdots + a^{\phi(n) - 1} \equiv 0 \pmod{n}.$$

   [*Hint:* Recall that

$$a^{\phi(n)} - 1 = (a - 1)(a^{\phi(n) - 1} + \cdots + a^2 + a + 1).]$$

5. If m and n are relatively prime positive integers, prove that

$$m^{\phi(n)} + n^{\phi(m)} \equiv 1 \pmod{mn}.$$

6. Fill in any missing details in the following proof of Euler's Theorem: Let p be a prime divisor of n and $\gcd(a, p) = 1$. By Fermat's Theorem, $a^{p - 1} \equiv 1 \pmod{p}$, so that $a^{p - 1} = 1 + tp$ for some t. Therefore $a^{p(p - 1)} = (1 + tp)^p = 1 + \binom{p}{1}(tp) + \cdots + (tp)^p \equiv 1 \pmod{p^2}$ and, by induction, $a^{p^{k - 1}(p - 1)} \equiv 1 \pmod{p^k}$ where $k = 1, 2, \ldots$. Raise both sides of this congruence to the $\phi(n)/p^{k - 1}(p - 1)$ power to get $a^{\phi(n)} \equiv 1 \pmod{p^k}$. Thus $a^{\phi(n)} \equiv 1 \pmod{n}$.

7. Find the units digit of 3^{100} by means of Euler's Theorem.

8. (a) If $\gcd(a, n) = 1$, show that the linear congruence $ax \equiv b \pmod{n}$ has the solution $x \equiv ba^{\phi(n) - 1} \pmod{n}$.
   (b) Use part (a) to solve the linear congruences $3x \equiv 5 \pmod{26}$, $13x \equiv 2 \pmod{40}$ and $10x \equiv 21 \pmod{49}$.

9. Use Euler's Theorem to evaluate $2^{100000} \pmod{77}$.

10. For any integer a, show that a and $a^{4n + 1}$ have the same last digit.

11. For any prime p, establish each of the assertions below:
    (a) $\tau(p!) = 2\tau((p - 1)!)$;
    (b) $\sigma(p!) = (p + 1)\sigma((p - 1)!)$;
    (c) $\phi(p!) = (p - 1)\phi((p - 1)!)$.

12. Given $n \geq 1$, a set of $\phi(n)$ integers which are relatively prime to n and which are incongruent modulo n is called a *reduced set of residues modulo n* (that is, a reduced set of residues are those members of a complete set of residues modulo n which are relatively prime to n).
    Verify that
    (a) the integers $-31, -16, -8, 13, 25, 80$ form a reduced set of residues modulo 9;
    (b) the integers $3, 3^2, 3^3, 3^4, 3^5, 3^6$ form a reduced set of residues modulo 14;
    (c) the integers $2, 2^2, 2^3, \ldots, 2^{18}$ form a reduced set of residues modulo 27.

13. If p is an odd prime, show that the integers

$$-\frac{p - 1}{2}, \cdots, -2, -1, 1, 2, \ldots, \frac{p - 1}{2}$$

   form a reduced set of residues modulo p.

7.4 SOME PROPERTIES OF THE PHI-FUNCTION

The next theorem points out a curious feature of the phi-function; namely, that the sum of the values of $\phi(d)$, as d ranges over the positive divisors of n, is equal to n itself. This was first noticed by Gauss.

THEOREM 7-6 (Gauss).

For each positive integer $n \geq 1$,

$$n = \sum_{d \mid n} \phi(d),$$

the sum being extended over all positive divisors of n.

Proof: The integers between 1 and n can be separated into classes as follows: if d is a positive divisor of n, we put the integer m in the class S_d provided that $\gcd(m, n) = d$. Stated in symbols,

$$S_d = \{m \mid \gcd(m, n) = d; 1 \leq m \leq n\}.$$

Now $\gcd(m, n) = d$ if and only if $\gcd(m/d, n/d) = 1$. Thus the number of integers in the class S_d is equal to the number of positive integers not exceeding n/d which are relatively prime to n/d; in other words, equal to $\phi(n/d)$. Since each of the n integers in the set $\{1, 2, \ldots, n\}$ lies in exactly one class S_d, we obtain the formula

$$n = \sum_{d \mid n} \phi(n/d).$$

But as d runs through all positive divisors of n, so does n/d; hence,

$$\sum_{d \mid n} \phi(n/d) = \sum_{d \mid n} \phi(d)$$

which proves the theorem.

Example 7-3

A simple numerical example of what we have just said is provided by $n = 10$. Here, the classes S_d are

$$S_1 = \{1, 3, 7, 9\},$$

$$S_2 = \{2, 4, 6, 8\},$$

$$S_5 = \{5\},$$

$$S_{10} = \{10\}.$$

These contain $\phi(10) = 4$, $\phi(5) = 4$, $\phi(2) = 1$, and $\phi(1) = 1$ integers, respectively. Therefore,

$$\sum_{d \mid 10} \phi(d) = \phi(10) + \phi(5) + \phi(2) + \phi(1)$$

$$= 4 + 4 + 1 + 1 = 10.$$

It is instructive to give a second proof of Theorem 7-6, this one depending on the fact that ϕ is multiplicative. The details are as follows: If $n = 1$, then clearly

$$\sum_{d \mid n} \phi(d) = \sum_{d \mid 1} \phi(d) = \phi(1) = 1 = n.$$

Assuming that $n > 1$, let us consider the number-theoretic function

$$F(n) = \sum_{d \mid n} \phi(d).$$

Since ϕ is known to be a multiplicative function, Theorem 6-4 asserts that F is also multiplicative. Hence, if $n = p_1^{k_1} p_2^{k_2} \cdots p_r^{k_r}$ is the prime factorization of n, then

$$F(n) = F(p_1^{k_1}) F(p_2^{k_2}) \cdots F(p_r^{k_r}).$$

For each value of i,

$$F(p_i^{k_i}) = \sum_{d \mid p_i^{k_i}} \phi(d)$$

$$= \phi(1) + \phi(p_i) + \phi(p_i^2) + \phi(p_i^3) + \cdots + \phi(p_i^{k_i})$$

$$= 1 + (p_i - 1) + (p_i^2 - p_i) + (p_i^3 - p_i^2) + \cdots + (p_i^{k_i} - p_i^{k_i - 1})$$

$$= p_i^{k_i},$$

since the terms in the foregoing expression cancel each other, save for the term $p_i^{k_i}$. Knowing this, we end up with

$$F(n) = p_1^{k_1} p_2^{k_2} \cdots p_r^{k_r} = n$$

and so

$$n = \sum_{d \mid n} \phi(d),$$

as desired.

We should mention in passing that there is another interesting identity which involves the phi-function.

THEOREM 7-7.

For $n > 1$, the sum of the positive integers less than n and relatively prime to n is $\frac{1}{2} n \phi(n)$; in symbols,

$$\frac{1}{2} n \phi(n) = \sum_{\substack{\gcd(k,n) = 1 \\ 1 \leq k < n}} k.$$

Proof: Let $a_1, a_2, \ldots, a_{\phi(n)}$ be the positive integers less than n and relatively prime to n. Now, since $\gcd(a, n) = 1$ if and only if $\gcd(n - a, n) = 1$, the numbers $n - a_1, n - a_2, \ldots, n - a_{\phi(n)}$ are equal in some order to $a_1, a_2, \ldots, a_{\phi(n)}$. Thus

$$a_1 + a_2 + \cdots + a_{\phi(n)} = (n - a_1) + (n - a_2) + \cdots + (n - a_{\phi(n)})$$

$$= \phi(n)n - (a_1 + a_2 + \cdots + a_{\phi(n)}).$$

Hence,

$$2(a_1 + a_2 + \cdots + a_{\phi(n)}) = \phi(n)n,$$

leading to the stated conclusion.

Example 7-4

Consider the case $n = 30$. The $\phi(30) = 8$ integers which are less than 30 and relatively prime to it are

$$1, 7, 11, 13, 17, 19, 23, 29.$$

In this setting, we find that the desired sum is

$$1 + 7 + 11 + 13 + 17 + 19 + 23 + 29 = 120 = \tfrac{1}{2} \cdot 30 \cdot 8.$$

Also note the pairings

$$1 + 29 = 30, 7 + 23 = 30, 11 + 19 = 30, 13 + 17 = 30.$$

This is a good point at which to give an application of the Möbius Inversion Formula.

THEOREM 7-8.

For any positive integer n,

$$\phi(n) = n \sum_{d \mid n} \mu(d)/d.$$

Proof: The proof is deceptively simple: If one applies the inversion formula to

$$F(n) = n = \sum_{d \mid n} \phi(d),$$

the result is

$$\phi(n) = \sum_{d \mid n} \mu(d)F(n/d) = \sum_{d \mid n} \mu(d)n/d.$$

Let us illustrate the situation with $n = 10$ again. As can easily be seen,

$$10 \sum_{d \mid 10} \mu(d)/d = 10[\mu(1) + \mu(2)/2 + \mu(5)/5 + \mu(10)/10]$$

$$= 10[1 + (-1)/2 + (-1)/5 + (-1)^2/10]$$

$$= 10[1 - 1/2 - 1/5 + 1/10] = 10 \cdot 2/5 = 4 = \phi(10).$$

Starting with Theorem 7-8, it is an easy matter to determine the value of the phi-function for any positive integer n. Suppose that the prime-power decomposition of n is $n = p_1^{k_1} p_2^{k_2} \cdots p_r^{k_r}$ and consider the product

$$P = \prod_{p_i \mid n} (\mu(1) + \mu(p_i)/p_i + \cdots + \mu(p_i^{k_i})/p_i^{k_i}).$$

Multiplying this out, we obtain a sum of terms of the form

$$\mu(1)\mu(p_1^{a_1})\mu(p_2^{a_2}) \cdots \mu(p_r^{a_r})/p_1^{a_1}p_2^{a_2} \cdots p_r^{a_r}, \qquad 0 \le a_i \le k_i$$

or, since μ is known to be multiplicative,

$$\mu(p_1^{a_1}p_2^{a_2} \cdots p_r^{a_r})/p_1^{a_1}p_2^{a_2} \cdots p_r^{a_r} = \mu(d)/d,$$

where the summation is over the set of divisors $d = p_1^{a_1}p_2^{a_2} \cdots p_r^{a_r}$ of n. Hence, $P = \sum_{d \mid n} \mu(d)/d$. It follows from Theorem 7-8 that

$$\phi(n) = n \sum_{d \mid n} \mu(d)/d = n \prod_{p_i \mid n} (\mu(1) + \mu(p_i)/p_i + \cdots + \mu(p_i^{k_i})/p_i^{k_i}).$$

But $\mu(p_i^{a_i}) = 0$ whenever $a_i \ge 2$. As a result, the last-written equation reduces to

$$\phi(n) = n \prod_{p_i \mid n} (\mu(1) + \mu(p_i)/p_i) = n \prod_{p_i \mid n} (1 - 1/p_i),$$

which agrees with the formula established earlier by different reasoning. What is significant about this argument is that no assumption is made concerning the multiplicative character of the phi-function, only of μ.

PROBLEMS 7.4

1. For a positive integer n, prove that

$$\sum_{d \mid n}(-1)^{n/d}\phi(d) = \begin{cases} 0 \text{ if } n \text{ is even} \\ -n \text{ if } n \text{ is odd} \end{cases}$$

[Hint: If $n = 2^k N$, where N is odd, then $\sum_{d \mid n}(-1)^{n/d}\phi(d) = \sum_{d \mid 2^{k-1}N}\phi(d) - \sum_{d \mid N}\phi(2^k d)$.]

2. Confirm that $\sum_{d \mid 36}\phi(d) = 36$ and $\sum_{d \mid 36}(-1)^{36/d}\phi(d) = 0$.

3. For a positive integer n, prove that $\sum_{d \mid n}\mu^2(d)/\phi(d) = n/\phi(n)$. [Hint: Both sides of the equation are multiplicative functions.]

4. Use Problem 3, Section 6.2, to give a different proof of the fact that

$$n \sum_{d \mid n}\mu(d)/d = \phi(n).$$

5. If the integer $n > 1$ has the prime factorization $n = p_1^{k_1}p_2^{k_2} \cdots p_r^{k_r}$, establish the following:

    (a) $\sum_{d \mid n}\mu(d)\phi(d) = (2 - p_1)(2 - p_2) \cdots (2 - p_r)$

    (b) $\sum_{d \mid n}d\phi(d) = \left(\dfrac{p_1^{2k_1 + 1} + 1}{p_1 + 1} \right) \left(\dfrac{p_2^{2k_2 + 1} + 1}{p_2 + 1} \right) \cdots \left(\dfrac{p_r^{2k_r + 1} + 1}{p_r + 1} \right)$

    (c) $\sum_{d \mid n}\phi(d)/d = \left(1 + \dfrac{k_1(p_1 - 1)}{p_1} \right) \left(1 + \dfrac{k_2(p_2 - 1)}{p_2} \right) \cdots \left(1 + \dfrac{k_r(p_r - 1)}{p_r} \right)$

[Hint: For part (a), use Problem 3, Section 6-2.]

6. Verify the formula $\sum_{d=1}^{n} \phi(d)[n/d] = n(n+1)/2$ for any positive integer n. [*Hint:* This is a direct application of Theorems 6-11 and 7-6.]

7. If n is a square-free integer, prove that $\sum_{d \mid n} \sigma(d^{k-1})\phi(d) = n^k$ for all integers $k \geq 2$.

8. For a square-free integer $n > 1$, show that $\tau(n^2) = n$ if and only if $n = 3$.

9. Prove that $3 \mid \sigma(3n+2)$ and $4 \mid \sigma(4n+3)$ for any positive integer n.

10. (a) Given $k > 0$, establish that there exists a sequence of k consecutive integers $n+1$, $n+2, \ldots, n+k$ satisfying

$$\mu(n+1) = \mu(n+2) = \cdots = \mu(n+k) = 0.$$

[*Hint:* Consider the system of linear congruences

$$x \equiv -1 \pmod{4}, \; x \equiv -2 \pmod{9}, \; \ldots, \; x \equiv -k \pmod{p_k{}^2}$$

where p_k is the kth prime.]
    (b) Find four consecutive integers for which $\mu(n) = 0$.

11. Modify the proof of Gauss's Theorem to establish that $\sum_{k=1}^{n} \gcd(k,n) = \sum_{d \mid n} d\phi(n/d) = n \sum_{d \mid n} \phi(d)/d$ for $n \geq 1$.

12. For $n > 2$, establish the inequality $\phi(n^2) + \phi((n+1)^2) \leq 2n^2$.

13. Given an integer n, prove that there exists at least one k for which $n \mid \phi(k)$.

14. Show that if n is a product of twin primes, say $n = p(p+2)$, then
$$\phi(n) \sigma(n) = (n+1)(n-3).$$

15. Prove that $\sum_{d \mid n} \sigma(d) \phi(n/d) = n \tau(n)$ and $\sum_{d \mid n} \tau(d) \phi(n/d) = \sigma(n)$.

16. If $a_1, a_2, \cdots, a_{\phi(n)}$ is a reduced set of residues modulo n, show that $a_1 + a_2 + \cdots + a_{\phi(n)} \equiv 0 \pmod{n}$ for $n > 2$.

7.5 AN APPLICATION TO CRYPTOGRAPHY

Classically, the making and breaking of secret codes has usually been confined to diplomatic and military practices. With the growing quantity of digital data stored and communicated by electronic data-processing systems, organizations in both the public and commercial sectors have felt the need to protect information from unwanted intrusion. Indeed, the widespread use of electronic funds transfers has made privacy a pressing concern in most financial transactions. There has thus been a recent surge of interest by mathematicians and computer scientists in *cryptography* (from the Greek *kryptos* meaning *hidden* and *graphein* meaning *to write*)—the science of making communications unintelligible to all except authorized parties. Cryptography is the only known practical means for protecting information transmitted through public communications networks, such as those using telephone lines, microwaves or satellites.

In the language of cryptography, where codes are called *ciphers,* the information to be concealed is called *plaintext.* After transformation to a secret form, a message is called *ciphertext.* The process of converting from plaintext to ciphertext is said to be *encrypting* (or *enciphering*), while the reverse process of changing from ciphertext back to plaintext is called *decrypting* (or *deciphering*).

One of the earliest cryptographic systems was used by the great Roman emperor Julius Caesar around 50 B.C. Caesar wrote to Marcus Cicero using a rudimentary substitution cipher in which each letter of the alphabet is replaced by the letter which occurs three places down the alphabet, with the last three letters cycled back to the first three letters. If we write the ciphertext equivalent underneath the plaintext letter, the substitution alphabet for the *Caesar cipher* is given by:

Plaintext: A B C D E F G H I J K L M N O P Q R S T U V W X Y Z

Ciphertext: D E F G H I J K L M N O P Q R S T U V W X Y Z A B C

For example, the plaintext message

(1)   CAESAR WAS GREAT

is transformed into the ciphertext

(2)   FDHVDU ZDV JUHDV.

The Caesar cipher can be described easily using congruence theory. Any plaintext is first expressed numerically by translating the characters of the text into digits by means of some correspondence like the following:

A	B	C	D	E	F	G	H	I	J	K	L	M
01	02	03	04	05	06	07	08	09	10	11	12	13

N	O	P	Q	R	S	T	U	V	W	X	Y	Z
14	15	16	17	18	19	20	21	22	23	24	25	26

If P is the digital equivalent of a plaintext letter and C is the digital equivalent of the corresponding ciphertext letter, then

$$C \equiv P + 3 \pmod{26}.$$

Thus, for instance, the letters of message (1) are converted to their equivalents

(3)   03 01 05 19 01 18 23 01 19 07 18 05 01 20

Using the congruence $C \equiv P + 3 \pmod{26}$, this becomes the ciphertext

(4)   06 04 08 22 04 21 26 04 22 10 21 08 04 23.

To recover the plaintext, the procedure is simply reversed by means of the congruence

$$P \equiv C - 3 \equiv C + 23 \pmod{26}.$$

The Caesar cipher is very simple and hence extremely insecure. Caesar himself soon abandoned this scheme, not only because of its insecurity, but also because he didn't trust Cicero, with whom he necessarily shared the secret of the cipher.

In conventional cryptographic systems, such as Caesar's cipher, the sender and receiver jointly have a secret *key*. The sender uses the key to encrypt the plaintext to be sent, while the receiver uses the same key to decrypt the ciphertext obtained. Public-key cryptography differs from conventional cryptography in that it uses two keys, an encryption key and a decryption key. Although the two keys effect inverse operations and are therefore related, there is no easily computed method of deriving the decryption key from the encryption key. Thus the encryption key can be made public without compromising the decryption key; each user can encrypt messages, but only the intended recipient (whose decryption key is kept secret) can decipher them. A major advantage of a public-key cryptosystem is that it is unnecessary for each sender and receiver to exchange a key in advance of their decision to communicate with each other.

In 1977, R. Rivest, A. Shamir, and L. Adleman proposed a public key cryptosystem which uses only elementary ideas from number theory. Their enciphering system is called RSA, after the initials of the algorithm's inventors. Its security depends on the assumption that in the currrent state of computer technology, the factorization of composite numbers with large prime factors is prohibitively time-consuming.

Each user of the RSA system chooses a pair of distinct primes, p and q, large enough that the factorization of their product $n = pq$, called the *enciphering modulus*, is beyond all current computational capabilities. For instance, one might pick p and q with 200 digits each, so that n has roughly 400 digits. Having selected n, the user then chooses a random positive integer k, the *enciphering exponent*, satisfying $\gcd(k, \phi(n)) = 1$. The pair (n, k) is placed in a public file, analogous to a telephone directory, as the user's personal encryption key. This will allow anyone else in the communication network to encrypt and send a message to that individual. Notice that while n is openly revealed, the listed public key does not mention the factors p and q of n.

The encryption process begins with the conversion of the message to be sent into an integer M by means of a "digital alphabet" in which each letter, number, or punctuation mark of the plaintext is replaced by a two-digit integer. One standard procedure is to use the assignment:

A = 01	K = 11	U = 21	1 = 31
B = 02	L = 12	V = 22	2 = 32
C = 03	M = 13	W = 23	3 = 33
D = 04	N = 14	X = 24	4 = 34
E = 05	O = 15	Y = 25	5 = 35
F = 06	P = 16	Z = 26	6 = 36
G = 07	Q = 17	, = 27	7 = 37
H = 08	R = 18	. = 28	8 = 38
I = 09	S = 19	? = 29	9 = 39
J = 10	T = 20	0 = 30	! = 40

with 00 indicating a space between words. In this scheme, the message

The brown fox is quick.

is transformed into the numerical string

$$M = 2008050002181523140006152400091900172109031128.$$

It is assumed that the plaintext number $M < n$, where n is the enciphering modulus. Otherwise it would be impossible to distinguish M from any larger integer congruent to it modulo n. If the message is too lengthy to be handled as a single number $M < n$, then M can be broken up into blocks of digits $M_1, M_2, \cdots, M_r$ of the appropriate size. Each block would be encrypted separately.

Looking up the intended recipient's encryption key (n, k) in the public directory, the sender disguises the plaintext number M as a ciphertext number r by raising M to the kth power and then reducing the result modulo n; that is,

$$M^k \equiv r \pmod{n}.$$

A 200-character message can be encrypted in seconds on a high-speed computer. Recall that the public enciphering exponent k was originally selected so that $\gcd(k, \phi(n)) = 1$. While there are many suitable choices for k, an obvious suggestion is to pick k to be any prime larger than both p and q.

At the other end, the authorized recipient deciphers the transmitted information by first determining the integer j, the secret recovery exponent, for which

$$kj \equiv 1 \pmod{\phi(n)}.$$

Since $\gcd(k, \phi(n)) = 1$, this linear congruence has a unique solution modulo $\phi(n)$. In fact, the Euclidean algorithm will produce j as a solution x to the equation

$$kx + \phi(n)y = 1.$$

The recovery exponent can only be calculated by someone who knows both k and $\phi(n) = (p - 1)(q - 1)$, hence knows the prime factors p and q of n. Thus, j is secure from an illegitimate third party whose knowledge is limited to the public key (n, k).

Matters have been arranged so that the recipient can now retrieve M from r by simply calculating r^j modulo n. Because $kj = 1 + \phi(n)t$ for some integer t, it follows that

$$r^j \equiv (M^k)^j \equiv M^{1 + \phi(n)t}$$

$$\equiv M(M^{\phi(n)})^t \equiv M \cdot 1^t \equiv M \pmod{n},$$

whenever $\gcd(M, n) = 1$. In other words, raising the ciphertext number to the jth power and reducing it modulo n recovers the original plaintext number M.

The assumption that $\gcd(M, n) = 1$ was made in order to use Euler's Theorem. In the unlikely event that M and n are not relatively prime, a similar argument establishes that $r^j \equiv M \pmod{p}$ and $r^j \equiv M \pmod{q}$, which then yields the desired congruence $r^j \equiv M \pmod{n}$. We omit the details.

The major advantage of this ingenious procedure is that the encryption of a message does not require the knowledge of the two primes p and q, but only their product n; there is no need for anyone other than the receiver of the message ever to know the prime factors critical to the decryption process.

Example 7-5

For the reader to gain familiarity with the RSA public-key algorithm, let us work an example in detail. We first select two primes

$$p = 29, q = 53$$

of an unrealistically small size, in order to get an easy-to-handle illustration. In practice, p and q would be large enough so that the factorization of the nonsecret $n = pq$ is not feasible. Our enciphering modulus is $n = 29 \cdot 53 = 1537$ and $\phi(n) = 28 \cdot 52 = 1456$. Since $\gcd(47,1456) = 1$, we may choose $k = 47$ to be the enciphering exponent. Then the recovery exponent, the unique integer j satisfying the congruence $kj \equiv 1 \pmod{\phi(n)}$, is $j = 31$. To encrypt the message

NO WAY,

first translate each letter into its digital equivalent using the substitution mentioned earlier; this yields the plaintext number

$$M = 141500230125.$$

We want each plaintext block to be an integer less than 1537. Given this restriction, it seems reasonable to split M into blocks of three digits each. The first block, 141, encrypts as the ciphertext number

$$141^{47} \equiv 658 \pmod{1537}.$$

These will be the first digits of the secret transmission. At the other end, knowing that the recovery exponent is $j = 31$, the authorized recipient would begin to recover the plaintext number by computing

$$658^{31} \equiv 141 \pmod{1537}.$$

The total ciphertext of our message is

$$0658 \qquad 1408 \qquad 1250 \qquad 1252.$$

For the RSA cryptosystem to be secure it must not be computationally feasible to recover the plaintext M from the information assumed to be known to a third party; namely, the listed public-key (n, k). The direct method of attack would be to attempt to factor n, an integer of huge magnitude; for once the factors are determined, the recovery exponent j can be calculated from $\phi(n) = (p - 1)(q - 1)$ and k. Our confidence in the RSA system rests on what is known as the "work factor," the expected amount of computer time needed to factor the product of two large primes. Factoring is computationally more difficult than distinguishing between primes and composites. On today's fastest computers, a 200-digit number can routinely be tested for primality in less than 10 minutes, whereas the running time required to factor a composite number of the same size is prohibitive. It has been estimated that the quickest factoring algorithm known can use approximately $(1.2)10^{23}$ computer operations to resolve an integer with 200 digits into its prime factors; assuming that each operation takes one microsecond (10^{-6} seconds), then the factorization time would be about $(3.8)10^{9}$ years.

Given unlimited computing time and some unimaginably efficient factoring algorithm, the RSA cryptosystem could be broken, but for the present it appears to be quite safe. Indeed, even the use of a large network of computers working in parallel poses no serious threat. All we need to do is choose larger primes p and q for the enciphering moduli, always staying ahead of the current state of the art in factoring integers.

A public-key cryptosystem can also be based on the classical problem in combinatorics known as the *knapsack problem,* or the subset sum problem. This problem may be stated as follows: Given a knapsack of volume V and n items of various volumes $a_1, a_2, \ldots, a_n$, can a subset of these items be found which will completely fill the knapsack? An alternative formulation is:

For positive integers $a_1, a_2, \ldots, a_n$ and a sum V, solve the equation

$$V = a_1 x_2 + a_2 x_2 + \cdots + a_n x_n,$$

where $x_i = 0$ or 1 for $i = 1, 2, \ldots, n$.

There might be no solution, or more than one solution, to the problem, depending on the choice of the sequence $a_1, a_2, \ldots, a_n$ and the integer V. For instance, the knapsack problem

$$22 = 3x_1 + 7x_2 + 9x_3 + 11x_4 + 20x_5$$

is not solvable; but

$$27 = 3x_1 + 7x_2 + 9x_3 + 11x_4 + 20x_5$$

has two distinct solutions, namely

$$x_2 = x_3 = x_4 = 1, \qquad x_1 = x_5 = 0 \qquad \text{and}$$

$$x_2 = x_5 = 1, \qquad x_1 = x_3 = x_4 = 0.$$

Finding a solution to a randomly chosen knapsack problem is notoriously difficult. None of the known methods for attacking the problem are substantially less time-consuming than conducting an exhaustive direct search; that is, by testing all the 2^n possibilities for $x_1, x_2, \ldots, x_n$. This is computationally impracticable for n greater than 100 or so.

However, if the sequence of integers $a_1, a_2, \ldots, a_n$ happens to have some special properties, the knapsack problem becomes much easier to solve. We call a sequence $a_1, a_2, \ldots, a_n$ *superincreasing* when each a_i is larger than the sum of all the preceding ones: in other words,

$$a_i > a_1 + a_2 + \cdots + a_{i-1} \qquad\qquad i = 2, 3, \ldots, n.$$

A simple illustration of a superincreasing sequence is $1, 2, 4, 8, \ldots, 2^n$, where $2^i > 2^i - 1 = 1 + 2 + 4 + \cdots + 2^{i-1}$. For the corresponding knapsack problem,

$$V = x_1 + 2x_2 + 4x_3 + \cdots + 2^n x_n, \qquad\qquad V < 2^{n+1},$$

the x_i are just the digits in the binary expansion of V.

Knapsack problems based on superincreasing sequences are uniquely solvable whenever they are solvable at all, as our next example shows.

Example 7–6

Let us solve the superincreasing knapsack problem

$$28 = 3x_1 + 5x_2 + 11x_3 + 20x_4 + 41x_5.$$

We start with the largest coefficient in this equation, namely 41. Since $41 > 28$, it cannot be part of our subset sum, hence $x_5 = 0$. The next-largest coefficient is 20, with $20 < 28$. Now the sum of the preceding coefficients is $3 + 5 + 11 < 28$, so that these cannot "fill" the knapsack; therefore 20 must be included in the sum, and so $x_4 = 1$. Knowing the values of x_4 and x_5, the original problem may be rewritten as

$$8 = 3x_1 + 5x_2 + 11x_3.$$

A repetition of our earlier reasoning now determines whether 11 should be in our knapsack sum. In fact, the inequality $11 > 8$ forces us to take $x_3 = 0$. To clinch matters we are reduced to solving the equation $8 = 3x_1 + 5x_2$, which has the obvious solution $x_1 = x_2 = 1$. This identifies a subset of 3, 5, 11, 20, 41 having the desired sum:

$$28 = 3 + 5 + 20.$$

It is not difficult to see how the procedure described in Example 7–6 operates in general. Suppose that we wish to solve the knapsack problem

$$V = a_1x_1 + a_2x_2 + \cdots + a_nx_n,$$

where $a_1, a_2, \ldots, a_n$ is a superincreasing sequence of integers. Assume that V can be obtained by using some subset of the sequence, so that V is not larger than the sum $a_1 + a_2 + \cdots + a_n$. Working from right to left in our sequence, we begin by letting $x_n = 1$ if $V \geq a_n$ and $x_n = 0$ if $V < a_n$. Then obtain $x_{n-1}, x_{n-2}, \ldots, x_1$, in turn, by choosing

$$x_i = \begin{cases} 1 & \text{if } V - (a_{i+1}x_{i+1} + \cdots + a_nx_n) \geq a_i \\ 0 & \text{if } V - (a_{i+1}x_{i+1} + \cdots + a_nx_n) < a_i. \end{cases}$$

With this algorithm, knapsack problems using superincreasing sequences can be solved quite readily.

A public-key cryptosystem based on the knapsack problem was devised by R. Merkle and M. Hellman in 1978. It works as follows: a typical user of the system starts by choosing a superincreasing sequence $a_1, a_2, \ldots, a_n$. Now select a modulus $m > 2a_n$ and a multiplier a, with $0 < a < m$ and $\gcd(a, m) = 1$. This insures that the congruence $ax \equiv 1 \pmod{m}$ has a unique solution, say, $x \equiv c \pmod{m}$. Finally, form the sequence of integers $b_1, b_2, \ldots, b_n$ defined by

$$b_i \equiv aa_i \pmod{m} \qquad\qquad i = 1, 2, \ldots, n$$

where $0 < b_i < m$. Carrying out this last transformation generally destroys the superincreasing property enjoyed by the a_i.

The user keeps secret the original sequence $a_1, a_2, \ldots, a_n$, as well as the numbers m and a; but publishes $b_1, b_2, \ldots, b_n$ in a public directory. Anyone wishing to send a message to the user will employ the publicly available sequence as the encryption key.

The sender begins by converting the plaintext message into a string M of 0's and 1's using the binary equivalent of letters:

Letter	Binary Equivalent	Letter	Binary Equivalent
A	00000	N	01101
B	00001	O	01110
C	00010	P	01111
D	00011	Q	10000
E	00100	R	10001
F	00101	S	10010
G	00110	T	10011
H	00111	U	10100
I	01000	V	10101
J	01001	W	10110
K	01010	X	10111
L	01011	Y	11000
M	01100	Z	11001

For example, the message

$$\text{First Place}$$

would be converted into the numerical representation

$$M = 00101 \quad 01000 \quad 10001 \quad 10010 \quad 10011 \quad 01111 \quad 01011 \quad 00000$$
$$00010 \quad 00100.$$

The string is then split into blocks of n binary digits, with the last block being filled out with 1's at the end if necessary. The public encrypting sequence $b_1, b_2, \ldots, b_n$ is next used to transform a given plaintext block, say $x_1 x_2 \cdots x_n$, into the sum

$$S = b_1 x_1 + b_2 x_2 + \cdots + b_n x_n.$$

The number S is the hidden information that the sender transmits over a communication channel which is presumed to be insecure.

Notice that since each x_i is either 0 or 1, the problem of recreating the plaintext block from S is equivalent to solving an apparently difficult knapsack problem ("difficult" because the sequence $b_1, b_2, \ldots, b_n$ is not necessarily superincreasing). On first impression, the intended recipient and any eavesdropper are faced with the same task. However, with the aid of the private decryption key, the recipient can change the difficult knapsack problem into an easy one. No one without the private key can make this change.

Knowing c and m, the recipient computes

$$S' \equiv cS \pmod{m} \qquad\qquad 0 \le S' < m$$

or, expanding this,

$$S' \equiv cb_1 x_1 + cb_2 x_2 + \cdots + cb_n x_n \pmod{m}$$
$$\equiv caa_1 x_1 + caa_2 x_2 + \cdots + caa_n x_n \pmod{m}$$

Now $ca \equiv 1 \pmod{m}$, so that the above congruence becomes

$$S' \equiv a_1x_1 + a_2x_2 + \cdots + a_nx_n \pmod{m}.$$

Since m was chosen to satisfy $m > 2a_n > a_1 + a_2 + \cdots + a_n$, we obtain $a_1x_1 + a_2x_2 + \cdots + a_nx_n < m$. In light of the condition $0 \leq S' < m$, the equality

$$S' = a_1x_1 + a_2x_2 + \cdots + a_nx_n$$

must hold. The solution to this superincreasing knapsack problem furnishes the solution to the difficult problem, and the plaintext block $x_1x_2 \cdots x_n$ of n digits is thereby recovered from S.

To help make the technique clearer, we consider a small-scale example with $n = 5$.

Example 7-7

Suppose that a typical user of this cryptosystem selects as a secret key the superincreasing sequence 3, 5, 11, 20, 41, the modulus $m = 75$, and the multiplier $a = 44$. Each member of the superincreasing sequence is multiplied by 44 and reduced modulo 75 to yield 57, 70, 34, 55, 4. This is the encryption key that the user submits to the public directory.

Someone who wants to send a plaintext message to the user, such as

HELP US,

first converts it into the following string of 0's and 1's:

$$M = 00111 \quad 00100 \quad 01011 \quad 01111 \quad 10100 \quad 10010$$

The string is then broken up into blocks of digits, in the current case blocks of length 5. Using the listed public key to encrypt, the sender transforms the successive blocks into

$$93 = 57 \cdot 0 + 70 \cdot 0 + 34 \cdot 1 + 55 \cdot 1 + 4 \cdot 1$$
$$34 = 57 \cdot 0 + 70 \cdot 0 + 34 \cdot 1 + 55 \cdot 0 + 4 \cdot 0$$
$$108 = 57 \cdot 0 + 70 \cdot 1 + 34 \cdot 1 + 55 \cdot 1 + 4 \cdot 1$$
$$129 = 57 \cdot 0 + 70 \cdot 1 + 34 \cdot 1 + 55 \cdot 1 + 4 \cdot 1$$
$$91 = 57 \cdot 1 + 70 \cdot 0 + 34 \cdot 1 + 55 \cdot 0 + 4 \cdot 0$$
$$112 = 57 \cdot 1 + 70 \cdot 0 + 34 \cdot 0 + 55 \cdot 1 + 4 \cdot 0$$

The transmitted ciphertext consists of the sequence of positive integers

$$93 \quad 34 \quad 108 \quad 129 \quad 91 \quad 112.$$

In order to read the message, the legitimate receiver first solves the congruence $44x \equiv 1 \pmod{75}$, yielding $x \equiv 29 \pmod{75}$. Then each ciphertext number is multiplied by 29 and reduced modulo 75, to produce a superincreasing knapsack problem. For instance, 93 is converted to 72, since $93 \cdot 29 \equiv 72 \pmod{75}$; and the corresponding knapsack problem is

$$72 = 3x_1 + 5x_2 + 11x_3 + 20x_4 + 41x_5.$$

The procedure for handling superincreasing knapsack problems quickly produces the solution $x_1 = x_2 = 0$, $x_3 = x_4 = x_5 = 1$. In this way, the first block 00111 of the binary equivalent of the plaintext is recovered.

The Merkle-Hellman cryptosystem aroused a great deal of interest when it was first proposed, since it was based on a provably difficult problem. However, in 1982 A. Shamir invented a "reasonably fast" algorithm for solving knapsack problems which involved sequences $b_1, b_2, \ldots, b_n$, where $b_i \equiv aa_i \pmod{m}$ and $a_1, a_2, \ldots, a_n$ is superincreasing. The weakness of the system is that the public encryption key $b_1, b_2, \ldots, b_n$ is too special; multiplying by a and reducing modulo m does not completely disguise the sequence $a_1, a_2, \ldots, a_n$. The system can be made somewhat more secure by iterating the modular multiplication method with different values of a and m, so that the public and private sequences differ by several transformations. But even this construction was successfully broken in 1985. While most variations of the Merkle-Hellman scheme have been shown to be insecure, there are a few that have so far resisted attack.

PROBLEMS 7.5

1. Encrypt the message *RETURN HOME* using the Caesar cipher.

2. If the Caesar cipher produced *KDSSB ELUWKGDB*, what is the plaintext message?

3. (a) A linear cipher is defined by the congruence $C \equiv aP + b \pmod{26}$, where a and b are integers with gcd $(a, 26) = 1$. Show that the corresponding decrypting congruence is $P \equiv a'(C - b) \pmod{26}$, where the integer a' satisfies $aa' \equiv 1 \pmod{26}$.
   (b) Using the linear cipher $C \equiv 5P + 11 \pmod{26}$, encrypt the message *NUMBER THEORY IS EASY.*
   (c) Decrypt the message *TZSVIW JQBVMIJ HL MVOOVI*, which was produced using the linear cipher $C \equiv 3P + 7 \pmod{26}$.

4. If $n = pq = 274279$ and $\phi(n) = 272376$, find the primes p and q. [*Hint:* Note that
$$p + q = n - \phi(n) + 1,$$
$$p - q = [(p + q)^2 - 4n]^{1/2}.]$$

5. When the RSA algorithm is based on the key $(n, k) = (3233, 37)$, what is the recovery exponent for the cryptosystem?

6. Encrypt the plaintext message *GOLD MEDAL* using the RSA algorithm with key $(n, k) = (2419, 3)$.

7. The ciphertext message produced by the RSA algorithm with key $(n, k) = (1643, 223)$ is

    1451   0103   1263   0560   0127   0897.

   Determine the original plaintext message. [*Hint:* The recovery exponent is $j = 7$.]

8. Decrypt the ciphertext

    1037   0431   0629   0690   0204   2267   0456

   that was encrypted using the RSA algorithm with key $(n, k) = (2419, 211)$. [*Hint:* The recovery exponent is 11. Note that it may be necessary to fill out a plaintext block by adding zeros on the left.]

9. Obtain all solutions of the knapsack problem

$$21 = 2x_1 + 3x_2 + 5x_3 + 7x_4 + 9x_5 + 11x_6 .$$

10. Determine which of the sequences below are superincreasing:
    (a) 3, 13, 20, 37, 81;
    (b) 5, 13, 25, 42, 90;
    (c) 7, 27, 47, 97, 197, 397.

11. Find the unique solution of each of the following superincreasing knapsack problems:
    (a) $118 = 4x_1 + 5x_2 + 10x_3 + 20x_4 + 41x_5 + 99x_6$;
    (b) $51 = 3x_1 + 5x_2 + 9x_3 + 18x_4 + 37x_5$;
    (c) $54 = x_1 + 2x_2 + 5x_3 + 9x_4 + 18x_5 + 40x_6$.

12. Consider a sequence of positive integers $a_1, a_2, \ldots, a_n$, where $a_{i+1} > 2a_i$ for $i = 1, 2,$ $\ldots, n - 1$. Show that the sequence is superincreasing.

13. A user of the knapsack cryptosystem has the sequence 49, 32, 30, 43 as a listed encryption key. If the user's private key involves the modulus $m = 50$ and multiplier $a = 33$, determine the secret superincreasing sequence.

14. The ciphertext message produced by the knapsack cryptosystem employing the superincreasing sequence 1, 3, 5, 11, 35, modulus $m = 53$ and multiplier $a = 5$ is

$$2, 15, 58, 43, 21.$$

Obtain the plaintext message. [*Hint:* Note that $5 \cdot 32 \equiv 1 \pmod{53}$.]

15. A user of the knapsack cryptosystem has a private key consisting of the superincreasing sequence 2, 3, 7, 13, 27, modulus $m = 40$ and multiplier $a = 7$.
    (a) Find the user's listed public key.
    (b) With the aid of the public key, encrypt the message SEND MONEY.

8

Primitive Roots and Indices

". . . mathematical proofs, like diamonds, are hard as well as clear, and will be touched with nothing but strict reasoning."

JOHN LOCKE

8.1 THE ORDER OF AN INTEGER MODULO n

In view of Euler's Theorem, we know that $a^{\phi(n)} \equiv 1 \pmod{n}$, whenever $\gcd(a, n) = 1$. However, there are often powers of a smaller than $a^{\phi(n)}$ which are congruent to 1 modulo n. This prompts the following definition:

DEFINITION 8-1.
  Let $n > 1$ and $\gcd(a, n) = 1$. The *order of a modulo n* (in older terminology: the *exponent to which a belongs modulo n*) is the smallest positive integer k such that $a^k \equiv 1 \pmod{n}$.

Consider the successive powers of 2 modulo 7. For this modulus, we obtain the congruences

$$2^1 \equiv 2, \ 2^2 \equiv 4, \ 2^3 \equiv 1, \ 2^4 \equiv 2, \ 2^5 \equiv 4, \ 2^6 \equiv 1, \ \ldots,$$

from which it follows that the integer 2 has order 3 modulo 7.

Observe that if two integers are congruent modulo n, then they have the same order modulo n. For if $a \equiv b \pmod{n}$ and $a^k \equiv 1 \pmod{n}$, Theorem 4-2 implies that $a^k \equiv b^k \pmod{n}$, whence $b^k \equiv 1 \pmod{n}$.

It should be emphasized that our definition of order modulo n concerns only integers a for which $\gcd(a, n) = 1$. Indeed, if $\gcd(a, n) > 1$, then we know from Theorem 4-7 that the linear congruence $ax \equiv 1 \pmod{n}$ has no solution; hence, the relation

$$a^k \equiv 1 \pmod{n}, \qquad\qquad k \geq 1$$

cannot hold, for this would imply that $x = a^{k-1}$ is a solution of $ax \equiv 1 \pmod{n}$. Thus, whenever there is reference to the order of a modulo n, it is to be assumed that $\gcd(a, n) = 1$, even if it is not explicitly stated.

In the example given above, we have $2^k \equiv 1 \pmod 7$ whenever k is a multiple of 3, where 3 is the order of 2 modulo 7. Our first theorem shows that this is typical of the general situation.

THEOREM 8-1.

Let the integer a have order k modulo n. Then $a^h \equiv 1 \pmod n$ if and only if $k \mid h$; in particular, $k \mid \phi(n)$.

> *Proof:* Suppose to begin with that $k \mid h$, so that $h = jk$ for some integer j. Since $a^k \equiv 1 \pmod n$, Theorem 4-2 yields $(a^k)^j \equiv 1^j \pmod n$ or $a^h \equiv 1 \pmod n$.
>
> Conversely, let h be any positive integer satisfying $a^h \equiv 1 \pmod n$. By the Division Algorithm, there exist q and r such that $h = qk + r$, where $0 \le r < k$. Consequently,
>
> $$a^h = a^{qk + r} = (a^k)^q a^r.$$
>
> By hypothesis, both $a^h \equiv 1 \pmod n$ and $a^k \equiv 1 \pmod n$, the implication of which is that $a^r \equiv 1 \pmod n$. Since $0 \le r < k$, we end up with $r = 0$; otherwise, the choice of k as the smallest positive integer such that $a^k \equiv 1 \pmod n$ is contradicted. Hence $h = qk$, and $k \mid h$.

Theorem 8-1 expedites the computation when we attempt to find the order of an integer a modulo n: instead of considering all powers of a, the exponents can be restricted to the divisors of $\phi(n)$. Let us obtain, by way of illustration, the order of 2 modulo 13. Since $\phi(13) = 12$, the order of 2 must be one of the integers 1, 2, 3, 4, 6, 12. From

$$2^2 \equiv 4,\ 2^3 \equiv 8,\ 2^4 \equiv 3,\ 2^6 \equiv 12,\ 2^{12} \equiv 1 \pmod{13},$$

it is seen that 2 has order 12 modulo 13.

For an arbitrarily selected divisor d of $\phi(n)$, it is not always true that there exists an integer a having order d modulo n. An example is $n = 12$. Here $\phi(12) = 4$, yet there is no integer which is of order 4 modulo 12; indeed, one finds that

$$1^2 \equiv 5^2 \equiv 7^2 \equiv 11^2 \equiv 1 \pmod{12}$$

and so the only choice for orders is 1 or 2.

Here is another basic fact regarding the order of an integer.

THEOREM 8-2.

If a has order k modulo n, then $a^i \equiv a^j \pmod n$ if and only if $i \equiv j \pmod k$.

> *Proof:* First, suppose that $a^i \equiv a^j \pmod n$, where $i \ge j$. Since a is relatively prime to n, we may cancel a power of a to obtain $a^{i-j} \equiv 1 \pmod n$. According to Theorem 8-1, this last congruence holds only if $k \mid i - j$, which is just another way of saying that $i \equiv j \pmod k$.
>
> Conversely, let $i \equiv j \pmod k$. Then we have $i = j + qk$ for some integer q. By the definition of k, $a^k \equiv 1 \pmod n$, so that
>
> $$a^i \equiv a^{j + qk} \equiv a^j(a^k)^q \equiv a^j \pmod n,$$
>
> which is the desired conclusion.

COROLLARY.

> *If a has order k modulo n, then the integers a, a^2, . . . , a^k are incongruent modulo n.*

> *Proof:* If $a^i \equiv a^j$ (mod n) for $1 \leq i \leq j \leq k$, then the theorem insures that $i \equiv j$ (mod k). But this is impossible unless $i = j$.

A fairly natural question presents itself: is it possible to express the order of any integral power of a in terms of the order of a? The answer is contained in

THEOREM 8-3.

> *If the integer a has order k modulo n and h > 0, then a^h has order k/gcd(h , k) modulo n.*

> *Proof:* Let $d = \gcd(h , k)$. Then we may write $h = h_1 d$ and $k = k_1 d$, with $\gcd(h_1 , k_1) = 1$. Clearly,

> $$(a^h)^{k_1} = (a^{h_1 d})^{k/d} = (a^k)^{h_1} \equiv 1 \;(\text{mod } n).$$

> If a^h is assumed to have order r modulo n, then Theorem 8-1 asserts that $r \mid k_1$. On the other hand, since a has order k modulo n, the congruence

> $$a^{hr} \equiv (a^h)^r \equiv 1 \;(\text{mod } n)$$

> indicates that $k \mid hr$; in other words, $k_1 d \mid h_1 dr$ or $k_1 \mid h_1 r$. But $\gcd(k_1 , h_1) = 1$ and therefore $k_1 \mid r$. This divisibility relation, when combined with the one obtained earlier, gives

> $$r = k_1 = k/d = k/\gcd(h , k),$$

> proving the theorem.

The preceding theorem has a corollary for which the reader may supply a proof.

COROLLARY.

> *Let a have order k modulo n. Then a^h also has order k if and only if* $\gcd(h , k) = 1$.

Let us see how all this works in a specific instance.

Example 8-1

The following table exhibits the orders modulo 13 of the positive integers less than 13:

integer	1	2	3	4	5	6	7	8	9	10	11	12
order	1	12	3	6	4	12	12	4	3	6	12	2

We observe that the order of 2 modulo 13 is 12, while the orders of 2^2 and 2^3 are 6 and 4, respectively; it is easy to verify that

$$6 = 12/\gcd(2 , 12) \text{ and } 4 = 12/\gcd(3 , 12)$$

in accordance with Theorem 8-3. The integers which also have order 12 modulo 13 are powers 2^k for which $\gcd(k, 12) = 1$; namely,

$$2^5 \equiv 6,\ 2^7 \equiv 11,\ 2^{11} \equiv 7 \pmod{13}.$$

If an integer a has the largest order possible, then we call it a primitive root of n.

DEFINITION 8-2.
If $\gcd(a, n) = 1$ and a is of order $\phi(n)$ modulo n, then a is a *primitive root* of the integer n.

To put it another way, n has a as a primitive root if $a^{\phi(n)} \equiv 1 \pmod{n}$, but $a^k \not\equiv 1 \pmod{n}$ for all positive integers $k < \phi(n)$.
It is easy to see that 3 is a primitive root of 7, for

$$3^1 \equiv 3,\ 3^2 \equiv 2,\ 3^3 \equiv 6,\ 3^4 \equiv 4,\ 3^5 \equiv 5,\ 3^6 \equiv 1 \pmod{7}.$$

More generally, one can prove that primitive roots exist for any prime modulus, which is a result of fundamental importance. While it is possible for a primitive root of n to exist when n is not a prime (for instance, 2 is a primitive root of 9), there is no reason to expect that every integer n will possess a primitive root; indeed, the existence of primitive roots is more often the exception than the rule.

Example 8-2

Let us show that if $F_n = 2^{2^n} + 1$, $n > 1$, is a prime, then 2 is not a primitive root of F_n. (Clearly, 2 is a primitive root of $5 = F_1$.) Since $2^{2^{n+1}} - 1 = (2^{2^n} + 1)(2^{2^n} - 1)$, we have

$$2^{2^{n+1}} \equiv 1 \pmod{F_n},$$

which implies that the order of 2 modulo F_n does not exceed 2^{n+1}. But if F_n is assumed to be prime,

$$\phi(F_n) = F_n - 1 = 2^{2^n}$$

and a straightforward induction argument confirms that $2^{2^n} > 2^{n+1}$, whenever $n > 1$. Thus the order of 2 modulo F_n is smaller than $\phi(F_n)$; referring to Definition 8-2 we see that 2 cannot be a primitive root of F_n.

One of the chief virtues of primitive roots lies in our next theorem.

THEOREM 8-4.

Let $\gcd(a, n) = 1$ and let $a_1, a_2, \ldots, a_{\phi(n)}$ be the positive integers less than n and relatively prime to n. If a is a primitive root of n, then

$$a, a^2, \ldots, a^{\phi(n)}$$

are congruent modulo n to $a_1, a_2, \ldots, a_{\phi(n)}$, in some order.

Proof: Since a is relatively prime to n, the same holds for all the powers of a; hence, each a^k is congruent modulo n to some one of the a_i. The $\phi(n)$ numbers in the set $\{a, a^2, \ldots , a^{\phi(n)}\}$ are incongruent by the corollary to Theorem 8-2, hence these powers must represent (not necessarily in order of appearance) the integers $a_1, a_2, \ldots , a_{\phi(n)}$.

One consequence of what has just been proved is that, in those cases in which a primitive root exists, we can now state exactly how many there are.

COROLLARY.

If n has a primitive root, then it has exactly $\phi(\phi(n))$ of them.

Proof: Suppose that a is a primitive root of n. By the theorem, any other primitive root of n is found among the members of the set $\{a, a^2, \ldots , a^{\phi(n)}\}$. But the number of powers a^k, $1 \le k \le \phi(n)$, which have order $\phi(n)$ is equal to the number of integers k for which $\gcd(k , \phi(n)) = 1$; there are $\phi(\phi(n))$ such integers, hence $\phi(\phi(n))$ primitive roots of n.

Theorem 8-4 can be illustrated by taking $a = 2$ and $n = 9$. Since $\phi(9) = 6$, the first six powers of 2 must be congruent modulo 9, in some order, to the positive integers less than 9 and relatively prime to it. Now the integers less than and relatively prime to 9 are 1, 2, 4, 5, 7, 8 and we see that

$$2^1 \equiv 2, \ 2^2 \equiv 4, \ 2^3 \equiv 8, \ 2^4 \equiv 7, \ 2^5 \equiv 5, \ 2^6 \equiv 1 \ (\text{mod } 9).$$

By virtue of the corollary, there are exactly $\phi(\phi(9)) = \phi(6) = 2$ primitive roots of 9, these being the integers 2 and 5.

PROBLEMS 8.1

1. Find the order of the integers 2, 3, and 5: (a) modulo 17, (b) modulo 19, and (c) modulo 23.

2. Establish each of the statements below:
    (a) If a has order hk modulo n, then a^h has order k modulo n.
    (b) If a has order $2k$ modulo the odd prime p, then $a^k \equiv -1 \ (\text{mod } p)$.
    (c) If a has order $n - 1$ modulo n, then n is a prime.

3. Prove that $\phi(2^n - 1)$ is a multiple of n for any $n > 1$. [*Hint:* The integer 2 has order n modulo $2^n - 1$.]

4. Assume that the order of a modulo n is h and the order of b modulo n is k. Show that the order of ab modulo n divides hk; in particular, if $\gcd(h , k) = 1$, then ab has order hk.

5. Given that a has order 3 modulo p, where p is an odd prime, show that $a + 1$ must have order 6 modulo p. [*Hint:* From $a^2 + a + 1 \equiv 0 \ (\text{mod } p)$, it follows that $(a + 1)^2 \equiv a$ (mod p) and $(a + 1)^3 \equiv -1 \ (\text{mod } p)$.]

6. Verify the following assertions:
    (a) The odd prime divisors of the integer $n^2 + 1$ are of the form $4k + 1$. [*Hint:* $n^2 \equiv -1 \ (\text{mod } p)$, where p is an odd prime, implies that $4 \mid \phi(p)$ by Theorem 8-1.]
    (b) The odd prime divisors of the integer $n^4 + 1$ are of the form $8k + 1$.
    (c) The odd prime divisors of the integer $n^2 + n + 1$ which are different from 3 are of the form $6k + 1$.

7. Establish that there are infinitely many primes of each of the forms $4k + 1$, $6k + 1$, and $8k + 1$. [*Hint:* Assume that there are only finitely many primes of the form $4k + 1$; call them $p_1, p_2, \ldots, p_r$. Consider the integer $(2p_1 p_2 \cdots p_r)^2 + 1$ and apply the previous problem.]

8. (a) Prove that if p and q are odd primes and $q \mid a^p - 1$, then either $q \mid a - 1$ or else $q = 2kp + 1$ for some integer k. [*Hint:* Since $a^p \equiv 1 \pmod{q}$, the order of a modulo q is either 1 or p; in the latter case, $p \mid \phi(q)$.]
   (b) Use part (a) to show that if p is an odd prime, then the prime divisors of $2^p - 1$ are of the form $2kp + 1$.
   (c) Find the smallest prime divisor of the integers $2^{17} - 1$ and $2^{29} - 1$.

9. Prove that there are infinitely many primes of the form $2kp + 1$, where p is an odd prime. [*Hint:* Assume that there are finitely many primes of the form $2kp + 1$, call them $q_1, q_2, \ldots, q_r$, and consider the integer $(q_1 q_2 \cdots q_r)^p - 1$.]

10. (a) Verify that 2 is a primitive root of 19, but not of 17.
    (b) Show that 15 has no primitive root by calculating the orders of 2, 4, 7, 8, 11, 13, and 14 modulo 15.

11. Let r be a primitive root of the integer n. Prove that r^k is a primitive root of n if and only if $\gcd(k, \phi(n)) = 1$.

12. (a) Find two primitive roots of 10.
    (b) Use the information that 3 is a primitive root of 17 to obtain the eight primitive roots of 17.

13. (a) Prove that if p and $q > 3$ are both odd primes and $q \mid R_p$, then $q = 2kp + 1$ for some integer k.
    (b) Find the smallest prime divisors of the repunits $R_5 = 11111$ and $R_7 = 1111111$.

14. (a) Let $p > 5$ be prime. If R_n is the smallest repunit for which $p \mid R_n$, establish that $n \mid p - 1$. For example, R_8 is the smallest repunit divisible by 73, and $8 \mid 72$. [*Hint:* The order of 10 modulo p is n.]
    (b) Find the smallest R_n divisible by 13.

8.2 PRIMITIVE ROOTS FOR PRIMES

Since primitive roots play a crucial role in many theoretical investigations, a problem exerting a natural appeal is that of describing all integers which possess primitive roots. We shall, over the course of the next few pages, prove the existence of primitive roots for all primes. Before doing this, let us turn aside briefly to establish Lagrange's Theorem, which deals with the number of solutions of a polynomial congruence.

THEOREM 8-5 (Lagrange).
   If p is a prime and
$$f(x) = a_n x^n + a_{n-1} x^{n-1} + \cdots + a_1 x + a_0, \quad a_n \not\equiv 0 \pmod{p}$$
is a polynomial of degree $n \geq 1$ with integral coefficients, then the congruence
$$f(x) \equiv 0 \pmod{p}$$
has at most n incongruent solutions modulo p.

Proof: We proceed by induction on *n,* the degree of $f(x)$. If $n = 1$, then our polynomial is of the form

$$f(x) = a_1x + a_0.$$

Since $\gcd(a_1, p) = 1$, we know by Theorem 4-7 that the congruence $a_1x \equiv -a_0$ (mod *p*) has a unique solution modulo *p*. Thus, the theorem holds for $n = 1$.

Now assume inductively that the theorem is true for polynomials of degree $k - 1$, and consider the case in which $f(x)$ has degree *k.* Either the congruence $f(x) \equiv 0$ (mod *p*) has no solutions (and we are finished), or it has at least one solution, call it *a.* If $f(x)$ is divided by $x - a$, the result is

$$f(x) = (x - a)q(x) + r,$$

in which $q(x)$ is a polynomial of degree $k - 1$ with integral coefficients and *r* is an integer. Substituting $x = a$, we obtain

$$0 \equiv f(a) = (a - a)q(a) + r = r \text{ (mod } p)$$

and so $f(x) \equiv (x - a)q(x)$ (mod *p*).

If *b* is another one of the incongruent solutions of $f(x) \equiv 0$ (mod *p*), then

$$0 \equiv f(b) = (b - a)q(b) \text{ (mod } p).$$

Since $b - a \not\equiv 0$ (mod *p*), we may cancel to conclude that $q(b) \equiv 0$ (mod *p*); in other words, any solution of $f(x) \equiv 0$ (mod *p*) which is different from *a* must satisfy $q(x) \equiv 0$ (mod *p*). By our induction assumption, the latter congruence can possess at most $k - 1$ incongruent solutions and so $f(x) \equiv 0$ (mod *p*) will have no more than *k* incongruent solutions. This completes the induction step and the proof.

From this theorem, we can pass easily to

COROLLARY.

If p is a prime number and $d \mid p - 1$, then the congruence

$$x^d - 1 \equiv 0 \text{ (mod } p)$$

has exactly d solutions.

Proof: Since $d \mid p - 1$, we have $p - 1 = dk$ for some *k.* Then

$$x^{p-1} - 1 = (x^d - 1)f(x),$$

where the polynomial $f(x) = x^{d(k-1)} + x^{d(k-2)} + \cdots + x^d + 1$ has integral coefficients and is of degree $d(k - 1) = p - 1 - d$. By Lagrange's Theorem, the congruence $f(x) \equiv 0$ (mod *p*) has at most $p - 1 - d$ solutions. We also know from Fermat's Theorem that $x^{p-1} - 1 \equiv 0$ (mod *p*) has precisely $p - 1$ incongruent solutions; namely, the integers $1, 2, \ldots, p - 1$.

Now any solution $x = a$ of $x^{p-1} - 1 \equiv 0$ (mod *p*) that is not a solution of $f(x) \equiv 0$ (mod *p*) must satisfy $x^d - 1 \equiv 0$ (mod *p*). For

$$0 \equiv a^{p-1} - 1 = (a^d - 1)f(a) \text{ (mod } p),$$

with $p \nmid f(a)$, implies that $p \mid a^d - 1$. It follows that $x^d - 1 \equiv 0 \pmod{p}$ must have at least

$$p - 1 - (p - 1 - d) = d$$

solutions. This last congruence can possess no more than d solutions (Lagrange's Theorem enters again), hence has exactly d solutions.

We take immediate advantage of this corollary to prove Wilson's Theorem in a different way: given a prime p, define the polynomial $f(x)$ by

$$f(x) = (x - 1)(x - 2) \cdots (x - (p - 1)) - (x^{p-1} - 1)$$
$$= a_{p-2}x^{p-2} + a_{p-3}x^{p-3} + \cdots + a_1 x + a_0,$$

which is of degree $p - 2$. Fermat's Theorem implies that the $p - 1$ integers 1, 2, . . . , $p - 1$ are incongruent solutions of the congruence

$$f(x) \equiv 0 \pmod{p}.$$

But this contradicts Lagrange's Theorem, unless

$$a_{p-2} \equiv a_{p-3} \equiv \cdots \equiv a_1 \equiv a_0 \equiv 0 \pmod{p}.$$

It follows that, for any choice of the integer x,

$$(x - 1)(x - 2) \cdots (x - (p - 1)) - (x^{p-1} - 1) \equiv 0 \pmod{p}.$$

Now substitute $x = 0$ to obtain

$$(-1)(-2) \cdots (-(p - 1)) + 1 \equiv 0 \pmod{p}$$

or $(-1)^{p-1}(p - 1)! + 1 \equiv 0 \pmod{p}$. Either $p - 1$ is even or else $p = 2$, in which case $-1 \equiv 1 \pmod{p}$; at any rate, we get

$$(p - 1)! \equiv -1 \pmod{p}.$$

Lagrange's Theorem has provided us with the entering wedge. We are now in a position to prove that, for any prime p, there exist integers with order corresponding to each divisor of $p - 1$. Stated more precisely:

THEOREM 8-6.

If p is a prime number and $d \mid p - 1$, then there are exactly $\phi(d)$ incongruent integers having order d modulo p.

Proof: Let $d \mid p - 1$ and $\psi(d)$ denote the number of integers k, $1 \le k \le p - 1$, that have order d modulo p. Since each integer between 1 and $p - 1$ has order d for some $d \mid p - 1$,

$$p - 1 = \sum_{d \mid p - 1} \psi(d).$$

At the same time, Gauss' Theorem tells us that

$$p - 1 = \sum_{d \mid p - 1} \phi(d)$$

and so, putting these together,

$$(1) \qquad \sum_{d \mid p - 1} \psi(d) = \sum_{d \mid p - 1} \phi(d).$$

Our aim is to show that $\psi(d) \le \phi(d)$ for each divisor d of $p - 1$, since this, in conjunction with equation (1), would produce the equality $\psi(d) = \phi(d) \ne 0$ (otherwise, the first sum would be strictly smaller than the second).

Given an arbitrary divisor d of $p - 1$, there are two possibilities: either we have $\psi(d) = 0$ or $\psi(d) > 0$. If $\psi(d) = 0$, then certainly $\psi(d) \le \phi(d)$. Suppose that $\psi(d) > 0$, so that there exists an integer a of order d. Then the d integers $a, a^2, \ldots, a^d$ are incongruent modulo p and each of them satisfies the polynomial congruence

$$(2) \qquad x^d - 1 \equiv 0 \pmod{p};$$

for, $(a^k)^d \equiv (a^d)^k \equiv 1 \pmod{p}$. By the corollary to Lagrange's Theorem, there can be no other solutions of equation (2). It follows that any integer that has order d modulo p must be congruent to one of $a, a^2, \ldots, a^d$. But only $\phi(d)$ of the just-mentioned powers have order d, namely those a^k for which the exponent k has the property $\gcd(k, d) = 1$. Hence, in the present situation, $\psi(d) = \phi(d)$, and the number of integers having order d modulo p is equal to $\phi(d)$. This establishes the result we set out to prove.

Taking $d = p - 1$ in Theorem 8-6, we arrive at

COROLLARY.

If p is a prime, then there are exactly $\phi(p - 1)$ incongruent primitive roots of p.

An illustration is afforded by the prime $p = 13$. For this modulus, 1 has order 1; 12 has order 2; 3 and 9 have order 3; 5 and 8 have order 4; 4 and 10 have order 6; and four integers, namely 2, 6, 7, 11, have order 12. Thus,

$$\sum_{d \mid 12} \psi(d) = \psi(1) + \psi(2) + \psi(3) + \psi(4) + \psi(6) + \psi(12)$$

$$= 1 + 1 + 2 + 2 + 2 + 4 = 12$$

as it should. Notice too that

$$\psi(1) = 1 = \phi(1), \quad \psi(4) = 2 = \phi(4)$$
$$\psi(2) = 1 = \phi(2), \quad \psi(6) = 2 = \phi(6)$$
$$\psi(3) = 2 = \phi(3), \quad \psi(12) = 4 = \phi(12).$$

Incidentally, there is a shorter and more elegant way of proving that $\psi(d) = \phi(d)$ for each $d \mid p - 1$. We simply subject the formula $d = \sum_{c \mid d} \psi(c)$ to Möbius inversion to deduce that

$$\psi(d) = \sum_{c \mid d} \mu(c)(d/c).$$

In light of Theorem 7-8, the right-hand side of the foregoing equation is equal to $\phi(d)$. Of course, the validity of this argument rests upon using the corollary on page 157 to show that $d = \sum_{c \mid d} \psi(c)$.

We can use this last theorem to give another proof of the fact that if p is a prime of the form $4k + 1$, then the quadratic congruence $x^2 \equiv -1 \pmod{p}$ admits a solution. Since $4 \mid p - 1$, Theorem 8-6 tells us that there is an integer a having order 4 modulo p; in other words,

$$a^4 \equiv 1 \pmod{p}$$

or equivalently,

$$(a^2 - 1)(a^2 + 1) \equiv 0 \pmod{p}.$$

Because p is a prime, it follows that either

$$a^2 - 1 \equiv 0 \pmod{p} \text{ or } a^2 + 1 \equiv 0 \pmod{p}.$$

If the first congruence held, then a would have order less than or equal to 2, a contradiction. Hence, $a^2 + 1 \equiv 0 \pmod{p}$, making the integer a a solution to the congruence $x^2 \equiv -1 \pmod{p}$.

Theorem 8-6, as proved, has an obvious drawback; while it does indeed imply the existence of primitive roots for a given prime p, the proof is nonconstructive. To find a primitive root, usually one must either proceed by brute force or else fall back on the extensive tables that have been constructed. The accompanying table lists the smallest positive primitive root for each prime below 200.

Prime	Least positive primitive root	Prime	Least positive primitive root
2	1	89	3
3	2	97	5
5	2	101	2
7	3	103	5
11	2	107	2
13	2	109	6
17	3	113	3
19	2	127	3
23	5	131	2
29	2	137	3
31	3	139	2
37	2	149	2
41	6	151	6
43	3	157	5
47	5	163	2
53	2	167	5
59	2	173	2
61	2	179	2
67	2	181	2
71	7	191	19
73	5	193	5
79	3	197	2
83	2	199	3

If $\chi(p)$ designates the smallest positive primitive root of the prime p, then the table presented shows that $\chi(p) \leq 19$ for all $p < 200$. In fact, $\chi(p)$ becomes arbitrarily large as p increases without bound. The table suggests, although the answer is not yet known, that there exist an infinite number of primes p for which $\chi(p) = 2$.

In most cases $\chi(p)$ is quite small. Among the first 19862 odd primes up to 223051, $\chi(p) \leq 6$ holds for about 80% of these primes; $\chi(p) = 2$ takes place for 7429 primes or approximately 37% of the time, while $\chi(p) = 3$ happens for 4515 primes, or 23% of the time.

In his *Disquisitiones Arithmeticae*, Gauss conjectured that there are infinitely many primes having 10 as a primitive root. In 1927 Emil Artin generalized this unresolved question as: For a not equal to 1, -1, or a perfect square, do there exist infinitely many primes having a as a primitive root? While there is little doubt that this conjecture is true, it has yet to be proved. Recent (1986) work has shown that there are infinitely many a for which Artin's conjecture is true, and at most two primes for which it fails.

The restrictions in Artin's conjecture are justified as follows. Let a be a perfect square, say $a = x^2$, and let p be an odd prime with $\gcd(a, p) = 1$. If $p \nmid x$, then Fermat's Theorem yields $x^{p-1} \equiv 1 \pmod{p}$, whence

$$a^{(p-1)/2} \equiv (x^2)^{(p-1)/2} \equiv 1 \pmod{p}.$$

Thus a cannot serve as a primitive root of p [if $p \mid x$, then $p \mid a$ and surely $a^{p-1} \not\equiv 1 \pmod{p}$]. Furthermore, since $(-1)^2 = 1$, -1 is not a primitive root of p whenever $p - 1 > 2$.

Example 8-3

Let us employ the various techniques of this section to find the $\phi(6) = 2$ integers having order 6 modulo 31. To start, we know that there are

$$\phi(\phi(31)) = \phi(30) = 8$$

primitive roots of 31. Obtaining one of them is a matter of trial and error. Since $2^5 \equiv 1 \pmod{31}$, the integer 2 is clearly ruled out. We need not search too far, since 3 turns out to be a primitive root of 31. Observe that in computing the integral powers of 3 it is not necessary to go beyond 3^{15}; for the order of 3 must divide $\phi(31) = 30$ and the calculation

$$3^{15} \equiv (27)^5 \equiv (-4)^5 \equiv (-64)(16) \equiv -2(16) \equiv -1 \not\equiv 1 \pmod{31}$$

shows that its order is greater than 15.

Because 3 is a primitive root of 31, any integer which is relatively prime to 31 is congruent modulo 31 to an integer of the form 3^k, where $1 \leq k \leq 30$. Theorem 8-3 asserts that the order of 3^k is $30/\gcd(k, 30)$; this will equal 6 if and only if $\gcd(k, 30) = 5$. The values of k for which the last equality holds are $k = 5$ and $k = 25$. Thus our problem is now reduced to evaluating 3^5 and 3^{25} modulo 31. A simple calculation gives

$$3^5 \equiv (27)9 \equiv (-4)9 \equiv -36 \equiv 26 \pmod{31},$$
$$3^{25} \equiv (3^5)^5 \equiv (26)^5 \equiv (-5)^5 \equiv (-125)(25) \equiv -1(25) \equiv 6 \pmod{31},$$

so that 6 and 26 are the only integers having order 6 modulo 31.

PROBLEMS 8.2

1. If p is an odd prime, prove that
   (a)   the only incongruent solutions of $x^2 \equiv 1 \pmod{p}$ are 1 and $p - 1$;
   (b)   the congruence $x^{p-2} + \cdots + x^2 + x + 1 \equiv 0 \pmod{p}$ has exactly $p - 2$ incongruent solutions and they are the integers $2, 3, \ldots, p - 1$.

2. Verify that each of the congruences $x^2 \equiv 1 \pmod{15}$, $x^2 \equiv -1 \pmod{65}$ and $x^2 \equiv -2 \pmod{33}$ has four incongruent solutions; hence, Lagrange's Theorem need not hold if the modulus is a composite number.

3. Determine all the primitive roots of the primes $p = 11, 19,$ and 23, expressing each as a power of some one of the roots.

4. Given that 3 is a primitive root of 43, find
   (a)   all positive integers less than 43 having order 6 modulo 43;
   (b)   all positive integers less than 43 having order 21 modulo 43.

5. Find all positive integers less than 61 having order 4 modulo 61.

6. Assuming that r is a primitive root of the odd prime p, establish the following facts:
   (a)   The congruence $r^{(p-1)/2} \equiv -1 \pmod{p}$ holds.
   (b)   If r' is any other primitive root of p, then rr' is not a primitive root of p. [*Hint:* By part (a), $(rr')^{(p-1)/2} \equiv 1 \pmod{p}$.]
   (c)   If the integer r' is such that $rr' \equiv 1 \pmod{p}$, then r' is a primitive root of p.

7. For a prime $p > 3$, prove that the primitive roots of p occur in incongruent pairs r, r' where $rr' \equiv 1 \pmod{p}$. [*Hint:* If r is a primitive root of p, consider the integer $r' = r^{p-2}$.]

8. Let r be a primitive root of the odd prime p. Prove that
   (a)   if $p \equiv 1 \pmod{4}$, then $-r$ is also a primitive root of p;
   (b)   if $p \equiv 3 \pmod{4}$, then $-r$ has order $(p - 1)/2$ modulo p.

9. Give a different proof of Theorem 5-5 by showing that if r is a primitive root of the prime $p \equiv 1 \pmod{4}$, then $r^{(p-1)/4}$ satisfies the quadratic congruence $x^2 + 1 \equiv 0 \pmod{p}$.

10. Use the fact that each prime p has a primitive root to give a different proof of Wilson's Theorem. [*Hint:* If p has a primitive root r, then Theorem 8-4 implies that $(p - 1)! \equiv r^{1+2+\cdots+(p-1)} \pmod{p}$.]

11. If p is a prime, show that the product of the $\phi(p - 1)$ primitive roots of p is congruent modulo p to $(-1)^{\phi(p-1)}$. [*Hint:* If r is a primitive root of p, then the integer r^k is a primitive root of p provided that $\gcd(k, p - 1) = 1$; now use Theorem 7-7.]

12. For an odd prime p, verify that the sum

$$1^n + 2^n + 3^n + \cdots + (p - 1)^n \equiv \begin{cases} 0 \pmod{p} & \text{if } (p - 1) \nmid n \\ -1 \pmod{p} & \text{if } (p - 1) \mid n \end{cases}$$

$$\left[\textit{Hint: If } (p - 1) \nmid n, \text{ and } r \text{ is a primitive root of } p, \text{ then the sum is congruent modulo } p \text{ to} \right.$$

$$\left. 1 + r^n + r^{2n} + \cdots + r^{(p-2)n} = \frac{r^{(p-1)n} - 1}{r^n - 1}. \right]$$

8.3 COMPOSITE NUMBERS HAVING PRIMITIVE ROOTS

We saw earlier that 2 is a primitive root of 9, so that composite numbers can also possess primitive roots. The next step of our program is to determine all composite numbers for which there exist primitive roots. Some information is available in the following two negative results.

THEOREM 8-7.

For $k \geq 3$, *the integer* 2^k *has no primitive roots.*

Proof: For reasons that will become clear later, we start by showing that if a is an odd integer, then for $k \geq 3$

$$a^{2^{k-2}} \equiv 1 \pmod{2^k}.$$

If $k = 3$, this congruence becomes $a^2 \equiv 1 \pmod 8$, which is certainly true (indeed, $1^2 \equiv 3^2 \equiv 5^2 \equiv 7^2 \equiv 1 \pmod 8$). For $k > 3$, we proceed by induction on k. Assume that the asserted congruence holds for the integer k; that is, $a^{2^{k-2}} \equiv 1 \pmod{2^k}$. This is equivalent to the equation

$$a^{2^{k-2}} = 1 + b2^k,$$

where b is an integer. Squaring both sides, we obtain

$$a^{2^{k-1}} = (a^{2^{k-2}})^2 = 1 + 2(b2^k) + (b2^k)^2$$
$$= 1 + 2^{k+1}(b + b^2 2^{k-1})$$
$$\equiv 1 \pmod{2^{k+1}},$$

so that the asserted congruence holds for $k + 1$ and hence for all $k \geq 3$.

Now the integers which are relatively prime to 2^k are precisely the odd integers; also, $\phi(2^k) = 2^{k-1}$. By what was just proved, if a is an odd integer and $k \geq 3$,

$$a^{\phi(2^k)/2} \equiv 1 \pmod{2^k}$$

and, consequently, there are no primitive roots of 2^k.

Another theorem in this same spirit is

THEOREM 8-8.

If $\gcd(m, n) = 1$, *where* $m > 2$ *and* $n > 2$, *then the integer* mn *has no primitive roots.*

Proof: Consider any integer a for which $\gcd(a, mn) = 1$; then $\gcd(a, m) = 1$ and $\gcd(a, n) = 1$. Put $h = \text{lcm}(\phi(m), \phi(n))$ and $d = \gcd(\phi(m), \phi(n))$.

Since $\phi(m)$ and $\phi(n)$ are both even (Theorem 7-4), surely $d \geq 2$. In consequence,

$$h = \frac{\phi(m)\phi(n)}{d} \leq \frac{\phi(mn)}{2}.$$

Now Euler's Theorem asserts that $a^{\phi(m)} \equiv 1 \pmod{m}$. Raising this equation to the $\phi(n)/d$ power, we get

$$a^h = (a^{\phi(m)})^{\phi(n)/d} \equiv 1^{\phi(n)/d} \equiv 1 \pmod{m}.$$

Similar reasoning leads to $a^h \equiv 1 \pmod{n}$. Together with the hypothesis $\gcd(m, n) = 1$, these congruences force the conclusion that

$$a^h \equiv 1 \pmod{mn}.$$

The point which we wish to make is that the order of any integer relatively prime to mn does not exceed $\phi(mn)/2$, whence there can be no primitive roots for mn.

Some special cases of Theorem 8-8 are of particular interest and we list these below.

COROLLARY.

The integer n fails to have a primitive root if either
(1) n is divisible by two odd primes, or
(2) n is of the form $n = 2^m p^k$, where p is an odd prime and $m \geq 2$.

The significant feature of this last series of results is that they restrict our search for primitive roots to the integers 2, 4, p^k and $2p^k$, where p is an odd prime. In this section, we shall prove that each of the numbers just mentioned has a primitive root, the major task being the establishment of the existence of primitive roots for powers of an odd prime. The argument is somewhat long-winded, but otherwise routine; for the sake of clarity, it is broken down into several steps.

LEMMA 1.

If p is an odd prime, then a primitive root r of p exists such that $r^{p-1} \not\equiv 1 \pmod{p^2}$.

Proof: From Theorem 8-6, it is known that p has primitive roots. Choose any one, call it r. If $r^{p-1} \not\equiv 1 \pmod{p^2}$, then we are finished. In the contrary case, replace r by $r' = r + p$, which is also a primitive root of p. Then employing the Binomial Theorem,

$$(r')^{p-1} \equiv (r + p)^{p-1} \equiv r^{p-1} + (p-1)pr^{p-2} \pmod{p^2}.$$

But we have assumed that $r^{p-1} \equiv 1 \pmod{p^2}$; hence

$$(r')^{p-1} \equiv 1 - pr^{p-2} \pmod{p^2}.$$

Since r is a primitive root of p, $\gcd(r, p) = 1$ and so $p \nmid r^{p-2}$. The outcome of all this is that $(r')^{p-1} \not\equiv 1 \pmod{p^2}$, which proves the lemma.

COROLLARY.

If p is an odd prime, then p^2 has a primitive root; in fact, for a primitive root r of p, either r or $r + p$ is a primitive root of p^2.

Proof: The assertion is almost obvious: If r is a primitive root of p, then the order of r modulo p^2 is either $p - 1$ or else $p(p-1) = \phi(p^2)$. The foregoing proof shows that if r has order $p - 1$ modulo p^2, then $r + p$ will be a primitive root of p^2.

To reach our goal, another somewhat technical lemma is needed.

LEMMA 2.

Let p be an odd prime and r be a primitive root of p such that $r^{p-1} \not\equiv 1 \pmod{p^2}$. Then for each positive integer $k \geq 2$,

$$r^{p^{k-2}(p-1)} \not\equiv 1 \pmod{p^k}.$$

Proof: The proof proceeds by induction on k. By hypothesis, the assertion holds for $k = 2$. Let us assume that it is true for some $k \geq 2$ and show that it is true for $k + 1$. Since $\gcd(r, p^{k-1}) = \gcd(r, p^k) = 1$, Euler's Theorem indicates that

$$r^{p^{k-2}(p-1)} = r^{\phi(p^{k-1})} \equiv 1 \pmod{p^{k-1}}.$$

Hence, there exists an integer a satisfying

$$r^{p^{k-2}(p-1)} = 1 + ap^{k-1},$$

where $p \nmid a$ by our induction hypothesis. Raise both sides of this last-written equation to the pth power and expand to obtain

$$r^{p^{k-1}(p-1)} = (1 + ap^{k-1})^p \equiv 1 + ap^k \pmod{p^{k+1}}.$$

Since the integer a is not divisible by p, we have

$$r^{p^{k-1}(p-1)} \not\equiv 1 \pmod{p^{k+1}}.$$

This completes the induction step, thereby proving the lemma.

The hard work, for the moment, is over. We now stitch the pieces together to prove that the powers of any odd prime have a primitive root.

THEOREM 8-9.

If p is an odd prime number and $k \geq 1$, then there exists a primitive root for p^k.

Proof: The two lemmas allow us to choose a primitive root r of p for which $r^{p^{k-2}(p-1)} \not\equiv 1 \pmod{p^k}$; in fact, any integer r satisfying the condition $r^{p-1} \not\equiv 1 \pmod{p^2}$ will do. We argue that such an r serves as a primitive root for all powers of p.

Let n be the order of r modulo p^k. In compliance with Theorem 8-1, n must divide $\phi(p^k) = p^{k-1}(p - 1)$. Since $r^n \equiv 1 \pmod{p^k}$ yields $r^n \equiv 1 \pmod{p}$, we also have $p - 1 \mid n$ (Theorem 8-1 serves again). Consequently, n assumes the form $n = p^m(p - 1)$, where $0 \leq m \leq k - 1$. If it happened that $n \neq p^{k-1}(p - 1)$, then $p^{k-2}(p - 1)$ would be divisible by n and we would arrive at

$$r^{p^{k-2}(p-1)} \equiv 1 \pmod{p^k},$$

contradicting the way in which r was initially picked. Therefore, $n = p^{k-1}(p - 1)$ and r is a primitive root for p^k.

This leaves only the case $2p^k$ for our consideration.

COROLLARY.

There are primitive roots for $2p^k$, where p is an odd prime and $k \geq 1$.

Proof: Let r be a primitive root for p^k. There is no harm in assuming that r is an odd integer; for, if it is even, then $r + p^k$ is odd and is still a primitive root for p^k. Then $\gcd(r, 2p^k) = 1$. The order n of r modulo $2p^k$ must divide

$$\phi(2p^k) = \phi(2)\phi(p^k) = \phi(p^k).$$

But $r^n \equiv 1 \pmod{2p^k}$ implies that $r^n \equiv 1 \pmod{p^k}$, and so $\phi(p^k) \mid n$. Together these divisibility conditions force $n = \phi(2p^k)$, making r a primitive root of $2p^k$.

The prime 5 has $\phi(4) = 2$ primitive roots, namely the integers 2 and 3. Since

$$2^{5-1} \equiv 16 \not\equiv 1 \pmod{25} \quad \text{and} \quad 3^{5-1} \equiv 6 \not\equiv 1 \pmod{25},$$

these also serve as primitive roots for 5^2, hence for all higher powers of 5. The proof of the last corollary guarantees that 3 is a primitive root for all numbers of the form $2 \cdot 5^k$.

We summarize what has been accomplished in

THEOREM 8-10.

An integer $n > 1$ has a primitive root if and only if

$$n = 2, 4, p^k, \text{ or } 2p^k,$$

where p is an odd prime.

Proof: By virtue of Theorems 8-7 and 8-8, the only positive integers with primitive roots are those mentioned in the statement of our theorem. It may be checked that 1 is a primitive root for 2, while 3 is a primitive root of 4. We have just finished proving that primitive roots exist for any power of an odd prime and for twice such a power.

This seems the opportune moment to mention that Euler gave an essentially correct (although incomplete) proof in 1773 of the existence of primitive roots for any prime p and listed all the primitive roots for $p \leq 37$. Legendre, using Lagrange's Theorem, managed to repair the deficiency and showed (1785) that there are $\phi(d)$ integers of order d for each $d \mid (p - 1)$. The greatest advances in this direction were made by Gauss when, in 1801, he published a proof that there exist primitive roots of n if and only if $n = 2, 4, p^k$, and $2p^k$, where p is an odd prime.

PROBLEMS 8.3

1. (a) Find the four primitive roots of 26 and the eight primitive roots of 25.
   (b) Determine all the primitive roots of 3^2, 3^3 and 3^4.

2. For an odd prime p, establish the following facts:
   (a) There are as many primitive roots of $2p^n$ as of p^n.
   (b) Any primitive root r of p^n is also a primitive root of p. [Hint: Let r have order k modulo p. Show that $r^{pk} \equiv 1 \pmod{p^2}$, . . . , $r^{p^{n-1}k} \equiv 1 \pmod{p^n}$, hence $\phi(p^n) \mid p^{n-1}k$.]
   (c) A primitive root of p^2 is also a primitive root of p^n for $n \geq 2$.

3. If r is a primitive root of p^2, p being an odd prime, show that the solutions of the congruence $x^{p-1} \equiv 1 \pmod{p^2}$ are precisely the integers $r^p, r^{2p}, \ldots, r^{(p-1)p}$.

4. (a) Prove that 3 is a primitive root of all integers of the form 7^k and $2 \cdot 7^k$.
    (b) Find a primitive root for any integer of the form 17^k.

5. Obtain all the primitive roots of 41 and 82.

6. (a) Prove that a primitive root r of p^k, where p is an odd prime, is a primitive root of $2p^k$ if and only if r is an odd integer.
    (b) Confirm that 3, 3^3, 3^5, and 3^9 are primitive roots of $578 = 2 \cdot 17^2$, but that 3^4 and 3^{17} are not.

7. Assume that r is a primitive root of the odd prime p and $(r + tp)^{p-1} \not\equiv 1 \pmod{p^2}$. Show that $r + tp$ is a primitive root of p^k for each $k \geq 1$.

8. If $n = 2^{k_0}p_1^{k_1}p_2^{k_2} \ldots p_r^{k_r}$ is the prime factorization of $n > 1$, define the *universal exponent* $\lambda(n)$ of n by

$$\lambda(n) = \text{lcm}(\lambda(2^{k_0}), \phi(p_1^{k_1}), \ldots, \phi(p_r^{k_r}))$$

where $\lambda(2) = 1$, $\lambda(2^2) = 2$, and $\lambda(2^k) = 2^{k-2}$ for $k \geq 3$. Prove the following statements concerning the universal exponent:
    (a) For $n = 2, 4, p^k, 2p^k$, where p is an odd prime, $\lambda(n) = \phi(n)$.
    (b) If $\gcd(a, 2^k) = 1$, then $a^{\lambda(2^k)} \equiv 1 \pmod{2^k}$. [*Hint:* For $k \geq 3$, use induction on k and the fact that $\lambda(2^{k+1}) = 2\lambda(2^k)$.]
    (c) If $\gcd(a, n) = 1$, then $a^{\lambda(n)} \equiv 1 \pmod{n}$. [*Hint:* For each prime power p^k occurring in n, $a^{\lambda(n)} \equiv 1 \pmod{p^k}$.]

9. Verify that, for $5040 = 2^4 \cdot 3^2 \cdot 5 \cdot 7$, $\lambda(5040) = 12$ and $\phi(5040) = 1152$.

10. Use Problem 8 to show that if $n \neq 2, 4, p^k, 2p^k$, where p is an odd prime, then n has no primitive root. [*Hint:* Except for the cases 2, 4, p^k, $2p^k$, we have $\lambda(n) \mid \frac{1}{2}\phi(n)$; hence, $a^{\phi(n)/2} \equiv 1 \pmod{n}$ whenever $\gcd(a, n) = 1$.]

11. (a) Prove that if $\gcd(a, n) = 1$, then the linear congruence $ax \equiv b \pmod{n}$ has the solution $x \equiv ba^{\lambda(n)-1} \pmod{n}$.
    (b) Use part (a) to solve the congruences $13x \equiv 2 \pmod{40}$ and $3x \equiv 13 \pmod{77}$.

8.4 THE THEORY OF INDICES

The remainder of the chapter is concerned with a new idea, the concept of index. This was introduced by Gauss in his *Disquisitiones Arithmeticae*.

Let n be any integer which admits a primitive root r. As we know, the first $\phi(n)$ powers of r,

$$r, r^2, \ldots, r^{\phi(n)}$$

are congruent modulo n, in some order, to those integers less than n and relatively prime to it. Hence, if a is an arbitrary integer relatively prime to n, then a can be expressed in the form

$$a \equiv r^k \pmod{n}$$

for a suitable choice of k, where $1 \leq k \leq \phi(n)$. This allows us to frame the following definition.

DEFINITION 8-3.

Let r be a primitive root of n. If $\gcd(a , n) = 1$, then the smallest positive integer k such that $a \equiv r^k \pmod{n}$ is called the *index of a relative to r.*

One customarily denotes the index of a relative to r by $\text{ind}_r a$ or, if no confusion is likely to occur, by $\text{ind } a$. Clearly, $1 \le \text{ind}_r a \le \phi(n)$ and

$$r^{\text{ind}_r a} \equiv a \pmod{n}.$$

The notation $\text{ind}_r a$ is meaningless unless $\gcd (a, n) = 1$; in the future, this will be tacitly assumed.

For example, the integer 2 is a primitive root of 5 and

$$2^1 \equiv 2, \; 2^2 \equiv 4, \; 2^3 \equiv 3, \; 2^4 \equiv 1 \pmod{5}.$$

It follows that

$$\text{ind}_2 1 = 4, \; \text{ind}_2 2 = 1, \; \text{ind}_2 3 = 3, \; \text{ind}_2 4 = 2.$$

Observe that indices of integers which are congruent modulo n are equal. Thus, when setting up tables of values for $\text{ind } a,$ it suffices to consider only those integers a less than and relatively prime to the modulus n. To see this, let $a \equiv b \pmod{n}$, where a and b are relatively prime to n. Since $r^{\text{ind } a} \equiv a \pmod{n}$ and $r^{\text{ind } b} \equiv b \pmod{n}$, we have

$$r^{\text{ind } a} \equiv r^{\text{ind } b} \pmod{n}.$$

Invoking Theorem 8-2, it may be concluded that $\text{ind } a \equiv \text{ind } b \pmod{\phi(n)}$. But, because of the restrictions on the size of $\text{ind } a$ and $\text{ind } b,$ this is only possible when $\text{ind } a = \text{ind } b$.

Indices obey rules which are reminiscent of those for logarithms, with the primitive root playing a role analogous to that of the base for the logarithm.

THEOREM 8-11.

If n has a primitive root r and $\text{ind } a$ denotes the index of a relative to r, then the properties

(1) $\text{ind } (ab) \equiv \text{ind } a + \text{ind } b \pmod{\phi(n)}$,
(2) $\text{ind } a^k \equiv k \, \text{ind } a \pmod{\phi(n)}$ for $k > 0$,
(3) $\text{ind } 1 \equiv 0 \pmod{\phi(n)}$, $\text{ind } r \equiv 1 \pmod{\phi(n)}$.

hold.

Proof: By the definition of index, $r^{\text{ind } a} \equiv a \pmod{n}$ and $r^{\text{ind } b} \equiv b \pmod{n}$. Multiplying these congruences together, we obtain

$$r^{\text{ind } a + \text{ind } b} \equiv ab \pmod{n}.$$

But $r^{\text{ind } (ab)} \equiv ab \pmod{n}$, so that

$$r^{\text{ind } a + \text{ind } b} \equiv r^{\text{ind } (ab)} \pmod{n}.$$

It may very well happen that ind a + ind b exceeds $\phi(n)$. This presents no problem, for Theorem 8-1 guarantees that the last equation holds if and only if the exponents are congruent modulo $\phi(n)$; that is,

$$\text{ind } a + \text{ind } b \equiv \text{ind } (ab) \ (\text{mod } \phi(n)),$$

which is property (1).

The proof of property (2) proceeds along much the same lines. For we have $r^{\text{ind } a^k} \equiv a^k \ (\text{mod } n)$ while, by the laws of exponents, $r^{k \ \text{ind } a} = (r^{\text{ind } a})^k \equiv a^k \ (\text{mod } n)$; hence,

$$r^{\text{ind } a^k} \equiv r^{k \ \text{ind } a} \ (\text{mod } n).$$

As above, the implication is that ind $a^k \equiv k$ ind $a \ (\text{mod } \phi(n))$. The two parts of property (3) should be fairly apparent.

The theory of indices can be used to solve certain types of congruences. For instance, consider the binomial congruence

$$x^k \equiv a \ (\text{mod } n), \qquad\qquad\qquad k \geq 2$$

where n is a positive integer having a primitive root and $\gcd(a, n) = 1$. By properties (1) and (2) of Theorem 8-11, this congruence is entirely equivalent to the linear congruence

$$k \text{ ind } x \equiv \text{ind } a \ (\text{mod } \phi(n))$$

in the unknown ind x. If $d = \gcd(k, \phi(n))$ and $d \nmid$ ind a, there is no solution. But, if $d \mid$ ind a, then there are exactly d values of ind x that will satisfy this last congruence, hence there are d incongruent solutions of $x^k \equiv a \ (\text{mod } n)$.

The case in which $k = 2$ and $n = p$, with p an odd prime, is particularly important. Since $\gcd(2, p - 1) = 2$, the foregoing remarks imply that the congruence quadratic $x^2 \equiv a \ (\text{mod } p)$ has a solution if and only if $2 \mid$ ind a; when this condition is fulfilled, there are exactly two solutions. If r is a primitive root of p, then $r^k (1 \leq k \leq p - 1)$ runs modulo p through the integers $1, 2, \ldots, p - 1$, in some order. The even powers of r produce the values of a for which the congruence $x^2 \equiv a \ (\text{mod } p)$ is solvable; there are precisely $(p - 1)/2$ such choices for a.

Example 8-4

For an illustration of these ideas, let us solve the congruence

$$4x^9 \equiv 7 \ (\text{mod } 13).$$

A table of indices can be constructed once a primitive root of 13 is fixed. Using the primitive root 2, we simply calculate the powers $2, 2^2, \ldots, 2^{12}$ modulo 13. Here,

$2^1 \equiv 2,$	$2^5 \equiv 6,$	$2^9 \equiv 5$
$2^2 \equiv 4,$	$2^6 \equiv 12,$	$2^{10} \equiv 10$
$2^3 \equiv 8,$	$2^7 \equiv 11,$	$2^{11} \equiv 7$
$2^4 \equiv 3,$	$2^8 \equiv 9,$	$2^{12} \equiv 1$

all modulo 13, and hence our table is

a	1	2	3	4	5	6	7	8	9	10	11	12
$\text{ind}_2\, a$	12	1	4	2	9	5	11	3	8	10	7	6

Taking indices, the congruence $4x^9 \equiv 7 \pmod{13}$ has a solution if and only if

$$\text{ind}_2\, 4 + 9\, \text{ind}_2\, x \equiv \text{ind}_2\, 7 \pmod{12}.$$

The table gives the values $\text{ind}_2\, 4 = 2$ and $\text{ind}_2\, 7 = 11$, so that the last congruence becomes $9\, \text{ind}_2\, x \equiv 11 - 2 \equiv 9 \pmod{12}$ which in its turn is equivalent to $\text{ind}_2\, x \equiv 1 \pmod{4}$. It follows that

$$\text{ind}_2\, x = 1, 5, \text{ or } 9.$$

Consulting the table of indices again, we find that the original congruence $4\, x^9 \equiv 7 \pmod{13}$ possesses the three solutions

$$x \equiv 2, 5, \text{ and } 6 \pmod{13}.$$

If a different primitive root is chosen, one obviously obtains a different value for the index of $a;$ but, for purposes of solving the given congruence, it does not really matter which index table is available. The $\phi(\phi(13)) = 4$ primitive roots of 13 are obtained from the powers $2^k(1 \leq k \leq 12)$, where

$$\gcd(k, \phi(13)) = \gcd(k, 12) = 1.$$

These are

$$2^1 \equiv 2,\ 2^5 \equiv 6,\ 2^7 \equiv 11,\ 2^{11} \equiv 7 \pmod{13}.$$

The index table for, say, the primitive root 6 is displayed below:

a	1	2	3	4	5	6	7	8	9	10	11	12
$\text{ind}_6\, a$	12	5	8	10	9	1	7	3	4	2	11	6

Employing this table, the congruence $4x^9 \equiv 7 \pmod{13}$ is replaced by

$$\text{ind}_6\, 4 + 9\, \text{ind}_6\, x \equiv \text{ind}_6\, 7 \pmod{12}$$

or rather,

$$9\, \text{ind}_6\, x \equiv 7 - 10 \equiv -3 \equiv 9 \pmod{12}.$$

Thus, $\text{ind}_6\, x = 1, 5, \text{ or } 9$, leading to the solutions

$$x \equiv 2, 5, \text{ and } 6 \pmod{13},$$

as before.

The following criterion for solvability is often useful.

THEOREM 8-12.

Let n be an integer possessing a primitive root and let $\gcd(a, n) = 1$. *Then the congruence* $x^k \equiv a \pmod{n}$ *has a solution if and only if*

$$a^{\phi(n)/d} \equiv 1 \pmod{n},$$

where $d = \gcd(k, \phi(n))$; *if it has a solution, there are exactly d solutions modulo n.*

Proof: Taking indices, the congruence $a^{\phi(n)/d} \equiv 1 \pmod{n}$ is equivalent to

$$\frac{\phi(n)}{d} \text{ ind } a \equiv 0 \pmod{\phi(n)}$$

which in turn holds if and only if $d \mid \text{ind } a$. But we have just seen that the latter is a necessary and sufficient condition for the congruence $x^k \equiv a \pmod{n}$ to be solvable.

COROLLARY (Euler).

Let p be a prime and $\gcd(a, p) = 1$. *Then the congruence* $x^k \equiv a \pmod{p}$ *has a solution if and only if* $a^{(p-1)/d} \equiv 1 \pmod{p}$, *where* $d = \gcd(k, p-1)$.

Example 8-5

Let us consider the congruence

$$x^3 \equiv 4 \pmod{13}.$$

In this setting, $d = \gcd(3, \phi(13)) = \gcd(3, 12) = 3$ and so $\phi(13)/d = 4$. Since $4^4 \equiv 9 \not\equiv 1 \pmod{13}$, Theorem 8-12 asserts that the given congruence is not solvable.

On the other hand, the same theorem guarantees that

$$x^3 \equiv 5 \pmod{13}$$

will possess a solution (in fact, there are three incongruent solutions modulo 13); for, in this case, $5^4 \equiv 625 \equiv 1 \pmod{13}$. These solutions can be found by means of the index calculus as follows: The congruence $x^3 \equiv 5 \pmod{13}$ is equivalent to

$$3 \text{ ind}_2 x \equiv 9 \pmod{12},$$

which becomes

$$\text{ind}_2 x \equiv 3 \pmod{4}.$$

This last equation admits three incongruent solutions modulo 12, namely

$$\text{ind}_2 x = 3, 7, \text{ or } 11.$$

The integers corresponding to these indices are, respectively, 8, 11, and 7, so that the solutions of the congruence $x^3 \equiv 5 \pmod{13}$ are

$$x \equiv 7, 8, \text{ and } 11 \pmod{13}.$$

PROBLEMS 8.4

1. Find the index of 5 relative to each of the primitive roots of 13.

2. Using a table of indices for a primitive root of 11, solve the congruences
    (a) $7x^3 \equiv 3 \pmod{11}$
    (b) $3x^4 \equiv 5 \pmod{11}$
    (c) $x^8 \equiv 10 \pmod{11}$

3. The following is a table of indices for the prime 17 relative to the primitive root 3:

a	1	2	3	4	5	6	7	8	9	10	11	12	13	14	15	16
ind $_3a$	16	14	1	12	5	15	11	10	2	3	7	13	4	9	6	8

    With the aid of this table, solve the congruences
    (a) $x^{12} \equiv 13 \pmod{17}$ (b) $8x^5 \equiv 10 \pmod{17}$
    (c) $9x^8 \equiv 8 \pmod{17}$ (d) $7^x \equiv 7 \pmod{17}$

4. Find the remainder when $3^{24} \cdot 5^{13}$ is divided by 17. [*Hint:* Use the theory of indices.]

5. If r and r' are both primitive roots of the odd prime p, show that for $\gcd(a , p) = 1$

    $$\text{ind}_{r'} a \equiv (\text{ind}_r a)(\text{ind}_{r'} r)\pmod{p - 1}.$$

    This corresponds to the rule for changing the base of logarithms.

6. (a) Construct a table of indices for the prime 17 with respect to the primitive root 5.
         [*Hint:* By the previous problem, ind$_5 a \equiv 13$ ind$_3 a \pmod{16}$.]
    (b) Using the table in part (a), solve the congruences in Problem 3.

7. If r is a primitive root of the odd prime p, verify that
    $$\text{ind}_r (-1) = \text{ind}_r (p - 1) = \tfrac{1}{2}(p - 1).$$

8. (a) Determine the integers a ($1 \le a \le 12$) such that the congruence $ax^4 \equiv b \pmod{13}$
         has a solution for $b = 2, 5$, and 6.
    (b) Determine the integers a ($1 \le a \le p - 1$) such that the congruence $x^4 \equiv a$
         $\pmod p$ has a solution for $p = 7, 11$, and 13.

9. Employ the corollary to Theorem 8-12 to establish that if p is an odd prime, then
    (a) $x^2 \equiv -1 \pmod p$ is solvable if and only if $p \equiv 1 \pmod 4$;
    (b) $x^4 \equiv -1 \pmod p$ is solvable if and only if $p \equiv 1 \pmod 8$.

10. Given the congruence $x^3 \equiv a \pmod p$, where $p \ge 5$ is a prime number and $\gcd(a , p)$
    $= 1$, prove that
    (a) if $p \equiv 1 \pmod 6$, then the congruence has either no solutions or three incongruent
         solutions modulo p;
    (b) if $p \equiv 5 \pmod 6$, then the congruence has a unique solution modulo p.

11. Show that the congruence $x^3 \equiv 3 \pmod{19}$ has no solutions, while $x^3 \equiv 11 \pmod{19}$ has
    three incongruent solutions.

12. Determine whether the two congruences $x^5 \equiv 13 \pmod{23}$ and $x^7 \equiv 15 \pmod{29}$ are
    solvable.

13. If p is a prime and $\gcd(k , p - 1) = 1$, prove that the integers
    $$1^k, 2^k, 3^k, \ldots , (p - 1)^k$$
    form a reduced set of residues modulo p.

14. Let r be a primitive root of the odd prime p, and let $d = \gcd(k, p - 1)$. Prove that the values of a for which the congruence $x^k \equiv a \pmod{p}$ is solvable are r^d, r^{2d}, . . . , $r^{[(p-1)/d]d}$.

15. If r is a primitive root of the odd prime p, show that

$$\text{ind}_r(p - a) \equiv \text{ind}_r a + (p - 1)/2 \pmod{p - 1};$$

and consequently that only one-half of an index table need be calculated in order to complete the table.

16. (a) Let r be a primitive root of the odd prime p. Establish that the exponential congruence

$$a^x \equiv b \pmod{p}$$

has a solution if and only if $d \mid \text{ind}_r b$, where the integer $d = \gcd(\text{ind}_r a, p - 1)$; in this case, there are d incongruent solutions modulo $p - 1$.

   (b) Solve the exponential congruences $4^x \equiv 13 \pmod{17}$ and $5^x \equiv 4 \pmod{19}$.

17. For which values of b is the exponential congruence $9^x \equiv b \pmod{13}$ solvable?

9

The Quadratic
Reciprocity Law

"The moving power of mathematical invention is not reasoning but imagination."
A. DeMorgan

9.1 EULER'S CRITERION

As the heading suggests, the present chapter has as its goal another major contribution of Gauss: the Quadratic Reciprocity Law. For those who consider the theory of numbers "the Queen of Mathematics," this is one of the jewels in her crown. The intrinsic beauty of the Quadratic Reciprocity Law has long exerted a strange fascination for mathematicians. Since Gauss' time, over a hundred proofs of it, all more or less different, have been published (in fact, Gauss himself eventually devised seven). Among the eminent mathematicians of the 19th century who contributed their proofs appear the names of Cauchy, Jacobi, Dirichlet, Eisenstein, Kronecker, and Dedekind.

Roughly speaking, the Quadratic Reciprocity Law deals with the solvability of quadratic congruences. Therefore, it seems appropriate to begin by considering the congruence

$$(1) \qquad ax^2 + bx + c \equiv 0 \ (\mathrm{mod}\ p),$$

where p is an odd prime and $a \not\equiv 0 \ (\mathrm{mod}\ p)$; that is, $\gcd(a, p) = 1$. The supposition that p is an odd prime implies that $\gcd(4a, p) = 1$. Thus, quadratic congruence (1) is equivalent to

$$4a(ax^2 + bx + c) \equiv 0 \ (\mathrm{mod}\ p).$$

By using the identity

$$4a(ax^2 + bx + c) = (2ax + b)^2 - (b^2 - 4ac),$$

the last-written quadratic congruence may be expressed as

$$(2ax + b)^2 \equiv (b^2 - 4ac) \ (\mathrm{mod}\ p).$$

Now put $y = 2ax + b$ and $d = b^2 - 4ac$ to get

(2) $$y^2 \equiv d \pmod{p}.$$

If $x \equiv x_0 \pmod{p}$ is a solution of quadratic congruence (1), then the integer $y \equiv 2ax_0 + b \pmod{p}$ satisfies quadratic congruence (2). Conversely, if $y \equiv y_0 \pmod{p}$ is a solution of quadratic congruence (2), then $2ax \equiv y_0 - b \pmod{p}$ can be solved to obtain a solution of (1).

Thus, the problem of finding a solution to the quadratic congruence (1) is equivalent to that of finding a solution to a linear congruence and a quadratic congruence of the form

(3) $$x^2 \equiv a \pmod{p}.$$

If $p \mid a$, then quadratic congruence (3) has $x \equiv 0 \pmod{p}$ as its only solution. To avoid trivialities, let us agree to assume hereafter that $p \nmid a$.

Granting this, whenever $x^2 \equiv a \pmod{p}$ admits a solution $x = x_0$, then there is also a second solution $x = p - x_0$. This second solution is not congruent to the first. For $x_0 \equiv p - x_0 \pmod{p}$ implies that $2x_0 \equiv 0 \pmod{p}$, or $x_0 \equiv 0 \pmod{p}$, which is impossible. By Lagrange's Theorem, these two solutions exhaust the incongruent solutions of $x^2 \equiv a \pmod{p}$. In short: $x^2 \equiv a \pmod{p}$ has exactly two solutions or no solutions.

A simple numerical example of what we have just said is provided by the quadratic congruence

$$5x^2 - 6x + 2 \equiv 0 \pmod{13}.$$

To obtain the solution, one replaces this congruence by the simpler one

$$y^2 \equiv 9 \pmod{13}$$

with solutions $y \equiv 3, 10 \pmod{13}$. Next, solve the linear congruences

$$10x \equiv 9 \pmod{13}, \quad 10x \equiv 16 \pmod{13}.$$

It is not difficult to see that $x \equiv 10, 12 \pmod{13}$ satisfy these equations and, by our previous remarks, also the original quadratic congruence.

The major effort in this presentation is directed towards providing a test for the existence of solutions of the quadratic congruence

(4) $$x^2 \equiv a \pmod{p}, \quad \gcd(a, p) = 1.$$

To put it differently, we wish to identify those integers a which are perfect squares modulo p. Some additional terminology will help us to discuss this situation in a concise way:

DEFINITION 9-1.
   Let p be an odd prime and $\gcd(a, p) = 1$. If the congruence $x^2 \equiv a \pmod{p}$ has a solution, then a is said to be a *quadratic residue of p*. Otherwise, a is called a *quadratic nonresidue of p*.

The point to bear in mind is that if $a \equiv b \pmod{p}$, then a is a quadratic residue of p if and only if b is a quadratic residue of p.

Thus, we only need to determine the quadratic character of those positive integers less than p in order to ascertain that of any integer.

Example 9-1

Consider the case of the prime $p = 13$. To find out how many of the integers $1, 2, 3, \ldots, 12$ are quadratic residues of 13, we must know which of the congruences

$$x^2 \equiv a \pmod{13}$$

are solvable when a runs through the set $\{1, 2, \ldots, 12\}$. Modulo 13, the squares of the integers $1, 2, 3, \ldots, 12$ are

$$1^2 \equiv 12^2 \equiv 1,$$
$$2^2 \equiv 11^2 \equiv 4,$$
$$3^2 \equiv 10^2 \equiv 9,$$
$$4^2 \equiv 9^2 \equiv 3,$$
$$5^2 \equiv 8^2 \equiv 12,$$
$$6^2 \equiv 7^2 \equiv 10.$$

Consequently, the quadratic residues of 13 are 1, 3, 4, 9, 10, 12, while the non-residues are 2, 5, 6, 7, 8, 11. Observe that the integers between 1 and 12 are divided equally among the quadratic residues and nonresidues; this is typical of the general situation.

Euler devised a simple criterion for deciding whether an integer a is a quadratic residue of a given prime p.

THEOREM 9-1.

(Euler's Criterion). *Let p be an odd prime and $\gcd(a, p) = 1$. Then a is a quadratic residue of p if and only if $a^{(p-1)/2} \equiv 1 \pmod{p}$.*

Proof: Suppose that a is a quadratic residue of p, so that $x^2 \equiv a \pmod{p}$ admits a solution, call it x_1. Since $\gcd(a, p) = 1$, evidently $\gcd(x_1, p) = 1$. We may therefore appeal to Fermat's Theorem to obtain

$$a^{(p-1)/2} \equiv (x_1^2)^{(p-1)/2} \equiv x_1^{p-1} \equiv 1 \pmod{p}.$$

For the opposite direction, assume that the congruence $a^{(p-1)/2} \equiv 1 \pmod{p}$ holds, and let r be a primitive root of p. Then $a \equiv r^k \pmod{p}$ for some integer k, with $1 \leq k \leq p - 1$. It follows that

$$r^{k(p-1)/2} \equiv a^{(p-1)/2} \equiv 1 \pmod{p}.$$

By Theorem 8-1, the order of r (namely, $p - 1$) must divide the exponent $k(p - 1)/2$. The implication is that k is an even integer, say $k = 2j$. Hence,

$$(r^j)^2 = r^{2j} = r^k \equiv a \pmod{p},$$

making the integer r^j a solution of the congruence $x^2 \equiv a \pmod{p}$. This proves that a is a quadratic residue of the prime p.

Now if p (as always) is an odd prime and gcd $(a, p) = 1$, then

$$(a^{(p - 1)/2} - 1)(a^{(p - 1)/2} + 1) = a^{p - 1} - 1 \equiv 0 \pmod{p},$$

the last congruence being justified by Fermat's Theorem. Hence either

$$a^{(p - 1)/2} \equiv 1 \pmod{p} \quad \text{or} \quad a^{(p - 1)/2} \equiv -1 \pmod{p},$$

but not both. For, if both congruences held simultaneously, then we would have $1 \equiv -1 \pmod{p}$, or equivalently, $p \mid 2$, which conflicts with our hypothesis. Since a quadratic nonresidue of p does not satisfy $a^{(p - 1)/2} \equiv 1 \pmod{p}$, it must therefore satisfy $a^{(p - 1)/2} \equiv -1 \pmod{p}$. This observation provides an alternate formulation of Euler's Criterion: the integer a is a quadratic nonresidue of the prime p if and only if $a^{(p - 1)/2} \equiv -1 \pmod{p}$.

Putting the various pieces together, we come up with

COROLLARY.

Let p be an odd prime and $\gcd(a, p) = 1$. Then a is a quadratic residue or nonresidue of p according to whether

$$a^{(p - 1)/2} \equiv 1 \pmod{p} \quad or \quad a^{(p - 1)/2} \equiv -1 \pmod{p}.$$

Example 9-2

In the case where $p = 13$, we find that

$$2^{(13 - 1)/2} = 2^6 = 64 \equiv 12 \equiv -1 \pmod{13}.$$

Thus, by virtue of the last corollary, the integer 2 is a quadratic nonresidue of 13. Since

$$3^{(13 - 1)/2} = 3^6 = (27)^2 \equiv 1^2 \equiv 1 \pmod{13},$$

the same result indicates that 3 is a quadratic residue of 13 and so the congruence $x^2 \equiv 3 \pmod{13}$ is solvable; in fact, its two incongruent solutions are $x \equiv 4$ and 9 $\pmod{13}$.

There is an alternative proof of Euler's Criterion (due to Dirichlet) which is longer, but perhaps more illuminating. The reasoning proceeds as follows: Let a be a quadratic nonresidue of p and let c be any one of the integers $1, 2, \ldots, p - 1$. By the theory of linear congruences, there exists a solution c' of $cx \equiv a \pmod{p}$, with c' also in the set $\{1, 2, \ldots, p - 1\}$. Note that $c' \neq c$, for otherwise we would have $c^2 \equiv a$

(mod p), which contradicts what we assumed. Thus, the integers between 1 and $p - 1$ can be divided into $(p - 1)/2$ pairs, c, c', where $cc' \equiv a$ (mod p). This leads to $(p - 1)/2$ congruences,

$$c_1 c'_1 \equiv a \text{ (mod } p),$$

$$c_2 c'_2 \equiv a \text{ (mod } p),$$

$$\vdots$$

$$c_{(p-1)/2} c'_{(p-1)/2} \equiv a \text{ (mod } p).$$

Multiplying them together and observing that the product

$$c_1 c'_1 c_2 c'_2 \cdots c_{(p-1)/2} \, c'_{(p-1)/2}$$

is simply a rearrangement of $1 \cdot 2 \cdot 3 \cdots (p - 1)$, we obtain

$$(p - 1)! \equiv a^{(p-1)/2} \text{ (mod } p).$$

At this point, Wilson's Theorem enters the picture; for, $(p - 1)! \equiv -1$ (mod p), so that

$$a^{(p-1)/2} \equiv -1 \text{ (mod } p),$$

which is Euler's Criterion when a is a quadratic nonresidue of p.

We next examine the case in which a is a quadratic residue of p. In this setting the congruence $x^2 \equiv a$ (mod p) admits two solutions $x = x_1$ and $x = p - x_1$, for some x_1 satisfying $1 \leq x_1 \leq p - 1$. If x_1 and $p - x_1$ are removed from the set $\{1, 2, \ldots, p - 1\}$, then the remaining $p - 3$ integers can be grouped into pairs c, c' (where $c \neq c'$) such that $cc' \equiv a$ (mod p). To these $(p - 3)/2$ congruences, add the congruence

$$x_1(p - x_1) \equiv -x_1^2 \equiv -a \text{ (mod } p).$$

Upon taking the product of all the congruences involved, we arrive at the relation

$$(p - 1)! \equiv -a^{(p-1)/2} \text{ (mod } p).$$

Wilson's Theorem plays its role once again to produce

$$a^{(p-1)/2} \equiv 1 \text{ (mod } p).$$

Summing up, we have shown that $a^{(p-1)/2} \equiv 1$ (mod p) or $a^{(p-1)/2} \equiv -1$ (mod p) according to whether a is a quadratic residue or nonresidue of p.

Euler's Criterion is not offered as a practical test for determining whether a given integer is or is not a quadratic residue; the calculations involved are too cumbersome unless the modulus is small. But as a crisp criterion, easily worked with for theoretical purposes, it leaves little to be desired. A more effective method of computation is embodied in the Quadratic Reciprocity Law, which we shall prove later in the chapter.

PROBLEMS 9.1

1. Solve the following quadratic congruences:
   (a) $x^2 + 7x + 10 \equiv 0 \pmod{11}$;
   (b) $3x^2 + 9x + 7 \equiv 0 \pmod{13}$;
   (c) $5x^2 + 6x + 1 \equiv 0 \pmod{23}$.

2. Prove that the quadratic congruence $6x^2 + 5x + 1 \equiv 0 \pmod{p}$ has a solution for every prime p, even though the equation $6x^2 + 5x + 1 = 0$ has no solution in the integers.

3. (a) For an odd prime p, prove that the quadratic residues of p are congruent modulo p to the integers

$$1^2, 2^2, 3^2, \ldots, \left(\frac{p-1}{2}\right)^2.$$

   (b) Verify that the quadratic residues of 17 are 1, 2, 4, 8, 9, 13, 15, 16.

4. Show that 3 is a quadratic residue of 23, but a nonresidue of 31.

5. Given that a is a quadratic residue of the odd prime p, prove that
   (a) a is not a primitive root of p;
   (b) $p - a$ is a quadratic residue or nonresidue of p according 25 $p \equiv 1 \pmod{4}$ or $p \equiv 3 \pmod{4}$;
   (c) if $p \equiv 3 \pmod{4}$, then $x \equiv \pm a^{(p+1)/4} \pmod{p}$ are the solutions of the congruence $x^2 \equiv a \pmod{p}$.

6. Let p be an odd prime and $\gcd(a, p) = 1$. Establish that the quadratic congruence $ax^2 + bx + c \equiv 0 \pmod{p}$ is solvable if and only if $b^2 - 4ac$ is either zero or a quadratic residue of p.

7. If $p = 2^k + 1$ is prime, verify that every quadratic nonresidue of p is a primitive root of p. [*Hint:* Apply Euler's Criterion.]

8. Assume that the integer r is a primitive root of the prime p, where $p \equiv 1 \pmod{8}$.
   (a) Show that the solutions of the quadratic congruence $x^2 \equiv 2 \pmod{p}$ are given by

$$x \equiv \pm (r^{7(p-1)/8} + r^{(p-1)/8}) \pmod{p}.$$

   [*Hint:* First confirm that $r^{3(p-1)/2} \equiv -1 \pmod{p}$.]
   (b) Use part (a) to find all solutions to the congruences $x^2 \equiv 2 \pmod{17}$ and $x^2 \equiv 2 \pmod{41}$.

9. (a) If $ab \equiv r \pmod{p}$, where r is a quadratic residue of the odd prime p, prove that a and b are both quadratic residues of p or both nonresidues of p.
   (b) If a and b are both quadratic residues of the odd prime p or both nonresidues of p, show that the congruence $ax^2 \equiv b \pmod{p}$ has a solution. [*Hint:* Multiply the given congruence by a' where $aa' \equiv 1 \pmod{p}$.]

10. Let p be an odd prime and $\gcd(a, p) = \gcd(b, p) = 1$. Prove that either all three of the quadratic congruences

$$x^2 \equiv a \pmod{p}, \ x^2 \equiv b \pmod{p}, \ x^2 \equiv ab \pmod{p}$$

   are solvable or exactly one of them admits a solution.

11. (a) Knowing that 2 is a primitive root of 19, find all the quadratic residues of 19. [*Hint:* See the Proof of Theorem 9-1.]
    (b) Find the quadratic residues of 29 and 31.

12. If $n > 2$ and $\gcd(a, n) = 1$, then a is called a quadratic residue of n whenever there exists an integer x such that $x^2 \equiv a \pmod{n}$. Prove that if a is a quadratic residue of $n > 2$, then $a^{\phi(n)/2} \equiv 1 \pmod{n}$.

13. Show that the result of the previous problem does not provide a sufficient condition for the existence of a quadratic residue of n; in other words, find relatively prime integers a and n, with $a^{\phi(n)/2} \equiv 1 \pmod{n}$, for which the congruence $x^2 \equiv a \pmod{n}$ is not solvable.

9.2 THE LEGENDRE SYMBOL AND ITS PROPERTIES

Euler's studies on quadratic residues were further developed by the French mathematician Adrien Marie Legendre (1752–1833). Legendre's memoir "Recherches d'Analyse Indéterminée" (1785) contains an account of the Quadratic Reciprocity Law and its many applications, a sketch of a theory of the representation of an integer as the sum of three squares and the statement of a theorem that was later to become famous: Every arithmetic progression $ax + b$, where $\gcd(a, b) = 1$, contains an infinite number of primes. The topics covered in "Recherches" were taken up in a more thorough and systematic fashion in his *Essai sur la Theorie des Nombres*, which appeared in 1798. This represented the first "modern" treatise devoted exclusively to number theory, its precursors being translations or commentaries on Diophantus. Legendre's *Essai* was subsequently expanded into his *Theorie des Nombres*. The results of his later research papers, inspired to a large extent by Gauss, were included in 1830 in a two-volume third edition of the *Theorie des Nombres*. This remained, together with the *Disquisitiones Arithmeticae* of Gauss, a standard work on the subject for many years. Although Legendre made no great innovations in number theory, he raised fruitful questions which provided subjects of investigation for the mathematicians of the 19th century.

Before leaving Legendre's mathematical contributions, we should mention that he is also known for his work on elliptic integrals and for his *Éléments de Géométrie* (1794). In this last book, he attempted a pedagogical improvement of Euclid's *Elements* by rearranging and simplifying many of the proofs without lessening the rigor of the ancient treatment. The result was so favorably received that it became one of the most successful textbooks ever written, dominating instruction in geometry for over a century through its numerous editions and translations. An English translation was made in 1824 by the famous Scottish essayist and historian Thomas Carlyle, who was in early life a teacher of mathematics; Carlyle's translation ran through 33 American editions, the last not appearing until 1890. In fact, Legendre's revision was used at Yale University as late as 1885, when Euclid was finally abandoned as a text.

Our future efforts will be greatly simplified by the use of the symbol (a/p); this notation was introduced by Legendre in his *Essai* and is called, naturally enough, the Legendre symbol.

DEFINITION 9-2.
Let p be an odd prime and $\gcd(a, p) = 1$. The *Legendre symbol* (a/p) is defined by

$$(a/p) = \begin{cases} 1 & \text{if } a \text{ is a quadratic residue of } p \\ -1 & \text{if } a \text{ is a quadratic nonresidue of } p \end{cases}$$

For the want of better terminology, we shall refer to a as the *numerator* and p as the *denominator* of the symbol (a/p). Other standard notations for the Legendre symbol are $\left(\dfrac{a}{p}\right)$ or $(a \mid p)$.

Example 9-3

Let us look at the prime $p = 13$, in particular. Using the Legendre symbol, the results of an earlier example may be expressed as

$$(1/13) = (3/13) = (4/13) = (9/13) = (10/13) = (12/13) = 1$$

and

$$(2/13) = (5/13) = (6/13) = (7/13) = (8/13) = (11/13) = -1.$$

REMARK: For $p \mid a$, we have purposely left the symbol (a/p) undefined. Some authors find it convenient to extend Legendre's definition to this case by setting $(a/p) = 0$. One advantage of this would be that the number of solutions of $x^2 \equiv a \pmod{p}$ can then be given by the simple formula $1 + (a/p)$.

The next theorem establishes certain elementary facts concerning the Legendre symbol.

THEOREM 9-2.

Let p be an odd prime and a and b be integers which are relatively prime to p. Then the Legendre symbol has the following properties:

(1) *If $a \equiv b \pmod{p}$, then $(a/p) = (b/p)$.*
(2) $(a^2/p) = 1$.
(3) $(a/p) \equiv a^{(p-1)/2} \pmod{p}$.
(4) $(ab/p) = (a/p)(b/p)$.
(5) $(1/p) = 1$ and $(-1/p) = (-1)^{(p-1)/2}$.

Proof: If $a \equiv b \pmod{p}$, then the two congruences $x^2 \equiv a \pmod{p}$ and $x^2 \equiv b \pmod{p}$ have exactly the same solutions, if any at all. Thus $x^2 \equiv a \pmod{p}$ and $x^2 \equiv b \pmod{p}$ are both solvable, or neither one has a solution. This is reflected in the statement that $(a/p) = (b/p)$.

Regarding property (2), observe that the integer a trivially satisfies the congruence $x^2 \equiv a^2 \pmod{p}$; hence, $(a^2/p) = 1$. Property (3) is just the corollary to Theorem 9-1 rephrased in terms of the Legendre symbol. We use (3) to establish property (4):

$$(ab/p) \equiv (ab)^{(p-1)/2} \equiv a^{(p-1)/2}b^{(p-1)/2} \equiv (a/p)(b/p) \pmod{p}.$$

Now the Legendre symbol assumes only the values 1 or -1. If $(ab/p) \neq (a/p)(b/p)$, we would have $1 \equiv -1 \pmod{p}$ or $2 \equiv 0 \pmod{p}$; this cannot occur, since $p > 2$. It follows that

$$(ab/p) = (a/p)(b/p).$$

Finally, we observe that the first equality in property (5) is a special case of property (2), while the second one is obtained from property (3) upon setting $a = -1$. Since the quantities $(-1/p)$ and $(-1)^{(p-1)/2}$ are either 1 or -1, the resulting congruence

$$(-1/p) \equiv (-1)^{(p-1)/2} \pmod{p}$$

implies that $(-1/p) = (-1)^{(p-1)/2}$.

From parts (2) and (4) of Theorem 9-2, we may also abstract the relation

(6) $$(ab^2/p) = (a/p)(b^2/p) = (a/p).$$

In other words, a square factor which is relatively prime to p can be deleted from the numerator of the Legendre symbol without affecting its value.

Since $(p-1)/2$ is even for a prime p of the form $4k+1$ and odd for p of the form $4k+3$, the equation $(-1/p) = (-1)^{(p-1)/2}$ permits us to add a small supplemental corollary to Theorem 9-2.

COROLLARY.

If p is an odd prime, then

$$(-1/p) = \begin{cases} 1 \ \textit{if } p \equiv 1 \pmod{4} \\ -1 \ \textit{if } p \equiv 3 \pmod{4} \end{cases}$$

This corollary may be viewed as asserting that the congruence $x^2 \equiv -1 \pmod{p}$ has a solution for an odd prime p if and only if p is of the form $4k+1$. The result is not new, of course; we have merely provided the reader with a different path to Theorem 5-5.

Example 9-4

Let us ascertain whether the congruence $x^2 \equiv -46 \pmod{17}$ is solvable. This can be done by evaluating the Legendre symbol $(-46/17)$. We first appeal to properties (4) and (5) of Theorem 9-2 to write

$$(-46/17) = (-1/17)(46/17) = (46/17).$$

Since $46 \equiv 12 \pmod{17}$, it follows that

$$(46/17) = (12/17).$$

Now property (6) above gives

$$(12/17) = (3 \cdot 2^2/17) = (3/17).$$

But

$$(3/17) \equiv 3^{(17-1)/2} \equiv 3^8 \equiv (81)^2 \equiv (-4)^2 \equiv -1 \pmod{17},$$

where we have made appropriate use of property (3) of Theorem 9-2; hence, $(3/17) = -1$. Inasmuch as $(-46/17) = -1$, the quadratic congruence $x^2 \equiv -46 \pmod{17}$ admits no solution.

The Corollary to Theorem 9-2 lends itself to an application concerning the distribution of primes.

THEOREM 9-3.

There are infinitely many primes of the form $4k + 1$.

Proof: Suppose that there are finitely many such primes; let us call them p_1, p_2, . . . , p_n and consider the integer

$$N = (2p_1p_2 \cdots p_n)^2 + 1.$$

Clearly N is odd, so that there exists some odd prime p with $p \mid N$. To put it another way,

$$(2p_1p_2 \cdots p_n)^2 \equiv -1 \ (\text{mod } p)$$

or, if one prefers to phrase this in terms of the Legendre symbol, $(-1/p) = 1$. But the relation $(-1/p) = 1$ holds only if p is of the form $4k + 1$. Hence, p is one of the primes p_i. This implies that p_i divides $N - (2p_1p_2 \cdots p_n)^2$, or $p_i \mid 1$, which is a contradiction. The conclusion: there must exist infinitely many primes of the form $4k + 1$.

We dig deeper into the properties of quadratic residues with

THEOREM 9-4.

If p is an odd prime, then

$$\sum_{a=1}^{p-1} (a/p) = 0.$$

Hence, there are precisely $(p - 1)/2$ *quadratic residues and* $(p - 1)/2$ *quadratic nonresidues of p.*

Proof: Let r be a primitive root of p. We know that, modulo p, the powers r, r^2, . . . , r^{p-1} are just a permutation of the integers $1, 2, . . . , p - 1$. Thus for any a lying between 1 and $p - 1$, inclusive, there exists a unique positive integer k ($1 \le k \le p - 1$), such that $a \equiv r^k \ (\text{mod } p)$. By appropriate use of Euler's Criterion, we have

(1) $(a/p) = (r^k/p) \equiv (r^k)^{(p-1)/2} = (r^{(p-1)/2})^k \equiv (-1)^k \ (\text{mod } p),$

where, since r is a primitive root of p, $r^{(p-1)/2} \equiv -1 \ (\text{mod } p)$. But (a/p) and $(-1)^k$ are equal to either 1 or -1, so that equality holds in equation (1). Now add up the Legendre symbols in question to obtain

$$\sum_{a=1}^{p-1} (a/p) = \sum_{k=1}^{p-1} (-1)^k = 0,$$

which is the desired conclusion.

The proof of Theorem 9-4 serves to bring out the following point, which we record as

COROLLARY.

The quadratic residues of an odd prime p are congruent modulo p to the even powers of a primitive root r of p; the quadratic nonresidues are congruent to the odd powers of r.

For an illustration of the idea just introduced, we again fall back on the prime $p = 13$. Since 2 is a primitive root of 13, the quadratic residues of 13 are given by the even powers of 2, namely,

$$
\begin{array}{ll}
2^2 \equiv 4 & 2^8 \equiv 9 \\
2^4 \equiv 3 & 2^{10} \equiv 10 \\
2^6 \equiv 12 & 2^{12} \equiv 1
\end{array}
$$

all congruences being modulo 13. Similarly, the nonresidues occur as the odd powers of 2:

$$
\begin{array}{ll}
2^1 \equiv 2 & 2^7 \equiv 11 \\
2^3 \equiv 8 & 2^9 \equiv 5 \\
2^5 \equiv 6 & 2^{11} \equiv 7.
\end{array}
$$

Most proofs of the Quadratic Reciprocity Law, and ours as well, rest ultimately upon what is known as Gauss' Lemma. While this lemma gives the quadratic character of an integer, it is more useful from a theoretical point of view than as a computational device. We state and prove it below.

THEOREM 9-5.

(Gauss' Lemma). *Let p be an odd prime and let* $\gcd(a, p) = 1$. *If n denotes the number of integers in the set*

$$
S = \left\{ a, \, 2a, \, 3a, \, \ldots, \, \left(\frac{p-1}{2} \right) a \right\}
$$

whose remainders upon division by p exceed p/2, then

$$
(a/p) = (-1)^n.
$$

Proof: Since $\gcd(a, p) = 1$, none of the $(p-1)/2$ integers in S is congruent to zero and no two are congruent to each other modulo p. Let $r_1, \ldots, r_m$ be those remainders upon division by p such that $0 < r_i < p/2$ and let $s_1, \ldots, s_n$ be those remainders such that $p > s_i > p/2$. Then $m + n = (p-1)/2$, and the integers

$$
r_1, \, \ldots, \, r_m, \, p - s_1, \, \ldots, \, p - s_n
$$

are all positive and less than $p/2$.

To prove that these integers are all distinct, it suffices to show that no $p - s_i$ is equal to any r_j. Assume to the contrary that

$$
p - s_i = r_j
$$

for some choice of i and j. Then there exist integers u and v, with $1 \leq u, v \leq (p-1)/2$, satisfying $s_i \equiv ua \pmod{p}$ and $r_j \equiv va \pmod{p}$. Hence,

$$(u+v)a \equiv s_i + r_j \equiv p \equiv 0 \pmod{p}$$

which says that $u + v \equiv 0 \pmod{p}$. But the latter congruence cannot take place, since $1 < u + v \leq p - 1$.

The point we wish to bring out is that the $(p-1)/2$ numbers

$$r_1, \ldots, r_m, p - s_1, \ldots, p - s_n$$

are simply the integers $1, 2, \ldots, (p-1)/2$, not necessarily in order of appearance. Thus, their product is $[(p-1)/2]!$:

$$\left(\frac{p-1}{2}\right)! = r_1 \cdots r_m (p - s_1) \cdots (p - s_n)$$

$$\equiv r_1 \cdots r_m (-s_1) \cdots (-s_n) \pmod{p}$$

$$\equiv (-1)^n r_1 \cdots r_m s_1 \cdots s_n \pmod{p}.$$

But we know that $r_1, \ldots, r_m, s_1, \ldots, s_n$ are congruent modulo p to a, $2a, \ldots, [(p-1)/2]a$, in some order, so that

$$\left(\frac{p-1}{2}\right)! \equiv (-1)^n a \cdot 2a \cdots \left(\frac{p-1}{2}\right) a \pmod{p}$$

$$\equiv (-1)^n a^{(p-1)/2} \left(\frac{p-1}{2}\right)! \pmod{p}.$$

Since $[(p-1)/2]!$ is relatively prime to p, it may be cancelled from both sides of this congruence to give

$$1 \equiv (-1)^n a^{(p-1)/2} \pmod{p}$$

or, upon multiplying by $(-1)^n$,

$$a^{(p-1)/2} \equiv (-1)^n \pmod{p}.$$

Use of Euler's Criterion now completes the argument:

$$(a/p) \equiv a^{(p-1)/2} \equiv (-1)^n \pmod{p},$$

which implies that

$$(a/p) = (-1)^n.$$

By way of illustration, let $p = 13$ and $a = 5$. Then $(p-1)/2 = 6$, so that

$$S = \{5, 10, 15, 20, 25, 30\}.$$

Modulo 13, the members of S are the same as the integers

$$5, 10, 2, 7, 12, 4.$$

Three of these are greater than $13/2$; hence, $n = 3$ and Theorem 9-5 says that

$$(5/13) = (-1)^3 = -1.$$

Gauss' Lemma allows us to proceed to a variety of interesting results. For one thing, it provides a means for determining which primes have 2 as a quadratic residue.

THEOREM 9-6.

If p is an odd prime, then

$$(2/p) = \begin{cases} 1 \ \ if \ p \equiv 1 \ (\text{mod } 8) \ or \ p \equiv 7 \ (\text{mod } 8); \\ -1 \ \ if \ p \equiv 3 \ (\text{mod } 8) \ or \ p \equiv 5 \ (\text{mod } 8). \end{cases}$$

Proof: According to Gauss' Lemma, $(2/p) = (-1)^n$, where n is the number of integers in the set

$$S = \left\{ 2, 2 \cdot 2, 3 \cdot 2, \ldots, \left(\frac{p-1}{2} \right) \cdot 2 \right\}$$

which, upon division by p, have remainders greater than $p/2$. The members of S are all less than p, so that it suffices to count the number that exceed $p/2$. For $1 \le k \le (p-1)/2$, $2k < p/2$ if and only if $k < p/4$. If [ ] denotes the greatest integer function, then there are $[p/4]$ integers in S less than $p/2$, hence

$$n = \frac{p-1}{2} - [p/4]$$

is the number of integers which are greater than $p/2$.

Now we have four possibilities; for, any odd prime has one of the forms $8k + 1, 8k + 3, 8k + 5$, or $8k + 7$. A simple calculation shows that

if $p = 8k + 1$, then $n = 4k - [2k + \frac{1}{4}] = 4k - 2k = 2k,$

if $p = 8k + 3$, then $n = 4k + 1 - [2k + \frac{3}{4}] = 4k + 1 - 2k = 2k + 1,$

if $p = 8k + 5$, then $n = 4k + 2 - [2k + 1 + \frac{1}{4}]$
$$= 4k + 2 - (2k + 1) = 2k + 1,$$

if $p = 8k + 7$, then $n = 4k + 3 - [2k + 1 + \frac{3}{4}]$
$$= 4k + 3 - (2k + 1) = 2k + 2.$$

Thus, when p is of the form $8k + 1$ or $8k + 7$, n is even and $(2/p) = 1$; on the other hand, when p assumes the form $8k + 3$ or $8k + 5$, n is odd and $(2/p) = -1$.

Notice that if the odd prime p is of the form $8k \pm 1$ (equivalently, $p \equiv 1$ (mod 8) or $p \equiv 7$ (mod 8)), then

$$\frac{p^2 - 1}{8} = \frac{(8k \pm 1)^2 - 1}{8} = \frac{64k^2 \pm 16k}{8} = 8k^2 \pm 2k,$$

which is an even integer; in this situation, $(-1)^{(p^2 - 1)/8} = 1 = (2/p)$. On the other hand, if p is of the form $8k \pm 3$ (equivalently, $p \equiv 3 \pmod 8$ or $p \equiv 5 \pmod 8$), then

$$\frac{p^2 - 1}{8} = \frac{(8k \pm 3)^2 - 1}{8} = \frac{64k^2 \pm 48k + 8}{8} = 8k^2 \pm 6k + 1,$$

which is odd; here, we have $(-1)^{(p^2 - 1)/8} = -1 = (2/p)$. These observations are incorporated in the statement of the following corollary to Theorem 9-6.

COROLLARY.

If p is an odd prime, then

$$(2/p) = (-1)^{(p^2 - 1)/8}.$$

It is time for another look at primitive roots. As we have remarked, there is no general technique for obtaining a primitive root of an odd prime p; the reader might, however, find the next theorem useful on occasion.

THEOREM 9-7.

If p and $2p + 1$ are both odd primes, then the integer $(-1)^{(p - 1)/2}2$ is a primitive root of $2p + 1$.

Proof: For ease of discussion, let us put $q = 2p + 1$. We distinguish two cases: $p \equiv 1 \pmod 4$ and $p \equiv 3 \pmod 4$.

If $p \equiv 1 \pmod 4$, then $(-1)^{(p - 1)/2}2 = 2$. Since $\phi(q) = q - 1 = 2p$, the order of 2 modulo q is one of the numbers 1, 2, p, or $2p$. Taking note of property (3) of Theorem 9-2, we have

$$(2/q) \equiv 2^{(q - 1)/2} = 2^p \pmod q.$$

But, in the present setting, $q \equiv 3 \pmod 8$; whence, the Legendre symbol $(2/q) = -1$. It follows that $2^p \equiv -1 \pmod q$ and so 2 cannot have order p modulo q. The order of 2 being neither 1, 2, ($2^2 \equiv 1 \pmod q$ implies that $q \mid 3$, which is an impossibility) nor p, we are forced to conclude that the order of 2 modulo q is $2p$. This makes 2 a primitive root of q.

We now deal with the case $p \equiv 3 \pmod 4$. This time, $(-1)^{(p - 1)/2}2 = -2$ and

$$(-2)^p \equiv (-2/q) = (-1/q)(2/q) \pmod q.$$

Since $q \equiv 7 \pmod 8$, the corollary to Theorem 9-2 asserts that $(-1/q) = -1$, while once again we have $(2/q) = 1$. This leads to the congruence $(-2)^p \equiv -1 \pmod q$. From here on, the argument duplicates that of the last paragraph. Without analyzing further, we announce the decision: -2 is a primitive root of the prime q.

Theorem 9-7 indicates, for example, that the primes 11, 59, 107, and 179 have 2 as a primitive root. Likewise, the integer -2 serves as a primitive root for 7, 23, 47, and 167.

Before retiring from the field, we should mention another result of the same character: if p and $4p + 1$ are both primes, then 2 is a primitive root of $4p + 1$. Thus, to the list of prime numbers having 2 for a primitive root, one could add, say, 13, 29, 53, and 173.

There is an attractive proof of the infinitude of primes of the form $8k - 1$ which can be based on Theorem 9-6.

THEOREM 9-8.

There are infinitely many primes of the form $8k - 1$.

Proof: As usual, suppose that there are only a finite number of such primes. Let these be $p_1, p_2, \cdots, p_n$ and consider the integer

$$N = (4p_1p_2 \cdots p_n)^2 - 2.$$

There exists at least one odd prime divisor p of N, so that

$$(4p_1p_2 \cdots p_n)^2 \equiv 2 \pmod{p}$$

or $(2/p) = 1$. In view of Theorem 9-6, $p \equiv \pm 1 \pmod 8$. If all the odd prime divisors of N were of the form $8k + 1$, then N would be of the form $16a + 2$; this is clearly impossible, since N is of the form $16a - 2$. Thus, N must have a prime divisor q of the form $8k - 1$. But $q \mid N$ and $q \mid (4p_1p_2 \cdots p_n)^2$ leads to the contradiction that $q \mid 2$.

The next result, which allows us to effect the passage from Gauss' Lemma to the Quadratic Reciprocity Law (Theorem 9-9), has some independent interest.

LEMMA.

If p is an odd prime and a an odd integer, with $\gcd(a, p) = 1$, then

$$(a/p) = (-1)^{\sum_{k=1}^{(p-1)/2} [ka/p]}.$$

Proof: We shall employ the same notation as in the proof of Gauss' Lemma. Consider the set of integers

$$S = \left\{ a, 2a, \ldots, \left(\frac{p-1}{2} \right) a \right\}.$$

Divide each of these multiples of a by p to obtain

$$ka = q_k p + t_k, \qquad\qquad 1 \le t_k \le p - 1.$$

Then $ka/p = q_k + t_k/p$, so that $[ka/p] = q_k$. Thus for $1 \le k \le (p-1)/2$, we may write ka in the form

(1) $$ka = [ka/p]p + t_k.$$

If the remainder $t_k < p/2$, then it is one of the integers $r_1, \ldots, r_m$; on the other hand if $t_k > p/2$, then it is one of the integers $s_1, \ldots, s_n$.

Taking the sum of the $(p - 1)/2$ equations in (1), we get the relation

(2)
$$\sum_{k=1}^{(p-1)/2} ka = \sum_{k=1}^{(p-1)/2} [ka/p]\, p + \sum_{k=1}^{m} r_k + \sum_{k=1}^{n} s_k.$$

It was learned in proving Gauss' Lemma that the $(p - 1)/2$ numbers

$$r_1, \ldots, r_m, p - s_1, \ldots, p - s_n$$

are just a rearrangement of the integers $1, 2, \ldots, (p - 1)/2$. Hence,

(3)
$$\sum_{k=1}^{(p-1)/2} k = \sum_{k=1}^{m} r_k + \sum_{k=1}^{n} (p - s_k) = pn + \sum_{k=1}^{m} r_k - \sum_{k=1}^{n} s_k.$$

Subtracting equation (3) from equation (2) gives

(4)
$$(a - 1) \sum_{k=1}^{(p-1)/2} k = p \left(\sum_{k=1}^{(p-1)/2} [ka/p] - n \right) + 2 \sum_{k=1}^{n} s_k.$$

Let us use the fact that $p \equiv a \equiv 1 \pmod 2$ and translate this last equation into a congruence modulo 2:

$$0 \cdot \sum_{k=1}^{(p-1)/2} k \equiv 1 \cdot \left(\sum_{k=1}^{(p-1)/2} [ka/p] - n \right) \pmod 2$$

or

$$n \equiv \sum_{k=1}^{(p-1)/2} [ka/p] \pmod 2.$$

The rest follows from Gauss' Lemma; for,

$$(a/p) = (-1)^n = (-1)^{\sum_{k=1}^{(p-1)/2} [ka/p]}$$

as we wished to show.

For an example of this last result, again consider $p = 13$ and $a = 5$. Since $(p - 1)/2 = 6$, it is necessary to calculate $[ka/p]$ for $k = 1, \ldots, 6$:

$$[5/13] \ = [10/13] = 0;$$
$$[15/13] = [20/13] = [25/13] = 1;$$
$$[30/13] = 2.$$

By the lemma just proven, we have

$$(5/13) = (-1)^{1+1+1+2} = (-1)^5 = -1,$$

confirming what was earlier seen.

PROBLEMS 9.2

1. Find the value of the following Legendre symbols:
   (a) $(19/23)$,   (b) $(-23/59)$,   (c) $(20/31)$,   (d) $(18/43)$,   (e) $(-72/131)$.

2. Use Gauss's Lemma to compute each of the Legendre symbols below (that is, in each case obtain the integer n for which $(a/p) = (-1)^n$):
   (a) $(8/11)$,   (b) $(7/13)$,   (c) $(5/19)$,   (d) $(11/23)$,   (e) $(6/31)$.

3. For an odd prime p, prove that there are $(p - 1)/2 - \phi(p)$ quadratic nonresidues of p which are not primitive roots of p.

4. (a) Let p be an odd prime. Show that the Diophantine equation

   $$x^2 + py + a = 0, \ \gcd(a, p) = 1,$$

   has an integral solution if and only if $(-a/p) = 1$.
   (b) Determine whether $x^2 + 7y - 2 = 0$ has a solution in the integers.

5. Prove that 2 is not a primitive root of any prime of the form $p = 3 \cdot 2^n + 1$, except when $p = 13$. [*Hint:* Use Theorem 9-6.]

6. (a) If p is an odd prime and $\gcd(ab, p) = 1$, prove that at least one of a, b, or ab is a quadratic residue of p.
   (b) Given a prime p, show that p divides

   $$(n^2 - 2)(n^2 - 3)(n^2 - 6)$$

   for some choice of $n > 0$.

7. If p is an odd prime, show that

   $$\sum_{a=1}^{p-2} (a(a + 1)/p) = -1$$

   [*Hint:* If a' is defined by $aa' \equiv 1 \pmod{p}$, then $(a(a + 1)/p) = ((1 + a')/p)$. Note that $1 + a'$ runs through a complete set of residues modulo p, except for the integer 1.]

8. Prove the statements below:
   (a) If p and $q = 2p + 1$ are both odd primes, then -4 is a primitive root of q.
   (b) If $p \equiv 1 \pmod 4$ is a prime, then -4 and $(p - 1)/4$ are both quadratic residues of p.

9. For a prime $p \equiv 7 \pmod 8$, show that $p \mid 2^{(p-1)/2} - 1$. [*Hint:* Use Theorem 9-6.]

10. Use Problem 9 to confirm that the numbers $2^n - 1$ are composite for $n = 11, 23, 83, 131, 179, 183, 239, 251$.

11. Given that p and $q = 4p + 1$ are both primes, prove the following:
    (a) Any quadratic nonresidue of q is either a primitive root of q or has order 4 modulo q. [*Hint:* If a is a quadratic nonresidue of q, then $-1 \equiv (a/q) \equiv a^{2p} \pmod q$; hence a has order 1, 2, 4, p, 2p, or $4p$ modulo q.]
    (b) The integer 2 is a primitive root of q; in particular, 2 is a primitive root of 13, 29, 53 and 173.

12. If r is a primitive root of the odd prime p, prove that the product of the quadratic residues of p is congruent modulo p to $r^{(p^2 - 1)/4}$ while the product of the nonresidues of p is congruent modulo p to $r^{(p-1)^2/4}$. [*Hint:* Apply the Corollary to Theorem 9-4.]

13. Establish that the product of the quadratic residues of the odd prime p is congruent modulo p to 1 or -1 according as $p \equiv 3 \pmod 4$ or $p \equiv 1 \pmod 4$. [*Hint:* Use Problem 12 and the fact that $r^{(p-1)/2} \equiv -1 \pmod p$. Or, Problem 3(a) of Section 9.1 and the proof of Theorem 5-5.]

14. (a) If the prime $p > 3$, show that p divides the sum of its quadratic residues.
    (b) If the prime $p > 5$, show that p divides the sum of the squares of its quadratic non-residues.

15. Prove that for any prime $p > 5$ there exist integers $1 \le a, b \le p - 1$ for which

$$(a/p) = ((a + 1)/p) = 1 \quad \text{and} \quad (b/p) = ((b + 1)/p) = -1;$$

that is, there are consecutive quadratic residues of p and consecutive nonresidues.

16. (a) Let p be an odd prime and $\gcd(a, p) = \gcd(k, p) = 1$. Show that if the equation $x^2 - ay^2 = kp$ admits a solution, then $(a/p) = 1$; for example, $(2/7) = 1$, since $6^2 - 2 \cdot 2^2 = 4 \cdot 7$. [*Hint:* If x_0, y_0 satisfy the given equation, then $(x_0 y_0^{p-2})^2 \equiv a \pmod p$.]
    (b) By considering the equation $x^2 + 5y^2 = 7$, demonstrate that the converse of the result in part (a) need not hold.
    (c) Show that, for any prime $p \equiv \pm 3 \pmod 8$, the equation $x^2 - 2y^2 = p$ has no solution.

17. Prove that the odd prime divisors p of the integers $9^n + 1$ are of the form $p \equiv 1 \pmod 4$.

18. For a prime $p \equiv 1 \pmod 4$, verify that the sum of the quadratic residues of p is equal to $p(p - 1)/4$. [*Hint:* If $a_1, \cdots, a_r$ are the quadratic residues of p less than $p/2$, then $p - a_1, \cdots, p - a_r$ are those greater than $p/2$.]

9.3 QUADRATIC RECIPROCITY

Let p and q be distinct odd primes, so that both of the Legendre symbols (p/q) and (q/p) are defined. It is natural to inquire whether the value of (p/q) can be determined if that of (q/p) is known. To put the question more generally, is there any connection at all between the values of these two symbols? The basic relationship was conjectured experimentally by Euler in 1783 and imperfectly proved by Legendre two years thereafter. Using his symbol, Legendre stated this relationship in the elegant form that has since become known as the Quadratic Reciprocity Law:

$$(p/q)(q/p) = (-1)^{\frac{p-1}{2} \frac{q-1}{2}}.$$

Legendre went amiss in assuming a result which is as difficult to prove as the law itself, namely, that for any prime $p \equiv 1 \pmod 8$, there exists another prime $q \equiv 3 \pmod 4$ for which p is a quadratic residue. Undaunted, he attempted another proof in his *Essai sur la Théorie des Nombres* (1798); this one too contained a gap, since Legendre took for granted that there are an infinite number of primes in certain arithmetical progressions (a fact eventually proved by Dirichlet in 1837, using in the process very subtle arguments from complex variable theory).

At the age of eighteen, Gauss (in 1795), apparently unaware of the work of either Euler or Legendre, rediscovered this reciprocity law and, after a year's unremitting labor, obtained the first complete proof. "It tortured me," says Gauss, "for the whole year and eluded my most strenuous efforts before, finally, I got the proof explained in the fourth section of the *Disquisitiones Arithmeticae.*" In the *Disquisitiones Arith- meticae*—which was published in 1801, although finished in 1798—Gauss attributed the Quadratic Reciprocity Law to himself, taking the view that a theorem belongs to the one who gives the first rigorous demonstration. The indignant Legendre was led to complain: "This excessive impudence is unbelievable in a man who has sufficient per- sonal merit not to have the need of appropriating the discoveries of others." All dis- cussion of priority between the two was futile; since each clung to the correctness of his position, neither took heed of the other. Gauss went on to publish five different demonstrations of what he called "the gem of higher arithmetic," while another was found among his papers. The version presented below, a variant of one of Gauss' own arguments, is due to his student, Ferdinand Eisenstein (1823–1852). The proof is com- plicated (and it would perhaps be unreasonable to expect an easy proof), but the un- derlying idea is simple enough.

THEOREM 9-9 (Quadratic Reciprocity Law).
   If p and q are distinct odd primes, then

$$(p/q)(q/p) = (-1)^{\frac{p-1}{2}\frac{q-1}{2}}.$$

Proof: Consider the rectangle in the xy coordinate plane whose vertices are $(0,0)$, $(p/2, 0)$, $(0, q/2)$, and $(p/2, q/2)$. Let R denote the region within this rectangle, not including any of the bounding lines. The general plan of attack is to count the number of lattice points (that is, the points whose coordinates are integers) inside R in two different ways. Since p and q are both odd, the lattice points in R consist of all points (n, m), where $1 \leq n \leq (p - 1)/2$ and $1 \leq m \leq (q - 1)/2$; the number of such points is clearly

$$\frac{p-1}{2} \cdot \frac{q-1}{2}.$$

   Now the diagonal D from $(0, 0)$ to $(p/2, q/2)$ has the equation $y = (q/p)x$, or equivalently, $py = qx$. Since $\gcd(p, q) = 1$, none of the lattice points inside R will lie on D. For p must divide the x coordinate of any lattice point on the line $py = qx$, and q must divide its y coordinate; there are no such points in R. Suppose that T_1 denotes the portion of R which is below the diagonal D, and T_2 the portion above. By what we have just seen, it suffices to count the lattice points inside each of these triangles.

   The number of integers in the interval $0 < y < kq/p$ is equal to $[kq/p]$. Thus, for $1 \leq k \leq (p - 1)/2$, there are precisely $[kq/p]$ lattice points in T_1 directly above the point $(k, 0)$ and below D; in other words, lying on the vertical line segment from $(k, 0)$ to $(k, kq/p)$. It follows that the total number of lattice points contained in T_1 is

$$\sum_{k=1}^{(p-1)/2} [kq/p].$$

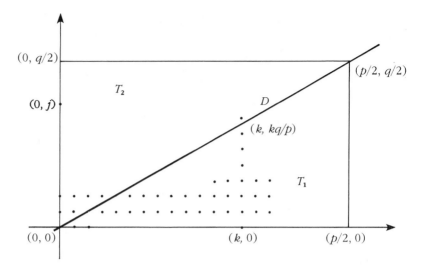

A similar calculation, with the roles of p and q interchanged, shows that the number of lattice points within T_2 is

$$\sum_{j=1}^{(q-1)/2} [jp/q].$$

This accounts for all of the lattice points inside R, so that

$$\frac{p-1}{2} \cdot \frac{q-1}{2} = \sum_{k=1}^{(p-1)/2} [kq/p] + \sum_{j=1}^{(q-1)/2} [jp/q].$$

The time has come for Gauss' Lemma to do its duty:

$$(p/q)(q/p) = (-1)^{\sum_{j=1}^{(q-1)/2} [jp/q]} \cdot (-1)^{\sum_{k=1}^{(p-1)/2} [kq/p]}$$

$$= (-1)^{\sum_{j=1}^{(q-1)/2} [jp/q] + \sum_{k=1}^{(p-1)/2} [kq/p]}$$

$$= (-1)^{\frac{p-1}{2}\frac{q-1}{2}}.$$

The proof of the Quadratic Reciprocity Law is now complete.

An immediate consequence of this is

COROLLARY 1.

   If p and q are distinct odd primes, then

$$(p/q)(q/p) = \begin{cases} 1 \text{ if } p \equiv 1 \text{ (mod 4) } or \text{ } q \equiv 1 \text{ (mod 4)} \\ -1 \text{ if } p \equiv q \equiv 3 \text{ (mod 4)}. \end{cases}$$

Proof: The number $(p - 1)/2 \cdot (q - 1)/2$ is even if and only if at least one of the integers p and q is of the form $4k + 1$; if both are of the form $4k + 3$, then $(p - 1)/2 \cdot (q - 1)/2$ is odd.

Multiplying each side of the Quadratic Reciprocity equation by (q/p) and using the fact that $(q/p)^2 = 1$, we could also formulate this as

COROLLARY 2.

If p and q are distinct odd primes, then

$$(p/q) = \begin{cases} (q/p) & \textit{if } p \equiv 1 \ (\mathrm{mod}\ 4) \textit{ or } q \equiv 1 \ (\mathrm{mod}\ 4) \\ -(q/p) & \textit{if } p \equiv q \equiv 3 \ (\mathrm{mod}\ 4). \end{cases}$$

Let us see what this last series of results accomplishes. Take p to be an odd prime and $a \neq \pm 1$ to be an integer not divisible by p. Suppose further that a has the factorization

$$a = \pm 2^{k_0} p_1^{k_1} p_2^{k_2} \cdots p_r^{k_r},$$

where the p_i are distinct odd primes. Since the Legendre symbol is multiplicative,

$$(a/p) = (\pm 1/p)(2/p)^{k_0}(p_1/p)^{k_1} \cdots (p_r/p)^{k_r}.$$

To evaluate (a/p), we have only to calculate each of the symbols $(-1/p)$, $(2/p)$, and (p_i/p). The values of $(-1/p)$ and $(2/p)$ were discussed earlier, so that the one stumbling block is (p_i/p), where p_i and p are distinct odd primes; this is where the Quadratic Reciprocity Law enters. For Corollary 2 allows us to replace (p_i/p) by a new Legendre symbol having a smaller denominator. Through continued inversion and division, the computation can be reduced to that of the known quantities

$$(-1/q), (1/q), \text{ and } (2/q).$$

This is all somewhat vague, of course, so let us look at a concrete example.

Example 9-5

Consider the Legendre symbol $(29/53)$, for instance. Since both $29 \equiv 1$ (mod 4) and $53 \equiv 1$ (mod 4), we see that

$$(29/53) = (53/29) = (24/29) = (2/29)(3/29)(4/29) = (2/29)(3/29).$$

With reference to Theorem 9-6, $(2/29) = -1$, while inverting again,

$$(3/29) = (29/3) = (2/3) = -1,$$

where we used the congruence $29 \equiv 2$ (mod 3). The net effect is that

$$(29/53) = (2/29)(3/29) = (-1)(-1) = 1.$$

The Quadratic Reciprocity Law provides a very satisfactory answer to the problem of finding odd primes $p \neq 3$ for which 3 is a quadratic residue. Since $3 \equiv 3 \pmod 4$, Corollary 2 above implies that

$$(3/p) = \begin{cases} (p/3) \text{ if } p \equiv 1 \pmod 4 \\ -(p/3) \text{ if } p \equiv 3 \pmod 4. \end{cases}$$

Now $p \equiv 1 \pmod 3$ or $p \equiv 2 \pmod 3$. By Theorems 9-2 and 9-6,

$$(p/3) = \begin{cases} 1 \text{ if } p \equiv 1 \pmod 3 \\ -1 \text{ if } p \equiv 2 \pmod 3 \end{cases}$$

the implication of which is that $(3/p) = 1$ if and only if

(1) $p \equiv 1 \pmod 4$ and $p \equiv 1 \pmod 3$,

or

(2) $p \equiv 3 \pmod 4$ and $p \equiv 2 \pmod 3$.

The restrictions in congruencies (1) are equivalent to requiring that $p \equiv 1 \pmod{12}$ while those in congruencies (2) are equivalent to $p \equiv 11 \equiv -1 \pmod{12}$. The upshot of all this:

THEOREM 9-10.

   If $p \neq 3$ is an odd prime, then

$$(3/p) = \begin{cases} 1 \text{ if } p \equiv \pm 1 \pmod{12} \\ -1 \text{ if } p \equiv \pm 5 \pmod{12}. \end{cases}$$

Example 9-6

For an example of the solution of a quadratic congruence with a composite modulus, consider

$$x^2 \equiv 196 \pmod{1357}.$$

Since $1357 = 23 \cdot 59$, the given congruence is solvable if and only if both

$$x^2 \equiv 196 \pmod{23} \quad \text{and} \quad x^2 \equiv 196 \pmod{59}$$

are solvable. Our procedure is to find the values of the Legendre symbols $(196/23)$ and $(196/59)$.

   The evaluation of $(196/23)$ requires the use of Theorem 9-10:

$$(196/23) = (12/23) = (3/23) = 1.$$

Thus, the congruence $x^2 \equiv 196 \pmod{23}$ admits a solution. As regards the symbol $(196/59)$, the Quadratic Reciprocity Law enables us to write

$$(196/59) = (19/59) = -(59/19) = -(2/19) = -(-1) = 1.$$

Therefore, it is possible to solve $x^2 \equiv 196 \pmod{59}$ and, in consequence, the congruence $x^2 \equiv 196 \pmod{1357}$ as well.

To arrive at an actual solution, notice that the congruence $x^2 \equiv 196 \equiv 12$ (mod 23) is satisfied by $x \equiv 9, 14$ (mod 23), while $x^2 \equiv 196 \equiv 19$ (mod 59) has solutions $x \equiv 14, 45$ (mod 59). We may now use the Chinese Remainder Theorem to obtain the simultaneous solutions of the four systems:

$$x \equiv 14 \text{ (mod 23)} \quad \text{and} \quad x \equiv 14 \text{ (mod 23)},$$

$$x \equiv 14 \text{ (mod 23)} \quad \text{and} \quad x \equiv 45 \text{ (mod 59)},$$

$$x \equiv 9 \ \text{ (mod 23)} \quad \text{and} \quad x \equiv 14 \text{ (mod 59)},$$

$$x \equiv 9 \ \text{ (mod 23)} \quad \text{and} \quad x \equiv 45 \text{ (mod 59)}.$$

The resulting values $x \equiv 14, 635, 722, 1343$ (mod 1357) are the desired solutions of the original congruence $x^2 \equiv 196$ (mod 1357).

Example 9-7

Let us turn to a quite different application of these ideas. At an earlier stage, it was observed that if $F_n = 2^{2^n} + 1, n > 1$, is a prime, then 2 is not a primitive root of F_n. We now possess the means to show that the integer 3 serves as a primitive root of any prime of this type.

As a first step in this direction, note that any F_n is of the form $12k + 5$. A simple induction argument confirms that $4^m \equiv 4$ (mod 12) for $m = 1, 2, \ldots$; hence, we must have

$$F_n = 2^{2^n} + 1 = 2^{2m} + 1 = 4^m + 1 \equiv 5 \text{ (mod 12)}.$$

If F_n happens to be prime, then Theorem 9-10 permits the conclusion

$$(3/F_n) = -1,$$

or, using Euler's Criterion,

$$3^{(F_n - 1)/2} \equiv -1 \text{ (mod } F_n).$$

Switching to the phi-function, the last congruence says that

$$3^{\phi(F_n)/2} \equiv -1 \text{ (mod } F_n).$$

From this, it may be inferred that 3 has order $\phi(F_n)$ modulo F_n, and so 3 is a primitive root of F_n.

PROBLEMS 9.3

1. Evaluate the following Legendre symbols:
   (a) $(71/73)$, (b) $(-219/383)$, (c) $(461/773)$,
   (d) $(1234/4567)$, (e) $(3658/12703)$. [*Hint:* $3658 = 2 \cdot 31 \cdot 59$.]

2. Prove that 3 is a quadratic nonresidue of all primes of the form $2^{2^n} + 1$, as well as all primes of the form $2^p - 1$, where p is an odd prime. [*Hint:* For all n, $4^n \equiv 4$ (mod 12).]

3. Determine whether the following quadratic congruences are solvable:
   (a) $x^2 \equiv 219 \pmod{419}$.
   (b) $3x^2 + 6x + 5 \equiv 0 \pmod{89}$.
   (c) $2x^2 + 5x - 9 \equiv 0 \pmod{101}$.

4. Verify that if p is an odd prime, then

$$(-2/p) = \begin{cases} 1 \text{ if } p \equiv 1 \pmod 8 & \text{or} \quad p \equiv 3 \pmod 8 \\ -1 \text{ if } p \equiv 5 \pmod 8 & \text{or} \quad p \equiv 7 \pmod 8. \end{cases}$$

5. (a) Prove that if $p > 3$ is an odd prime, then

$$(-3/p) = \begin{cases} 1 \text{ if } p \equiv 1 \pmod 6 \\ -1 \text{ if } p \equiv 5 \pmod 6. \end{cases}$$

   (b) Using part (a), show that there are infinitely many primes of the form $6k + 1$. [*Hint:* Assume that $p_1, p_2, \ldots, p_r$ are all the primes of the form $6k + 1$ and consider the integer $N = (2p_1p_2 \cdots p_r)^2 + 3$.]

6. Use Theorem 9-2 and Problems 4 and 5 to determine which primes can divide each of $n^2 + 1, n^2 + 2, n^2 + 3$ for some value of n.

7. Prove that there exist infinitely many primes of the form $8k + 3$. [*Hint:* Assume that there are only finitely many primes of the form $8k + 3$, say $p_1, p_2, \ldots, p_r$, and consider the integer $N = (p_1p_2 \cdots p_r)^2 + 2$.]

8. Find a prime number p which is simultaneously expressible in the forms $x^2 + y^2$, $u^2 + 2v^2$, and $r^2 + 3s^2$. [*Hint:* $(-1/p) = (-2/p) = (-3/p) = 1$.]

9. If p and q are odd primes satisfying $p = q + 4a$ for some a, establish that

$$(a/p) = (a/q)$$

   and, in particular, that $(6/37) = (6/13)$. [*Hint:* Note that $(a/p) = (-q/p)$ and use the Quadratic Reciprocity Law.]

10. Establish each of the following assertions:
    (a) $(5/p) = 1$ if and only if $p \equiv 1, 9, 11,$ or $19 \pmod{20}$;
    (b) $(6/p) = 1$ if and only if $p \equiv 1, 5, 19,$ or $23 \pmod{24}$;
    (c) $(7/p) = 1$ if and only if $p \equiv 1, 3, 9, 19, 25,$ or $27 \pmod{28}$.

11. Prove that there are infinitely many primes of the form $5k - 1$. [*Hint:* For any $n > 1$, the integer $5(n!)^2 - 1$ has a prime divisor $p > n$ which is not of the form $5k + 1$; hence, $(5/p) = 1$.]

12. Verify the following:
    (a) The prime divisors $p \neq 3$ of the integer $n^2 - n + 1$ are of the form $6k + 1$. [*Hint:* If $p | n^2 - n + 1$, then $(2n - 1)^2 \equiv -3 \pmod p$.]
    (b) The prime divisors $p \neq 5$ of the integer $n^2 + n - 1$ are of the form $10k + 1$ or $10k + 9$.
    (c) The prime divisors p of the integer $2n(n + 1) + 1$ are of the form $p \equiv 1 \pmod 4$. [*Hint:* If $p | 2n(n + 1) + 1$, then $(2n + 1)^2 \equiv -1 \pmod p$.]
    (d) The prime divisors p of the integer $3n(n + 1) + 1$ are of the form $p \equiv 1 \pmod 6$.

13. (a) Show that if p is a prime divisor of $839 = 38^2 - 5 \cdot 11^2$, then $(5/p) = 1$. Use this fact to conclude that 839 is a prime number. [*Hint:* It suffices to consider those primes $p < 29$.]
    (b) Prove that $397 = 20^2 - 3$ and $733 = 29^2 - 3 \cdot 6^2$ are both primes.

14. Solve the quadratic congruence $x^2 \equiv 11 \pmod{35}$. [*Hint:* After solving $x^2 \equiv 11 \pmod 5$ and $x^2 \equiv 11 \pmod 7$, use the Chinese Remainder Theorem.]

15. Establish that 7 is a primitive root of any prime of the form $p = 2^{4n} + 1$. [*Hint:* Since $p \equiv 3$ or $5 \pmod 7$, $(7/p) = (p/7) = -1$.]

16. Let a and $b > 1$ be relatively prime integers, with b odd. If $b = p_1 p_2 \cdots p_r$ is the decomposition of b into odd primes (not necessarily distinct) then the *Jacobi symbol* (a/b) is defined by

$$(a/b) = (a/p_1)(a/p_2) \cdots (a/p_r),$$

where the symbols on the right-hand side of the equality sign are Legendre symbols. Evaluate the Jacobi symbols

$$(21/221), (215/253), \text{ and } (631/1099).$$

17. Under the hypothesis of the previous problem, show that if a is a quadratic residue of b, then $(a/b) = 1$; but, the converse is false.

18. Prove that the following properties of the Jacobi symbol hold: If b and b' are positive odd integers and $\gcd(aa', bb') = 1$, then
   (a) $a \equiv a' \pmod b$ implies that $(a/b) = (a'/b)$;
   (b) $(aa'/b) = (a/b)(a'/b)$;
   (c) $(a/bb') = (a/b)(a/b')$;
   (d) $(a^2/b) = (a/b^2) = 1$;
   (e) $(1/b) = 1$;
   (f) $(-1/b) = (-1)^{(b-1)/2}$; [*Hint:* Whenever u and v are odd integers, then $(u-1)/2 + (v-1)/2 \equiv (uv-1)/2 \pmod 2$.]
   (g) $(2/b) = (-1)^{(b^2-1)/8}$. [*Hint:* Whenever u and v are odd integers, then $(u^2-1)/8 + (v^2-1)/8 \equiv [(uv)^2 - 1]/8 \pmod 2$.]

19. Derive the Generalized Quadratic Reciprocity Law: If a and b are relatively prime positive odd integers, each greater than 1, then

$$(a/b)(b/a) = (-1)^{\frac{a-1}{2}\frac{b-1}{2}}.$$

[*Hint:* See the hint in Problem 18(f).]

20. Using the Generalized Quadratic Reciprocity Law, determine whether the congruence $x^2 \equiv 231 \pmod{1105}$ is solvable.

9.4 QUADRATIC CONGRUENCES WITH COMPOSITE MODULI

So far in the proceedings, quadratic congruences with (odd) prime moduli have been of paramount importance. The remaining theorems broaden the horizon by allowing a composite modulus. To start, let us consider the situation where the modulus is a power of a prime.

THEOREM 9-11.

If p is an odd prime and $\gcd(a, p) = 1$, then the congruence

$$x^2 \equiv a \pmod{p^n}, \qquad\qquad n \geq 1$$

has a solution if and only if $(a/p) = 1$.

Proof: As is common with many "if and only if" theorems, one half of the proof is trivial while the other half requires considerable effort: If $x^2 \equiv a \pmod{p^n}$ has a solution, then so does $x^2 \equiv a \pmod{p}$—in fact, the same solution—whence $(a/p) = 1$.

For the converse, suppose that $(a/p) = 1$. We argue that $x^2 \equiv a \pmod{p^n}$ is solvable by inducting on n. If $n = 1$ there is really nothing to prove; indeed, $(a/p) = 1$ is just another way of saying that $x^2 \equiv a \pmod{p}$ can be solved. Assume that the result holds for $n = k \geq 1$, so that $x^2 \equiv a \pmod{p^k}$ admits a solution x_0. Then

$$x_0{}^2 = a + bp^k$$

for an appropriate choice of b. In passing from k to $k + 1$, we shall use x_0 and b to write down explicitly a solution to the congruence $x^2 \equiv a \pmod{p^{k+1}}$.

Towards this end, we first solve the linear congruence

$$2x_0 y \equiv -b \pmod{p},$$

obtaining a unique solution y_0 modulo p (this is certainly possible, since $\gcd(2x_0, p) = 1$). Next, consider the integer

$$x_1 = x_0 + y_0 p^k.$$

Upon squaring this integer, we get

$$(x_0 + y_0 p^k)^2 = x_0{}^2 + 2x_0 y_0 p^k + y_0{}^2 p^{2k}$$
$$= a + (b + 2x_0 y_0)p^k + y_0{}^2 p^{2k}.$$

But $p \mid (b + 2x_0 y_0)$, from which it follows that

$$x_1{}^2 = (x_0 + y_0 p^k)^2 \equiv a \pmod{p^{k+1}}.$$

Thus, the congruence $x^2 \equiv a \pmod{p^n}$ has a solution for $n = k + 1$ and, by induction, for all positive integers n.

Let us run through a specific example in detail. The first step in obtaining a solution of, say, the quadratic congruence

$$x^2 \equiv 23 \pmod{7^2}$$

is to solve $x^2 \equiv 23 \pmod{7}$, or what amounts to the same thing, the congruence

$$x^2 \equiv 2 \pmod{7}.$$

Since $(2/7) = 1$, a solution surely exists; in fact $x_0 = 3$ is an obvious choice. Now $x_0{}^2$ can be represented as

$$3^2 = 9 = 23 + (-2)7,$$

so that $b = -2$ (in our special case, the integer 23 plays the role of a). Following the proof of Theorem 9-11, we next determine y so that

$$6y \equiv 2 \pmod{7};$$

that is, $3y \equiv 1 \pmod 7$. This linear congruence is satisfied by $y_0 = 5$. Hence,

$$x_0 + 7y_0 = 3 + 7 \cdot 5 = 38$$

serves as a solution to the original congruence $x^2 \equiv 23 \pmod{49}$. It should be noted that $-38 \equiv 11 \bmod (49)$ is the only other solution.

If, instead, the congruence

$$x^2 \equiv 23 \pmod{7^3}$$

were proposed for solution, we would start with

$$x^2 \equiv 23 \pmod{7^2},$$

obtaining a solution $x_0 = 38$. Since

$$38^2 = 23 + 29 \cdot 7^2,$$

the integer $b = 29$. We would then find the unique solution $y_0 = 1$ of the linear congruence

$$76y \equiv -29 \pmod 7.$$

Then $x^2 \equiv 23 \pmod{7^3}$ is satisfied by

$$x_0 + y_0 \cdot 7^2 = 38 + 1 \cdot 49 = 87,$$

as well as $-87 \equiv 256 \pmod{7^3}$.

Having dwelt at length on odd primes, let us now take up the case $p = 2$. The next theorem supplies the pertinent information.

THEOREM 9-12.

Let a be an odd integer. Then

(1) $x^2 \equiv a \pmod 2$ *always has a solution;*
(2) $x^2 \equiv a \pmod 4$ *has a solution if and only if* $a \equiv 1 \pmod 4$;
(3) $x^2 \equiv a \pmod{2^n}$, *for* $n \geq 3$, *has a solution if and only if* $a \equiv 1 \pmod 8$.

Proof: The first assertion is obvious. The second depends on the observation that the square of any odd integer is congruent to 1 modulo 4. Consequently, $x^2 \equiv a \pmod 4$ can be solved only when a is of the form $4k + 1$; in this event, there are two solutions modulo 4, namely $x = 1$ and $x = 3$.

Now consider the case in which $n \geq 3$. Since the square of any odd integer is congruent to 1 modulo 8, we see that for the congruence $x^2 \equiv a \pmod{2^n}$ to be solvable it is necessary that a should be of the form $8k + 1$. To go the other way, let us suppose that $a \equiv 1 \pmod 8$ and proceed by induction on n. When $n = 3$, the congruence $x^2 \equiv a \pmod{2^n}$ is certainly solvable; indeed, each of the integers 1, 3, 5, 7 satisfies $x^2 \equiv 1 \pmod 8$. Fix a value of $n > 3$ and assume, for the induction hypothesis, that the congruence $x^2 \equiv a \pmod{2^n}$ admits a solution x_0. Then there exists an integer b for which

$$x_0^2 = a + b2^n.$$

Since a is odd, so is the integer x_0. It is therefore possible to find a unique solution y_0 of the linear congruence

$$x_0 y \equiv -b \pmod{2}.$$

We argue that the integer

$$x_1 = x_0 + y_0 2^{n-1}$$

satisfies the congruence $x^2 \equiv a \pmod{2^{n+1}}$. Squaring yields

$$(x_0 + y_0 2^{n-1})^2 = x_0^2 + x_0 y_0 2^n + y_0^2 2^{2n-2}$$
$$= a + (b + x_0 y_0) 2^n + y_0^2 2^{2n-2}.$$

By the way y_0 was chosen, $2 \mid (b + x_0 y_0)$, hence

$$x_1^2 = (x_0 + y_0 2^{n-1})^2 \equiv a \pmod{2^{n+1}}$$

(one also uses the fact that $2n - 2 = n + 1 + (n - 3) > n + 1$). Thus $x^2 \equiv a \pmod{2^{n+1}}$ is solvable, completing the induction step and the proof.

To illustrate: the congruence $x^2 \equiv 5 \pmod{4}$ has a solution, but $x^2 \equiv 5 \pmod{8}$ does not; on the other hand, $x^2 \equiv 17 \pmod{16}$ and $x^2 \equiv 17 \pmod{32}$ are both solvable.

In theory, we can now completely settle the question of when there exists an integer x such that

$$x^2 \equiv a \pmod{n}, \quad \gcd(a, n) = 1, \qquad\qquad n > 1.$$

For suppose that n has the prime-power decomposition

$$n = 2^{k_0} p_1^{k_1} p_2^{k_2} \cdots p_r^{k_r}, \qquad\qquad k_0 \geq 0, k_i > 0$$

where the p_i are distinct odd primes. Since the problem of solving the quadratic congruence $x^2 \equiv a \pmod{n}$ is equivalent to that of solving the system of congruences

$$x^2 \equiv a \pmod{2^{k_0}},$$
$$x^2 \equiv a \pmod{p_1^{k_1}},$$

$$\vdots$$

$$x^2 \equiv a \pmod{p_r^{k_r}},$$

our last two results may be combined to give the following general conclusion.

THEOREM 9-13.
    Let $n = 2^{k_0} p_1^{k_1} \cdots p_r^{k_r}$ be the prime factorization of $n > 1$ and let $\gcd(a, n) = 1$. Then $x^2 \equiv a \pmod{n}$ is solvable if and only if

(1) $(a/p_i) = 1$ for $i = 1, 2, \ldots, r$;
(2) $a \equiv 1 \pmod{4}$ if $4 \mid n$, but $8 \nmid n$; $a \equiv 1 \pmod{8}$ if $8 \mid n$.

PROBLEMS 9.4

1. (a) Show that 7 and 18 are the only incongruent solutions of $x^2 \equiv -1 \pmod{5^2}$.
   (b) Use part (a) to find the solutions of $x^2 \equiv -1 \pmod{5^3}$.

2. Solve each of the following quadratic congruences:
   (a) $x^2 \equiv 7 \pmod{3^3}$;
   (b) $x^2 \equiv 14 \pmod{5^3}$;
   (c) $x^2 \equiv 2 \pmod{7^3}$.

3. Solve the congruence $x^2 \equiv 31 \pmod{11^4}$.

4. Find the solutions of $x^2 + 5x + 6 \equiv 0 \pmod{5^3}$ and $x^2 + x + 3 \equiv 0 \pmod{3^3}$.

5. Prove that if the congruence $x^2 \equiv a \pmod{2^n}$, where a is odd and $n \geq 3$, has a solution, then it has exactly four incongruent solutions. [*Hint:* If x_0 is any solution, then the four integers $x_0, -x_0, x_0 + 2^{n-1}, -x_0 + 2^{n-1}$ are incongruent modulo 2^n and comprise all the solutions.]

6. From $23^2 \equiv 17 \pmod{2^7}$, find three other solutions of the congruence $x^2 \equiv 17 \pmod{2^7}$.

7. First determine the values of a for which the congruences below are solvable and then find the solutions of these congruences:
   (a) $x^2 \equiv a \pmod{2^4}$;
   (b) $x^2 \equiv a \pmod{2^5}$;
   (c) $x^2 \equiv a \pmod{2^6}$.

8. For fixed $n > 1$, show that all the solvable congruences $x^2 \equiv a \pmod{n}$ with $\gcd(a, n) = 1$ have the same number of solutions.

9. (a) Without actually finding them, determine the number of solutions of the congruences $x^2 \equiv 3 \pmod{11^2 \cdot 23^2}$ and $x^2 \equiv 9 \pmod{2^3 \cdot 3 \cdot 5^2}$.
   (b) Solve the congruence $x^2 \equiv 9 \pmod{2^3 \cdot 3 \cdot 5^2}$.

10. (a) For an odd prime p, prove that the congruence $2x^2 + 1 \equiv 0 \pmod{p}$ has a solution if and only if $p \equiv 1$ or $3 \pmod 8$.
    (b) Solve the congruence $2x^2 + 1 \equiv 0 \pmod{11^2}$. [*Hint:* Consider integers of the form $x_0 + 11k$, where x_0 is a solution of $2x^2 + 1 \equiv 0 \pmod{11}$.]

10

Perfect Numbers

"In most sciences one generation tears down what another has built and what one has established another undoes. In Mathematics alone each generation builds a new story to the old structure."
HERMANN HANKEL

10.1 THE SEARCH FOR PERFECT NUMBERS

The history of the theory of numbers abounds with famous conjectures and open questions. The present chapter focuses on some of the intriguing conjectures associated with perfect numbers. A few of these have been satisfactorily answered, but most remain unresolved; all have stimulated the development of the subject as a whole.

The Pythagoreans considered it rather remarkable that the number 6 is equal to the sum of its positive divisors, other than itself:

$$6 = 1 + 2 + 3.$$

The next number after 6 having this feature is 28; for the positive divisors of 28 are found to be 1, 2, 4, 7, 14, and 28, and

$$28 = 1 + 2 + 4 + 7 + 14.$$

In line with their philosophy of attributing mystical qualities to numbers, the Pythagoreans called such numbers "perfect." Stated precisely:

DEFINITION 10-1.
A positive integer n is said to be *perfect* if n is equal to the sum of all its positive divisors, excluding n itself.

The sum of the positive divisors of an integer n, each of them less than n, is given by $\sigma(n) - n$. Thus, the condition "n is perfect" amounts to asking that $\sigma(n) - n = n$, or equivalently, that

$$\sigma(n) = 2n.$$

For example, we have

$$\sigma(6) = 1 + 2 + 3 + 6 = 2 \cdot 6$$

and
$$\sigma(28) = 1 + 2 + 4 + 7 + 14 + 28 = 2 \cdot 28$$

so that 6 and 28 are both perfect numbers.

For many centuries, philosophers were more concerned with the mystical or religious significance of perfect numbers than with their mathematical properties. Saint Augustine explains that although God could have created the world all at once, He preferred to take six days because the perfection of the work is symbolized by the (perfect) number 6. Early commentators on the Old Testament argued that the perfection of the Universe is represented by 28, the number of days it takes the moon to circle the earth. In the same vein, the 8th century theologian Alcuin of York observed that the whole human race is descended from the eight souls on Noah's Ark and that this second Creation is less perfect than the first, 8 being an imperfect number.

Only four perfect numbers were known to the ancient Greeks. Nicomachus in his *Introductio Arithmeticae* (circa 100 A.D.) lists

$$P_1 = 6, \ P_2 = 28, \ P_3 = 496, \ P_4 = 8128.$$

He says that they are formed in an "orderly" fashion, one among the units, one among the tens, one among the hundreds, and one among the thousands (that is, less than 10,000). Based on this meager evidence, it was conjectured that

1. the nth perfect number P_n contains exactly n digits; and
2. the even perfect numbers end, alternately, in 6 and 8.

Both assertions are wrong. There is no perfect number with 5 digits; the next perfect number (first given correctly in an anonymous 15th century manuscript) is

$$P_5 = 33,550,336.$$

While the final digit of P_5 is 6, the succeeding perfect number, namely

$$P_6 = 8,589,869,056$$

ends in 6 also, not 8 as conjectured. To salvage something in the positive direction, we shall show later that the even perfect numbers do always end in 6 or 8—but not necessarily alternately.

If nothing else, the magnitude of P_6 should convince the reader of the rarity of perfect numbers. It is not yet known whether there are finitely many or infinitely many of them.

The problem of determining the general form of all perfect numbers dates back almost to the beginning of mathematical time. It was partially solved by Euclid when in Book IX of the *Elements* he proved that if the sum

$$1 + 2 + 2^2 + 2^3 + \cdots + 2^{k-1} = p$$

is a prime number, then $2^{k-1}p$ is a perfect number (of necessity even). For instance, $1 + 2 + 4 = 7$ is a prime; hence $4 \cdot 7 = 28$ is a perfect number. Euclid's argument makes use of the formula for the sum of a geometric progression

$$1 + 2 + 2^2 + 2^3 + \cdots + 2^{k-1} = 2^k - 1,$$

which is found in various Pythagorean texts. In this notation, the result reads as follows: If $2^k - 1$ is prime ($k > 1$), then $n = 2^{k-1}(2^k - 1)$ is a perfect number. About 2000 years after Euclid, Euler took a decisive step in proving that all even perfect numbers must be of this type. We incorporate both these statements in our first theorem.

THEOREM 10-1.

If $2^k - 1$ is prime ($k > 1$), then $n = 2^{k-1}(2^k - 1)$ is perfect and every even perfect number is of this form.

Proof: Let $2^k - 1 = p$, a prime, and consider the integer $n = 2^{k-1}p$. In as much gcd $(2^{k-1}, p) = 1$, the multiplicativity of σ (as well as Theorem 6-2) entails that

$$\sigma(n) = \sigma(2^{k-1}p) = \sigma(2^{k-1})\sigma(p)$$
$$= (2^k - 1)(p + 1)$$
$$= (2^k - 1)2^k = 2n,$$

making n a perfect number.

For the converse, assume that n is an even perfect number. We may write n as $n = 2^{k-1}m$, where m is an odd integer and $k \geq 2$. It follows from gcd$(2^{k-1}, m) = 1$ that

$$\sigma(n) = \sigma(2^{k-1}m) = \sigma(2^{k-1})\sigma(m) = (2^k - 1)\sigma(m),$$

while the requirement for a number to be perfect gives

$$\sigma(n) = 2n = 2^k m.$$

Together, these relations yield

$$2^k m = (2^k - 1)\sigma(m),$$

which is simply to say that $(2^k - 1) \mid 2^k m$. But $2^k - 1$ and 2^k are relatively prime, whence $(2^k - 1) \mid m$; say, $m = (2^k - 1)M$. Now the result of substituting this value of m into the last-displayed equation and cancelling $2^k - 1$ is that $\sigma(m) = 2^k M$. Since m and M are both divisors of m (with $M < m$), we have

$$2^k M = \sigma(m) \geq m + M = 2^k M,$$

leading to $\sigma(m) = m + M$. The implication of this equality is that m has only two positive divisors, to wit, M and m itself. It must be that m is prime and $M = 1$; in other words, $m = (2^k - 1)M = 2^k - 1$ is a prime number, completing the present proof.

Since the problem of finding even perfect numbers is reduced to the search for primes of the form $2^k - 1$, a closer look at these integers might be fruitful. One thing that can be proved is that if $2^k - 1$ is a prime number, then the exponent k must itself be prime. More generally:

LEMMA.

If $a^k - 1$ is prime ($a > 0, k \geq 2$), then $a = 2$ and k is also prime.

Proof: It can be verified without difficulty that

$$a^k - 1 = (a - 1)(a^{k-1} + a^{k-2} + \cdots + a + 1),$$

where, in the present setting,

$$a^{k-1} + a^{k-2} + \cdots + a + 1 \geq a + 1 > 1.$$

Since by hypothesis $a^k - 1$ is prime, the other factor must be 1; that is, $a - 1 = 1$ so that $a = 2$.

If k were composite, then we could write $k = rs$, with $1 < r$ and $1 < s$. Thus,

$$a^k - 1 = (a^r)^s - 1$$
$$= (a^r - 1)(a^{r(s-1)} + a^{r(s-2)} + \cdots + a^r + 1)$$

and each factor on the right is plainly greater than 1. But this violates the primality of $a^k - 1$, so that k must by contradiction be prime.

For $p = 2, 3, 5, 7$, the values 3, 7, 31, 127 of $2^p - 1$ are primes, so that

$$2(2^2 - 1) = 6,$$
$$2^2(2^3 - 1) = 28,$$
$$2^4(2^5 - 1) = 496,$$
$$2^6(2^7 - 1) = 8128$$

are all perfect numbers.

Many early writers erroneously believed that $2^p - 1$ is prime for every choice of the prime number p. But in 1536, Hudalrichus Regius in a work entitled *Utriusque Arithmetices* exhibits the correct factorization

$$2^{11} - 1 = 2047 = 23 \cdot 89.$$

If this seems a small accomplishment, it should be realized that his calculations were in all likelihood carried out in Roman numerals, with the aid of an abacus (not until the late 16th century did the Arabic numeral system win complete ascendancy over the Roman one). Regius also gave $p = 13$ as the next value of p for which the expression $2^p - 1$ is a prime. From this, one obtains the fifth perfect number

$$2^{12}(2^{13} - 1) = 33,550,336.$$

One of the difficulties in finding further perfect numbers was the unavailability of tables of primes. In 1603, Pietro Cataldi, who is remembered chiefly for this invention of the notation for continued fractions, published a list of all primes less than 5150. By the direct procedure of dividing by all primes not exceeding the square root of a number, Cataldi determined that $2^{17} - 1$ was prime and, in consequence, that

$$2^{16}(2^{17} - 1) = 8,589,869,056$$

is the sixth perfect number.

A question which immediately springs to mind is whether there are infinitely many primes of the type $2^p - 1$, with p a prime. If the answer were in the affirmative, then there would exist an infinitude of (even) perfect numbers. Unfortunately this is another famous unresolved problem.

This appears to be as good a place as any at which to prove our theorem on the final digits of even perfect numbers.

THEOREM 10-2.

An even perfect number n ends in the digit 6 or 8; that is, either $n \equiv 6$ (mod 10) or $n \equiv 8$ (mod 10).

Proof: Being an even perfect number, the integer n may be represented as $n = 2^{k-1}(2^k - 1)$, where $2^k - 1$ is a prime. According to the last lemma, the exponent k must also be prime. If $k = 2$, then $n = 6$ and the asserted result holds. We may therefore confine our attention to the case $k > 2$. The proof falls into two parts, according as k takes the form $4m + 1$ or $4m + 3$.

If k is of the form $4m + 1$, then

$$n = 2^{4m}(2^{4m+1} - 1)$$
$$= 2^{8m+1} - 2^{4m} = 2 \cdot 16^{2m} - 16^m.$$

A straightforward induction argument will make it clear that $16^t \equiv 6$ (mod 10) for any positive integer t. Utilizing this congruence, we get

$$n \equiv 2 \cdot 6 - 6 \equiv 6 \pmod{10}.$$

Now, in the case in which $k = 4m + 3$,

$$n = 2^{4m+2}(2^{4m+3} - 1)$$
$$= 2^{8m+5} - 2^{4m+2} = 2 \cdot 16^{2m+1} - 4 \cdot 16^m.$$

Falling back on the fact that $16^t \equiv 6$ (mod 10), we see that

$$n \equiv 2 \cdot 6 - 4 \cdot 6 \equiv -12 \equiv 8 \pmod{10}.$$

Consequently, every even perfect number has a last digit equal to 6 or to 8.

A little more argument establishes a sharper result, namely that any even perfect number $n = 2^{k-1}(2^k - 1)$ always ends in the digits 6 or 28. Since an integer is congruent modulo 100 to its last two digits, it suffices to prove that, if k is of the form $4m + 3$, then $n \equiv 28$ (mod 100). To see this, note that

$$2^{k-1} = 2^{4m+2} = 16^m \cdot 4 \equiv 6 \cdot 4 \equiv 4 \pmod{10}.$$

Moreover, for $k > 2$, we have $4 | 2^{k-1}$ and so the number formed by the last two digits of 2^{k-1} is divisible by 4. The situation is this: the last digit of 2^{k-1} is 4, while 4 divides the last two digits. Modulo 100, the various possibilities are

$$2^{k-1} \equiv 4, 24, 44, 64, \text{ or } 84.$$

But this implies that

$$2^k - 1 = 2 \cdot 2^{k-1} - 1 \equiv 7, 47, 87, 27, \text{ or } 67 \pmod{100},$$

whence

$$n = 2^{k-1}(2^k - 1)$$
$$\equiv 4 \cdot 7, 24 \cdot 47, 44 \cdot 87, 64 \cdot 27, \text{ or } 84 \cdot 67 \pmod{100}.$$

It is a modest exercise, which we bequeath to the reader, to verify that each of the products on the right-hand side of the last congruence is congruent to 28 modulo 100.

PROBLEMS 10.1

1. Prove that the integer $n = 2^{10}(2^{11} - 1)$ is not a perfect number by showing that $\sigma(n) \neq 2n$. [*Hint:* $2^{11} - 1 = 23 \cdot 89$.]

2. Verify each of the statements below:
   (a) No power of a prime can be a perfect number.
   (b) A perfect square cannot be a perfect number.
   (c) The product of two odd primes is never a perfect number. [*Hint:* Expand the inequality $(p - 1)(q - 1) > 2$ to get $pq > p + q + 1$.]

3. If n is a perfect number, prove that $\sum_{d \mid n} 1/d = 2$.

4. Prove that every even perfect number is a triangular number.

5. Given that n is an even perfect number, for instance $n = 2^{k-1}(2^k - 1)$, show that the integer $n = 1 + 2 + 3 + \cdots + (2^k - 1)$ and also that $\phi(n) = 2^{k-1}(2^{k-1} - 1)$.

6. For an even perfect number $n > 6$, show the following:
   (a) The sum of the digits of n is congruent to 1 modulo 9. [*Hint:* The congruence $2^6 \equiv 1 \pmod 9$ and the fact that any prime $p \geq 5$ is of the form $6k + 1$ or $6k + 5$ imply that $n = 2^{p-1}(2^p - 1) \equiv 1 \pmod 9$.)]
   (b) The integer n can be expressed as a sum of consecutive odd cubes. [*Hint:* Use Section 1.1, Problem 1(e) to establish the identity

$$1^3 + 3^3 + 5^3 + \cdots + (2^k - 1)^3 = 2^{2k-2}(2^{2k-1} - 1)$$

   for all $k \geq 1$.]

7. Show that no proper divisor of a perfect number can be perfect. [*Hint:* Apply the result of Problem 3.]

8. Find the last two digits of the perfect number

$$n = 2^{19936}(2^{19937} - 1).$$

9. If $\sigma(n) = kn$, where $k \geq 3$, then the positive integer n is called a *k-perfect number* (sometimes, *multiply perfect*). Establish the following assertions concerning k-perfect numbers:
   (a) $\quad\quad\quad 523{,}776 = 2^9 \cdot 3 \cdot 11 \cdot 31$ is 3-perfect;
   $\quad\quad\quad\quad 30{,}240 = 2^5 \cdot 3^3 \cdot 5 \cdot 7$ is 4-perfect;
   $\quad 14{,}182{,}439{,}040 = 2^7 \cdot 3^4 \cdot 5 \cdot 7 \cdot 11^2 \cdot 17 \cdot 19$ is 5-perfect.
   (b) If n is a 3-perfect number and $3 \nmid n$, then $3n$ is 4-perfect.
   (c) If n is a 5-perfect number and $5 \nmid n$, then $5n$ is 6-perfect.
   (d) If $3n$ is a $4k$-perfect number and $3 \nmid n$, then n is $3k$-perfect.

10. Show that 120 and 672 are the only 3-perfect numbers of the form $n = 2^k \cdot 3 \cdot p$, where p is an odd prime.

11. A positive integer n is *multiplicatively perfect* if n is equal to the product of all its positive divisors, excluding n itself; in other words, $n^2 = \prod_{d \mid n} d$. Find all multiplicatively perfect numbers. [*Hint:* Notice that $n^2 = n^{\tau(n)/2}$.]

12. (a) If $n > 6$ is an even perfect number, prove that $n \equiv 4 \pmod 6$. [*Hint:* $2^{p-1} \equiv 1 \pmod 3$ for an odd prime p.]

   (b) Prove that if $n \neq 28$ is an even perfect number, then $n \equiv 1$ or $-1 \pmod 7$.

13. For any even perfect number $n = 2^{k-1}(2^k - 1)$, show that $2^k \mid \sigma(n^2) + 1$.

14. Numbers n such that $\sigma(\sigma(n)) = 2n$ are called *superperfect numbers*.

   (a) If $n = 2^k$ with $2^{k+1} - 1$ a prime, prove that n is superperfect; hence, 16 and 64 are superperfect.

   (b) Find all even perfect numbers $n = 2^{k-1}(2^k - 1)$ which are also superperfect. [*Hint:* First establish the equality $\sigma(\sigma(n)) = 2^k (2^{k+1} - 1)$.]

15. The *harmonic mean* $H(n)$ of the divisors of a positive integer n is defined by the formula

$$\frac{1}{H(n)} = \frac{1}{\tau(n)} \sum_{d \mid n} \frac{1}{d} .$$

   Show that if n is a perfect number, then $H(n)$ must be an integer. [*Hint:* Observe that $H(n) = n\tau(n)/\sigma(n)$.]

16. The twin primes 5 and 7 are such that one-half their sum is a perfect number. Are there any other twin primes with this property? [*Hint:* Given the twin primes p and $p + 2$, with $p > 5, \frac{1}{2}(p + p + 2) = 6k$ for some $k > 1$.]

17. Prove that if $2^k - 1$ is prime, then the sum

$$2^{k-1} + 2^k + 2^{k+1} + \cdots + 2^{2k-2}$$

   will yield a perfect number. For instance, $2^3 - 1$ is prime and $2^2 + 2^3 + 2^4 = 28$, which is perfect.

18. Assuming that n is an even perfect number, say $n = 2^{k-1}(2^k - 1)$, prove that the product of the positive divisors of n is equal to n^k; in symbols,

$$\prod_{d \mid n} d = n^k.$$

19. If $n_1, n_2, \cdots, n_r$ are distinct even perfect numbers, establish that

$$\phi(n_1 n_2 \cdots n_r) = 2^{r-1} \phi(n_1)\phi(n_2) \cdots \phi(n_r).$$

   [*Hint:* See Problem 5.]

20. Given an even perfect number $n = 2^{k-1}(2^k - 1)$, show that

$$\phi(n) = n - 2^{2k-2}.$$

10.2   MERSENNE PRIMES

It has become traditional to call numbers of the form

$$M_n = 2^n - 1 \ (n \geq 1)$$

Mersenne numbers after a French monk, Father Marin Mersenne (1588–1648), who made an incorrect but provocative assertion concerning their primality. Those Mersenne numbers which happen to be prime are said to be *Mersenne primes*. By what we proved in Section 10.1, the determination of Mersenne primes M_n—and, in turn, of even perfect numbers—is narrowed down to the case in which n is itself prime.

In the preface of his *Cogitata Physica-Mathematica* (1644), Mersenne stated that M_p is prime for $p = 2, 3, 5, 7, 13, 17, 19, 31, 67, 127, 257$ and composite for all other primes $p < 257$. It was obvious to other mathematicians that Mersenne could not have tested for primality all the numbers he had announced; but neither could they. Euler verified (1772) that M_{31} was prime by examining all primes up to 46339 as possible divisors, but M_{67}, M_{127}, and M_{257} were beyond his technique; in any event, this yielded the eighth perfect number

$$2^{30}(2^{31} - 1) = 2,305,843,008,139,952,128.$$

It was not until 1947, after tremendous labor caused by unreliable desk calculators, that the examination of the prime or composite character of M_p for the 55 primes in the range $p \leq 257$ was completed. We know now that Mersenne made five mistakes. He erroneously concluded that M_{67} and M_{257} are prime and excluded M_{61}, M_{89}, and M_{107} from his predicted list of primes. It is rather astonishing that over 300 years were required to set the good friar straight.

All the composite numbers M_n with $n < 257$ have now been completely factored. The most difficult factorization, that of M_{251}, was obtained as recently as 1984 after a 32-hour search on a supercomputer.

An historical curiosity is that, in 1876, Edouard Lucas worked a test whereby he was able to prove that the Mersenne number M_{67} was composite; but he could not produce the actual factors. At the October 1903 meeting of the American Mathematical Society, the American mathematician Frank Nelson Cole had a paper on the program with the somewhat unassuming title "On the Factorization of Large Numbers." When called upon to speak, Cole walked to a board and, saying nothing, proceeded to raise the integer 2 to the 67th power; then he carefully subtracted 1 from the resulting number and let the figure stand. Without a word he moved to a clean part of the board and multiplied, longhand, the product

$$193,707,721 \times 761,838,257,287.$$

The two calculations agreed. The story goes that, for the first and only time on record, this venerable body rose to give the presenter of a paper a standing ovation. Cole took his seat without having uttered a word, and no one bothered to ask him a question. (Later, he confided to a friend that it took him twenty years of Sunday afternoons to find the factors of M_{67}.)

In the study of Mersenne numbers, one comes upon a strange fact: when each of the first four Mersenne primes (namely, 3, 7, 31, and 127) is substituted for n in the formula $2^n - 1$, a higher Mersenne prime is obtained. Mathematicians had hoped that this procedure would give rise to an infinite set of Mersenne primes; in other words, the conjecture was that if the number M_n is prime, then M_{M_n} is also a prime. Alas, in 1953 a high-speed computer found the next possibility

$$M_{M_{13}} = 2^{M_{13}} - 1 = 2^{8191} - 1$$

(a number with 2466 digits) to be composite.

There are various methods for determining whether certain special types of Mersenne numbers are prime or composite. One such test is presented on the following page.

THEOREM 10-3.

If p and $q = 2p + 1$ are primes, then either $q \mid M_p$ or $q \mid M_p + 2$, but not both.

Proof: With reference to Fermat's Theorem, we know that

$$2^{q-1} - 1 \equiv 0 \pmod{q}$$

and, factoring the left-hand side, that

$$(2^{(q-1)/2} - 1)(2^{(q-1)/2} + 1) = (2^p - 1)(2^p + 1)$$
$$\equiv 0 \pmod{q}.$$

What amounts to the same thing:

$$M_p(M_p + 2) \equiv 0 \pmod{q}.$$

The stated conclusion now follows directly from Theorem 3-1. One cannot have both $q \mid M_p$ and $q \mid M_p + 2$, for then $q \mid 2$, which is impossible.

A single application should suffice to illustrate Theorem 10-3: If $p = 23$, then $q = 2p + 1 = 47$ is also a prime, so that we may consider the case of M_{23}. The question reduces to one of whether $47 \mid M_{23}$ or, to put it differently, whether $2^{23} \equiv 1 \pmod{47}$. Now, we have

$$2^{23} = 2^3(2^5)^4 \equiv 2^3(-15)^4 \pmod{47}.$$

But

$$(-15)^4 = (225)^2 \equiv (-10)^2 \equiv 6 \pmod{47}.$$

Putting these two congruences together, it is seen that

$$2^{23} \equiv 2^3 \cdot 6 \equiv 48 \equiv 1 \pmod{47}$$

whence M_{23} is composite.

We might point out that Theorem 10-3 is of no help in testing the primality of M_{29}, say; in this instance, $59 \nmid M_{29}$, but instead $59 \mid M_{29} + 2$.

Of the two possibilities $q \mid M_p$ or $q \mid M_p + 2$, is it reasonable to ask: What conditions on q will ensure that $q \mid M_p$? The answer is to be found in

THEOREM 10-4.

If $q = 2n + 1$ is prime, then

(1) $q \mid M_n$, *provided that $q \equiv 1 \pmod{8}$ or $q \equiv 7 \pmod{8}$;*
(2) $q \mid M_n + 2$, *provided that $q \equiv 3 \pmod{8}$ or $q \equiv 5 \pmod{8}$.*

Proof: To say that $q \mid M_n$ is equivalent to asserting that

$$2^{(q-1)/2} = 2^n \equiv 1 \pmod{q}.$$

In terms of the Legendre symbol, the latter condition becomes the requirement that $(2/q) = 1$. But according to Theorem 9-6, $(2/q) = 1$ whenever we have $q \equiv 1 \pmod{8}$ or $q \equiv 7 \pmod{8}$. The proof of (2) proceeds along similar lines.

Let us consider an immediate consequence of Theorem 10-4.

COROLLARY.

If p and $q = 2p + 1$ are both odd primes, with $p \equiv 3$ (mod 4), then $q \mid M_p$.

Proof: An odd prime p is either of the form $4k + 1$ or $4k + 3$. If $p = 4k + 3$, then $q = 8k + 7$ and Theorem 10-4 yields $q \mid M_p$. In case $p = 4k + 1$, then $q = 8k + 3$ so that $q \nmid M_p$.

The following is a partial list of those prime numbers $p \equiv 3$ (mod 4) where $q = 2p + 1$ is also prime: $p = 11, 23, 83, 131, 179, 181, 239, 251$. In each instance, M_p is composite.

Exploring the matter a little further, we next tackle two results of Fermat which restrict the divisors of M_p. The first is

THEOREM 10-5.

If p is an odd prime, then any prime divisor of M_p is of the form $2kp + 1$.

Proof: Let q be any prime divisor of M_p, so that $2^p \equiv 1$ (mod q). If 2 has order k modulo q (that is, if k is the smallest positive integer which satisfies $2^k \equiv 1$ (mod q)), then Theorem 8-1 tells us that $k \mid p$. The case $k = 1$ cannot arise; for this would imply that $q \mid 1$, an impossible situation. Therefore, since both $k \mid p$ and $k > 1$, the primality of p forces $k = p$.

In compliance with Fermat's Theorem, we have $2^{q-1} \equiv 1$ (mod q) and so, thanks to Theorem 8-1 again, $k \mid q - 1$. Knowing that $k = p$, the net result is that $p \mid q - 1$. To be definite, let us put $q - 1 = pt$; then $q = pt + 1$. The proof is completed by noting that if t were an odd integer, then q would be even and a contradiction occurs. Hence, we must have $q = 2kp + 1$ for some choice of k, which gives q the required form.

As a further sieve to screen out possible divisors of M_p, we cite the following result.

THEOREM 10-6.

If p is an odd prime, then any prime divisor q of M_p is of the form $q \equiv \pm 1$ (mod 8).

Proof: Suppose that q is a prime divisor of M_p, so that $2^p \equiv 1$ (mod q). According to Theorem 10–5, q is of the form $q = 2kp + 1$ for some integer k. Thus, using Euler's Criterion, $(2/q) \equiv 2^{(q-1)/2} \equiv 1$ (mod q), whence $(2/q) = 1$. Theorem 9–6 can now be brought into play again to conclude that $q \equiv \pm 1$ (mod 8).

For an illustration of how these theorems can be used, one might look at M_{17}. Those integers of the form $34k + 1$ which are less than $362 < \sqrt{M_{17}}$ are

$$35, 69, 103, 137, 171, 205, 239, 273, 307, 341.$$

Since the smallest (nontrivial) divisor of M_{17} must be prime, we need only consider the primes among the foregoing ten numbers; namely,

$$103, 137, 239, 307.$$

The work can be shortened somewhat by noting that $307 \not\equiv \pm 1 \pmod 8$ and so 307 may be deleted from our list. Now either M_{17} is prime or one of the three remaining possibilities divides it. With a little calculation, one can check that M_{17} is divisible by none of 103, 137, and 239; the result: M_{17} is prime.

After giving the eighth perfect number $2^{30}(2^{31} - 1)$, Barlow in his book *Theory of Numbers* (published in 1811) concludes from its size that it "is the greatest that ever will be discovered; for as they are merely curious, without being useful, it is not likely that any person will ever attempt to find one beyond it." The very least that can be said is that Barlow underestimated obstinate human curiosity. While the subsequent search for larger perfect numbers provides us with one of the fascinating chapters in the history of mathematics, an extended discussion would be out of place here.

It is worth remarking however that the first twelve Mersenne primes (hence, twelve perfect numbers) have been known since 1914. The twelfth in order of discovery, namely M_{89}, was the last Mersenne prime disclosed by hand calculation; its primality was verified by both Powers and Cunningham in 1914, working independently and by different techniques. The prime M_{127} was found by Lucas in 1876 and for the next 75 years was the largest number actually known to be a prime.

Calculations whose mere size and tedium repel the mathematician are just grist for the mill of electronic computers. Starting in 1952, eighteen additional Mersenne primes (all huge) have come to light. The twenty-fifth Mersenne prime, M_{21701}, was discovered in 1978 by two eighteen-year-old high school students, Laura Nickel and Curt Noll, using 440 hours on a large computer. A few months later, Noll confirmed that M_{23209} is also prime. With the advent of much faster computers, even this record prime did not stand for long. The Mersenne prime M_{756839}, found in 1992, is currently the largest of the known prime numbers. It gives rise to the largest even perfect number, the 32nd one:

$$P_{32} = 2^{756838}(2^{756839} - 1),$$

an immense number of 455663 digits. The new perfect number, if printed out, would fill fourteen newspaper pages.

Except for brief interims, the largest known prime has always been a Mersenne prime. M_{216091} was a record holder for several years. But in the never ending pursuit of bigger and bigger prime numbers, the integer

$$391581 \cdot 2^{216193} - 1$$

was confirmed prime in 1989 and remained until recently the world's largest number verified prime. Its 65807 digits exceed those of M_{216091} by a mere 37 digits.

In 1989, a systematic computer search for Mersenne primes M_p with p in the interval $100000 < p < 139268$ resulted in the 31st discovery of a Mersenne prime, namely M_{110503}. Surprisingly, the newly revealed prime was lurking in the gap between two previously known Mersenne primes, so that M_{110503} is the 29th in order of size. Just as remarkable is that the search was run on a supercomputer in the incredibly fast time of 11 minutes. It is still unknown whether another Mersenne prime M_p exists in the interval $139268 < p < 216090$: that is, before the prime M_{216091}. Jumping ahead, however, there is no Mersenne prime in the range $216092 < p < 353620$.

An algorithm frequently used for testing the primality of M_p is the Lucas-Lehmer test. It relies on the inductively-defined sequence

$$S_1 = 4, \; S_{k+1} = S_k^2 - 2 \qquad (k \geq 1).$$

Thus, the sequence begins with the values 4, 14, 194, 37634, ... The basic theorem, as perfected by Derrick Lehmer in 1930 from the pioneering results of Lucas, is this: for $p > 2$, M_p is prime if and only if $S_{p-1} \equiv 0 \pmod{M_p}$. An equivalent formulation is that M_p is prime if and only if $S_{p-2} \equiv \pm 2^{(p+1)/2} \pmod{M_p}$.

A simple example is provided by the Mersenne number $M_7 = 2^7 - 1 = 127$. Working modulo 127, the computation runs as follows:

$$S_1 \equiv 4, \; S_2 \equiv 14, \; S_3 \equiv 67, \; S_4 \equiv 42, \; S_5 \equiv -16, \; S_6 \equiv 0.$$

This establishes that M_7 is prime.

For the reader's convenience, we list the 32 even perfect numbers, the number of digits in each, and its approximate date of discovery.

	Perfect number	Number of digits	Date of discovery
1	$2(2^2 - 1)$	1	unknown
2	$2^2(2^3 - 1)$	2	unknown
3	$2^4(2^5 - 1)$	3	unknown
4	$2^6(2^7 - 1)$	4	unknown
5	$2^{12}(2^{13} - 1)$	8	1456
6	$2^{16}(2^{17} - 1)$	10	1588
7	$2^{18}(2^{19} - 1)$	12	1588
8	$2^{30}(2^{31} - 1)$	19	1772
9	$2^{60}(2^{61} - 1)$	37	1883
10	$2^{88}(2^{89} - 1)$	54	1911
11	$2^{106}(2^{107} - 1)$	65	1914
12	$2^{126}(2^{127} - 1)$	77	1876
13	$2^{520}(2^{521} - 1)$	314	1952
14	$2^{606}(2^{607} - 1)$	366	1952
15	$2^{1278}(2^{1279} - 1)$	770	1952
16	$2^{2202}(2^{2203} - 1)$	1327	1952
17	$2^{2280}(2^{2281} - 1)$	1373	1952
18	$2^{3216}(2^{3217} - 1)$	1937	1957
19	$2^{4252}(2^{4253} - 1)$	2561	1961
20	$2^{4422}(2^{4423} - 1)$	2663	1961
21	$2^{9688}(2^{9689} - 1)$	5834	1963
22	$2^{9940}(2^{9941} - 1)$	5985	1963
23	$2^{11212}(2^{11213} - 1)$	6751	1963
24	$2^{19936}(2^{19937} - 1)$	12,003	1971
25	$2^{21700}(2^{21701} - 1)$	13,066	1978
26	$2^{23208}(2^{23209} - 1)$	13,973	1978
27	$2^{44496}(2^{44497} - 1)$	26,790	1979
28	$2^{86242}(2^{86243} - 1)$	51,924	1983
29	$2^{110502}(2^{110503} - 1)$	66,530	1989
30	$2^{132048}(2^{132049} - 1)$	79,502	1983
31	$2^{216090}(2^{216091} - 1)$	130,100	1985
32	$2^{756838}(2^{756839} - 1)$	455,663	1992

Most mathematicians believe that there are infinitely many Mersenne primes, but a proof of this seems hopelessly beyond reach. Known Mersenne primes M_p clearly become more scarce as p increases. It has been conjectured that about two primes M_p should be expected for all primes p in an interval $x < p < 2x$; the numerical evidence tends to support this.

The perfect numbers given above are the only ones which have been discovered. One of the celebrated problems of number theory is whether or not there exist any odd perfect numbers. While no odd perfect number has thus far been produced, it is nonetheless possible to find certain conditions for their existence. The oldest of these we owe to Euler, who proved that if n is an odd perfect number, then

$$n = p^\alpha q_1^{2\beta_1} q_2^{2\beta_2} \cdots q_r^{2\beta_r}$$

where $p, q_1, \ldots, q_r$ are distinct odd primes and $p \equiv \alpha \equiv 1 \pmod 4$. In 1937, Steuerwald showed that not all the β_i can be equal to 1; that is, if $n = p^\alpha q_1^2 q_2^2 \cdots q_r^2$ is an odd number with $p \equiv \alpha \equiv 1 \pmod 4$, then n is not perfect. Four years later, Kanold established that the β_i cannot all be equal to 2, nor is it possible to have one β_i equal to 2 and all the others equal to 1. The last few years have seen further progress: Hagis and McDaniel (1972) found that it is impossible to have $\beta_i = 3$ for all i.

With these comments out of the way, let us prove Euler's result.

THEOREM 10-7 (Euler).

If n is an odd perfect number, then

$$n = p_1^{k_1} p_2^{2j_2} \cdots p_r^{2j_r},$$

where the p_i are distinct odd primes and $p_1 \equiv k_1 \equiv 1 \pmod 4$.

Proof: Let $n = p_1^{k_1} p_2^{k_2} \cdots p_r^{k_r}$ be the prime factorization of n. Since n is perfect, we can write

$$2n = \sigma(n) = \sigma(p_1^{k_1})\sigma(p_2^{k_2}) \cdots \sigma(p_r^{k_r}).$$

Being an odd integer, $n \equiv 1 \pmod 4$ or $n \equiv 3 \pmod 4$; in either event, $2n \equiv 2 \pmod 4$. Thus, $\sigma(n) = 2n$ is divisible by 2, but not by 4. The implication is that one of the $\sigma(p_i^{k_i})$, say $\sigma(p_1^{k_1})$, must be an even integer (but not divisible by 4), while all the remaining $\sigma(p_i^{k_i})$ are odd integers.

For a given p_i, there are two cases to be considered: $p_i \equiv 1 \pmod 4$ and $p_i \equiv 3 \pmod 4$. If $p_i \equiv 3 \equiv -1 \pmod 4$, we would have

$$\sigma(p_i^{k_i}) = 1 + p_i + p_i^2 + \cdots + p_i^{k_i}$$

$$\equiv 1 + (-1) + (-1)^2 + \cdots + (-1)^{k_i} \pmod 4$$

$$\equiv \begin{cases} 0 \pmod 4 & \text{if } k_i \text{ is odd} \\ 1 \pmod 4 & \text{if } k \text{ is even} \end{cases}$$

Since $\sigma(p_1^{k_1}) \equiv 2 \pmod 4$, this tells us that $p_1 \not\equiv 3 \pmod 4$ or, to put it affirmatively, $p_1 \equiv 1 \pmod 4$. Furthermore, the congruence $\sigma(p_i^{k_i}) \equiv 0 \pmod 4$ signifies that 4 divides $\sigma(p_i^{k_i})$, which is not possible. The conclusion: if $p_i \equiv 3 \pmod 4$, where $i = 2, \ldots, r$, then its exponent k_i is an even integer.

Should it happen that $p_i \equiv 1 \pmod 4$—which is certainly true for $i = 1$—then

$$\sigma(p_i^{k_i}) = 1 + p_i + p_i^2 + \cdots + p_i^{k_i}$$
$$\equiv 1 + 1^1 + 1^2 + \cdots + 1^{k_i} \pmod 4$$
$$\equiv k_i + 1 \pmod 4.$$

The condition $\sigma(p_1^{k_1}) \equiv 2 \pmod 4$ forces $k_1 \equiv 1 \pmod 4$. For the other values of i, we know that $\sigma(p_i^{k_i}) \equiv 1$ or $3 \pmod 4$ and so $k_i \equiv 0$ or $2 \pmod 4$; in any case, k_i will be an even integer. The crucial point is that, regardless of whether $p_i \equiv 1 \pmod 4$ or $p_i \equiv 3 \pmod 4$, k_i is always even for $i \neq 1$. Our proof is now complete.

In view of the preceding theorem, any odd perfect number n can be expressed as

$$n = p_1^{k_1} p_2^{2j_2} \cdots p_r^{2j_r} = p_1^{k_1} (p_2^{j_2} \cdots p_r^{j_r})^2$$
$$= p_1^{k_1} m^2.$$

This leads directly to

COROLLARY.

If n is an odd perfect number, then n is of the form

$$n = p^k m^2,$$

where p is a prime, $p \nmid m$, and $p \equiv k \equiv 1 \pmod 4$; in particular, $n \equiv 1 \pmod 4$.

Proof: Only the last assertion is not obvious. Because $p \equiv 1 \pmod 4$, we have $p^k \equiv 1 \pmod 4$. Notice that m must be odd; hence, $m \equiv 1$ or $3 \pmod 4$ and so, upon squaring, $m^2 \equiv 1 \pmod 4$. It follows that

$$n = p^k m^2 \equiv 1 \cdot 1 \equiv 1 \pmod 4,$$

establishing our corollary.

Another line of investigation involves estimating the size of an odd perfect number n. The classical lower bound was obtained by Turcaninov in 1908: n has at least five distinct prime factors and exceeds $2 \cdot 10^6$. With the advent of electronic computers, the lower bound has been improved to $n > 10^{300}$. Recent investigations have shown that n must be divisible by at least eight distinct primes, the largest of which is greater than 100129, while the next largest exceeds 1009; if $3 \nmid n$, then the number of distinct prime factors of n is at least eleven.

While all of this lends support to the belief that there are no odd perfect numbers, only a proof of their nonexistence would be conclusive. We would then be in the curious position of having built up a whole theory for a class of numbers that didn't exist. "It must always," wrote the mathematician Joseph Sylvester in 1888, "stand to the credit of the Greek geometers that they succeeded in discovering a class of perfect numbers which in all probability are the only numbers which are perfect."

Another group of numbers that has had a continuous history extending from the early Greeks to the present time comprises the *amicable numbers.* Two numbers like

220 and 284 are called *amicable,* or friendly, because they have the remarkable property that each number is "contained" within the other, in the sense that each number is equal to the sum of all the positive divisors of the other, not counting the number itself. Thus as regards the divisors of 220,

$$1 + 2 + 4 + 5 + 10 + 11 + 20 + 22 + 44 + 55 + 110 = 284$$

while for 284,

$$1 + 2 + 4 + 71 + 142 = 220.$$

In terms of the σ function, amicable numbers m and n (or an *amicable pair*) are defined by the equations

$$\sigma(m) - m = n, \quad \sigma(n) - n = m,$$

or what amounts to the same thing,

$$\sigma(m) = m + n = \sigma(n).$$

Down through their quaint history, amicable numbers have been important in magic and astrology, and in casting horoscopes, making talismans, and concocting love potions. The Greeks believed that these numbers had a particular influence in establishing friendships between individuals. The philosopher Iamblichus of Chalcis (ca. A.D. 250–A.D.330) ascribed a knowledge of the pair 220 and 284 to the Pythagoreans. He wrote:

> They [the Pythagoreans] call certain numbers amicable numbers, adopting virtues and social qualities to numbers, as 284 and 220; for the parts of each have the power to generate the other. . . .

Biblical commentators spotted 220, the lesser of the classical pair, in Genesis 32:14 as numbering Jacob's present to Esau of 200 she-goats and 20 he-goats. According to one commentator, Jacob wisely counted out his gift (a "hidden secret arrangement") in order to secure the friendship of Esau. An Arab of the eleventh century, El Madschriti of Madrid, related that he had put to the test the erotic effect of these numbers by giving someone a confection in the shape of the smaller number, 220, to eat, while he himself ate the larger, 284. He failed, however, to describe whatever success the ceremony brought.

It is a mark of the slow development of number theory that until the 1630s no one had been able to add to the original pair of amicable numbers discovered by the Greeks. The first explicit rule described for finding certain types of amicable pairs is due to Thabit ibn Kurrah, an Arabian mathematician of the 9th century. In a manuscript composed at that time, he indicated:

> If the three numbers $p = 3 \cdot 2^{n-1} - 1$, $q = 3 \cdot 2^{n-1}$, and $r = 9 \cdot 2^{2n-1} - 1$ are all prime and $n \geq 2$, then $2^n pq$ and $2^n r$ are amicable numbers.

It was not until its rediscovery centuries later by Fermat and Descartes that Thabit's rule produced the second and third pairs of amicable numbers. In a letter to Mersenne in 1636, Fermat announced that 17,296 and 18,416 were an amicable pair, and Descartes wrote to Mersenne in 1638 that he had found the pair 9,363,584 and 9,437,056. Fermat's pair resulted from taking $n = 4$ in Thabit's rule ($p = 23$, $q = 47$, $r = 1151$ are all prime) and Descartes's from $n = 7$ ($p = 191$, $q = 383$, $r = 73,727$ are all prime).

In the 1700s, Euler drew up at one clip a list of sixty-four amicable pairs; two of these new pairs were later found to be "unfriendly," one in 1909 and one in 1914. Adrien Marie Legendre in 1830 found another pair, 2,172,649,216 and 8,520,191.

Extensive computer searches have currently revealed more than 7500 amicable pairs, some of them running to 282 digits; these include all those with values less than 10^{10}. It has not yet been established whether the number of amicable pairs is finite or infinite, nor has a pair been produced in which the numbers are relatively prime. What has been proved is that each integer in a pair of relatively prime amicable numbers must be greater than 10^{25}; and their product must be divisible by at least 22 distinct primes. Part of the difficulty is that in contrast with the single formula for generating (even) perfect numbers, there is no known rule for finding all amicable pairs of numbers.

PROBLEMS 10.2

1. Prove that the Mersenne number M_{13} is a prime; hence the integer $n = 2^{12}(2^{13} - 1)$ is perfect. [*Hint:* Since $\sqrt{M_{13}} < 91$, Theorem 10-5 implies that the only candidates for prime divisors of M_{13} are 53 and 79.]

2. Prove that the Mersenne number M_{19} is a prime; hence the integer $n = 2^{18}(2^{19} - 1)$ is perfect. [*Hint:* By Theorems 10-5 and 10-6, the only prime divisors to test are 191, 457, and 647.]

3. Prove that the Mersenne number M_{29} is composite.

4. A positive integer n is said to be a *deficient number* if $\sigma(n) < 2n$ and an *abundant number* if $\sigma(n) > 2n$. Prove each of the following:
   (a) There are infinitely many deficient numbers. [*Hint:* Consider the integers $n = p^k$, where p is an odd prime and $k \geq 1$.]
   (b) There are infinitely many even abundant numbers. [*Hint:* Consider the integers $n = 2^k \cdot 3$, where $k > 1$.]
   (c) There are infinitely many odd abundant numbers. [*Hint:* Consider the integers $n = 945 \cdot k$, where k is any positive integer not divisible by 2, 3, 5, or 7. Since $945 = 3^3 \cdot 5 \cdot 7$, it follows that $\gcd(945, k) = 1$ and so $\sigma(n) = \sigma(945)\sigma(k)$.]

5. Assuming that n is an even perfect number and $d \mid n$, where $1 < d < n$, show that d is deficient.

6. Prove that any multiple of a perfect number is abundant.

7. Confirm that the pairs of integers listed below are amicable:
   (a) $220 = 2^2 \cdot 5 \cdot 11$ and $284 = 2^2 \cdot 71$ (Pythagoras, 500 B.C.);
   (b) $17296 = 2^4 \cdot 23 \cdot 47$ and $18416 = 2^4 \cdot 1151$ (Fermat, 1636);
   (c) $9363584 = 2^7 \cdot 191 \cdot 383$ and $9437056 = 2^7 \cdot 73727$ (Descartes, 1638).

8. For a pair of amicable numbers m and n, prove that

$$\left(\sum_{d \mid m} 1/d \right)^{-1} + \left(\sum_{d \mid n} 1/d \right)^{-1} = 1.$$

9. Establish the following statements concerning amicable numbers:
   (a) Neither p nor p^2 can be one of an amicable pair, where p is a prime.
   (b) The larger integer in any amicable pair is a deficient number.
   (c) If m and n are an amicable pair, with m even and n odd, then n is a perfect square. [*Hint:* If p is an odd prime, then $1 + p + p^2 + \cdots + p^k$ is odd only when k is an even integer.]

10. In 1886, a 16-year-old Italian boy announced that $1184 = 2^5 \cdot 37$ and $1210 = 2 \cdot 5 \cdot 11^2$ form an amicable pair of numbers, but gave no indication of the method of discovery. Verify his assertion.

11. Prove "Thabit's rules" for amicable pairs: If $p = 3 \cdot 2^{n-1} - 1$, $q = 3 \cdot 2^n - 1$ and $r = 9 \cdot 2^{2n-1} - 1$ are all prime numbers, where $n \geq 2$, then $2^n pq$ and $2^n r$ are an amicable pair of numbers. This rule produces amicable numbers for $n = 2$, 4 and 7, but for no other $n \leq 20{,}000$.

12. By an *amicable triple of numbers* is meant three integers such that the sum of any two is equal to the sum of the divisors of the remaining integer, excluding the number itself. Verify that $2^5 \cdot 3 \cdot 13 \cdot 293 \cdot 337$, $2^5 \cdot 3 \cdot 5 \cdot 13 \cdot 16561$ and $2^5 \cdot 3 \cdot 13 \cdot 99371$ are an amicable triple.

13. A finite sequence of positive integers is said to be a *sociable chain* if each is the sum of the positive divisors of the preceding integer, excluding the number itself (the last integer is considered as preceding the first integer in the chain). Show that the following integers form a sociable chain:

$$14288, \ 15472, \ 14536, \ 14264, \ 12496.$$

Only two sociable chains were known until 1970, when nine chains of four integers apiece were found.

14. Prove that
   (a) any odd perfect number n can be represented in the form $n = pa^2$, where p is a prime;
   (b) if $n = pa^2$ is an odd perfect number, then $n \equiv p \pmod 8$.

15. If n is an odd perfect number, prove that n has at least three distinct prime factors. [*Hint:* Assume $n = p^k q^{2j}$, where $p \equiv k \equiv 1 \pmod 4$. Use the inequality $2 \equiv \sigma(n)/n \leq [p/(p-1)][q/(q-1)]$ to reach a contradiction.]

16. If the integer $n > 1$ is a product of distinct Mersenne primes, show that $\sigma(n) = 2^k$ for some k.

10.3 FERMAT NUMBERS

To round out the picture, let us mention another class of numbers that provides a rich source of conjectures, the Fermat numbers. These may be considered as a special case of the integers of the form $2^m + 1$. We observe that if $2^m + 1$ is a prime, then $m = 2^n$ for some $n > 0$. Assume to the contrary that m had an odd divisor $2k + 1 > 1$, say $m = (2k+1)r$; then $2^m + 1$ would admit the nontrivial factorization

$$2^m + 1 = 2^{(2k+1)r} + 1 = (2^r)^{2k+1} + 1$$

$$= (2^r + 1)(2^{2kr} - 2^{(2k-1)r} + \cdots + 2^{2r} - 2^r + 1),$$

which is impossible. In brief, $2^m + 1$ can be prime only if m is a power of 2.

DEFINITION 10-2.
   A *Fermat number* is an integer of the form

$$F_n = 2^{2^n} + 1, \quad n \geq 0.$$

If F_n is prime, it is said to be a *Fermat prime*.

Fermat, whose mathematical intuition was usually reliable, observed that the integers

$$F_0 = 3, F_1 = 5, F_2 = 17, F_3 = 257, F_4 = 65,537$$

are all primes and expressed his belief that F_n is prime for each value of n. In writing to Mersenne, he confidently announced: "I have found that numbers of the form $2^{2^n} + 1$ are always prime numbers and have long since signified to analysts the truth of this theorem." However, Fermat bemoaned his inability to come up with a proof and, in subsequent letters, his tone of growing exasperation suggests that he was continually trying to do so. The question was resolved negatively by Euler in 1732 when he found

$$F_5 = 2^{2^5} + 1 = 4,294,967,297$$

to be divisible by 641. To us, such a number does not seem very large; but in Fermat's time, the investigation of its primality was difficult and he obviously did not carry it out.

The following elementary proof that $641 \mid F_5$ does not explicitly involve division and is due to G. Bennett.

THEOREM 10-8.

The Fermat number F_5 is divisible by 641.

Proof: We begin by putting $a = 2^7$ and $b = 5$, so that

$$1 + ab = 1 + 2^7 \cdot 5 = 641.$$

It is easily seen that

$$1 + ab - b^4 = 1 + (a - b^3)b = 1 + 3b = 2^4.$$

But this implies that

$$\begin{aligned}
F_5 = 2^{2^5} + 1 &= 2^{32} + 1 \\
&= 2^4 a^4 + 1 \\
&= (1 + ab - b^4)a^4 + 1 \\
&= (1 + ab)a^4 + (1 - a^4 b^4) \\
&= (1 + ab)[a^4 + (1 - ab)(1 + a^2 b^2)],
\end{aligned}$$

which gives $641 \mid F_n$.

To this day it is not known whether there are infinitely many Fermat primes or, for that matter, whether there is at least one Fermat prime beyond F_4. The best "guess" is that all Fermat numbers $F_n > F_4$ are composite.

Part of the interest in Fermat primes stems from the discovery that they have a remarkable connection with the ancient problem of determining all regular polygons that can be constructed with ruler and compass alone (where the former is used only to draw straight lines and the latter only to draw arcs). In the seventh and last section of the *Disquisitiones Arithmeticae,* Gauss proved that a regular polygon of n sides is so constructible if and only if either

$$n = 2^k \quad \text{or} \quad n = 2^k p_1 p_2 \cdots p_r,$$

where $k \geq 0$ and $p_1, p_2, \ldots, p_r$ are distinct Fermat primes. The construction of regular polygons of 2^k, $2^k \cdot 3$, $2^k \cdot 5$ and $2^k \cdot 15$ sides had been known since the time of the Greek geometers. In particular, they could construct regular n-sided polygons for $n = 3, 4, 5, 6, 8, 10, 12, 15,$ and 16. What no one suspected before Gauss was that a regular 17-sided polygon can also be constructed by ruler and compass. Gauss was so proud of his discovery that he requested that a regular polygon of 17 sides be engraved on his tombstone; for some reason, this wish was never fulfilled, but such a polygon is inscribed on the side of a monument to Gauss erected in Brunswick, Germany, his birthplace.

A useful property of Fermat numbers is that they are relatively prime to each other.

THEOREM 10-9.

For Fermat numbers F_n and F_m, where $m > n \geq 0$, $\gcd(F_m, F_n) = 1$.

Proof: Put $d = \gcd(F_m, F_n)$. Since Fermat numbers are odd integers, d must be odd. If we set $x = 2^{2^n}$ and $k = 2^{m-n}$, then

$$\frac{F_m - 2}{F_n} = \frac{(2^{2^n})^{2^{m-n}} - 1}{2^{2^n} + 1}$$

$$= \frac{x^k - 1}{x + 1} = x^{k-1} - x^{k-2} + \cdots - 1,$$

whence $F_n \mid (F_m - 2)$. From $d \mid F_n$, it follows that $d \mid (F_m - 2)$. Now use the fact that $d \mid F_m$ to obtain $d \mid 2$. But d is an odd integer, and so $d = 1$, establishing the result claimed.

This leads to a pleasant little proof of the infinitude of primes: We know that each of the Fermat numbers $F_0, F_1, \ldots, F_n$ is divisible by a prime which, according to Theorem 10-9, does not divide any of the other F_k. Thus there are at least $n + 1$ distinct primes not exceeding F_n. Since there are infinitely many Fermat numbers, the number of primes is also infinite.

In 1877, T. Pepin devised the practical test (Pepin's Test) for determining the primality of F_n that is embodied in the following theorem:

THEOREM 10-10 (Pepin's Test).

For $n \geq 1$, the Fermat number $F_n = 2^{2^n} + 1$ is prime if and only if

$$3^{(F_n - 1)/2} \equiv -1 \pmod{F_n}.$$

Proof: First let us assume that

$$3^{(F_n - 1)/2} \equiv -1 \pmod{F_n}.$$

Upon squaring both sides, we get

$$3^{F_n - 1} \equiv 1 \pmod{F_n}.$$

The same congruence will hold for any prime p which divides F_n:

$$3^{F_n - 1} \equiv 1 \pmod{p}.$$

Now let k to be the order of 3 modulo p. Theorem 8-1 indicates that $k \mid F_n - 1$, or in other words that $k \mid 2^{2^n}$; therefore k must be a power of 2.

It is not possible that $k = 2^r$ for any $r \leq 2^n - 1$. For if this were so, repeated squaring of the congruence $3^k \equiv 1 \pmod{p}$ would yield

$$3^{2^{2^n - 1}} \equiv 1 \pmod{p},$$

or, what is the same thing,

$$3^{(F_n - 1)/2} \equiv 1 \pmod{p}.$$

We would then arrive at $1 \equiv -1 \pmod{p}$, resulting in $p = 2$, which is a contradiction. Thus the only possibility open to us is that

$$k = 2^{2^n} = F_n - 1.$$

Fermat's Theorem tells us now that $k \leq p - 1$, which means in turn that $F_n = k + 1 \leq p$. Since $p \mid F_n$, we also have $p \leq F_n$. Together these inequalities mean that $F_n = p$, so that F_n is a prime.

On the other hand, suppose that F_n, $n \geq 1$, is prime. The Quadratic Reciprocity Law gives

$$(3/F_n) = (F_n/3) = (2/3) = -1,$$

when we use the fact that $F_n \equiv (-1)^{2^n} + 1 = 2 \pmod{3}$. Applying Euler's Criterion, we end up with

$$3^{(F_n - 1)/2} \equiv -1 \pmod{F_n}.$$

Let us demonstrate the primality of $F_3 = 257$ using Pepin's Test. Working modulo 257, we have

$$3^{(F_3 - 1)/2} = 3^{128} = 3^3 (3^5)^{25}$$

$$\equiv 27(-14)^{25}$$

$$\equiv 27 \cdot 14^{24}(-14)$$

$$\equiv 27(-17)(-14)$$

$$\equiv 27 \cdot 19 \equiv 513 \equiv -1 \pmod{257},$$

so that F_3 is prime.

In 1905, J. C. Morehead and A. E. Western independently performed Pepin's Test on F_7, and communicated its composite character almost simultaneously. It took 66 years, until 1971, before Brillhart and Morrison discovered the prime factorization

$$F_7 = 2^{128} + 1$$

$$= 59{,}649{,}589{,}127{,}497{,}217 \cdot 5{,}704{,}689{,}200{,}685{,}129{,}054{,}721.$$

(The possibility of arriving at such a factorization without recourse to fast computers with large memories is remote.) Morehead and Western carried out (in 1909) a similar calculation for the compositeness of F_8, each doing half the work; but the actual factors were not found until 1981, when Brent and Pollard showed the smallest prime divisor of F_8 to be

$$1,238,926,361,552,897.$$

The other factor of F_8 is 62 digits long, and was shortly afterwards shown to be prime. A large F_n to which Pepin's Test has been applied is F_{14}, a number of 4933 digits; this Fermat number was determined to be composite by Selfridge and Hurwitz in 1963, although at present no divisor is known.

Our final theorem, due to Euler and Lucas, is a valuable aid in determining the divisors of Fermat numbers. As early as 1747, Euler established that every prime factor of F_n must be of the form $k \cdot 2^{n+1} + 1$; over 100 years later, in 1879, the French number theorist Edouard Lucas improved upon this result by showing that k can be taken to be even. From this, we have the following theorem.

THEOREM 10-11.

Any prime divisor p of the Fermat number $F_n = 2^{2^n} + 1$, where $n \geq 2$, is of the form $p = k \cdot 2^{n+2} + 1$.

Proof: For a prime divisor p of F_n,

$$2^{2^n} \equiv -1 \pmod{p},$$

which is to say, upon squaring, that

$$2^{2^{n+1}} \equiv 1 \pmod{p}.$$

If h is the order of 2 modulo p, this congruence tells us that

$$h \mid 2^{n+1}.$$

We cannot have $h = 2^r$ where $1 \leq r \leq n$, for this would lead to

$$2^{2^n} \equiv 1 \pmod{p},$$

and in turn to the contradiction that $p = 2$. This lets us conclude that $h = 2^{n+1}$. Since the order of 2 modulo p divides $\phi(p) = p - 1$, we may further conclude that $2^{n+1} \mid p - 1$. The point is that for $n \geq 2$, $p \equiv 1 \pmod{8}$ and so, by Theorem 9-6, the Legendre symbol $(2/p) = 1$. Using Euler's Criterion, we immediately pass to

$$2^{(p-1)/2} \equiv (2/p) = 1 \pmod{p}.$$

An appeal to Theorem 8-1 finishes the proof; it asserts that $h \mid (p-1)/2$, or equivalently, $2^{n+1} \mid (p-1)/2$. This forces $2^{n+2} \mid p - 1$, and we obtain $p = k \cdot 2^{n+2} + 1$ for some integer k.

Theorem 10-11 enables us to determine quite easily the nature of $F_4 = 2^{16} + 1 = 65537$. The prime divisors of F_4 must take the form $2^6 k + 1 = 64k + 1$. There is only one prime of this kind which is less than or equal to $\sqrt{F_4}$, namely the prime 193. Because this trial divisor fails to be a factor of F_4, we may conclude that F_4 is itself a prime.

The increasing availability and speed of computing equipment has allowed the search for prime factors of the Fermat numbers to be extended significantly. In 1980, G. Gostin discovered the prime factor 31065037602817 of F_{17} (none had previously been determined). Another computational development was the successful application (1987) of Pepin's Test by Young and Buell to F_{20}, which had 315,653 digits; until this, F_{14} had been the largest Fermat number exposed to the test. It is now known that F_n is composite for $5 \leq n \leq 20$, as well as some 60 other values of n, the largest being $n = 23,471$. For F_{22}, the challenge remains: it is the smallest Fermat number whose character is in doubt. Considering the size of F_{22}, its nature is not likely to be determined for some time.

The factorization of the 155-digit F_9 was accomplished in 1990. This required two months of work by hundreds of mathematicians using a worldwide network of more than a thousand computers. The computations showed F_9 to be a product of a previously known seven-digit prime and two new primes having 49 and 99 digits. Currently topping the "most wanted" list is F_{10}, which has not yet been completely factored although two prime factors are known.

A resume of the current primality status for the Fermat numbers F_n, $0 \leq n \leq 30$, is given below.

n	Character of F_n
0, 1, 2, 3, 4	Prime
5, 6, 7, 8, 9, 11	Completely factored
10, 12, 13, 19, 30	Two or four factors known
15, 16, 17, 18, 21, 23, 25, 26, 27	Only one prime factor known
14, 20	Composite but no factor known
22, 24, 28, 29	Character unknown

The case for F_{16} was settled in 1953 and lays to rest the tantalizing conjecture that all the terms of the sequence

$$2 + 1, \ 2^2 + 1, \ 2^{2^2} + 1, \ 2^{2^{2^2}} + 1, \ 2^{2^{2^{2^2}}} + 1, \ \cdots$$

are prime numbers. What is interesting is that none of the known prime factors p of a Fermat number F_n gives rise to a square factor p^2; indeed, it is speculated that the Fermat numbers are square-free.

Numbers of the form $k \cdot 2^n + 1$, which occur in the search for prime factors of Fermat numbers, are of considerable interest in their own right. The smallest n for which $k \cdot 2^n + 1$ is prime may be quite large in some cases; for instance, the first time $47 \cdot 2^n + 1$ is prime is when $n = 583$. But there also exist values of k such that $k \cdot 2^n + 1$ is always composite. Indeed, in 1960 it was proved that there exist infinitely many odd integers k with $k \cdot 2^n + 1$ composite for all $n \geq 1$. The problem of determining the least such value of k remains unsolved. Up to now, $k = 78557$ is the smallest known k for which $k \cdot 2^n + 1$ is never prime for any n.

PROBLEMS 10.3

1. By taking fourth powers of the congruence $5 \cdot 2^7 \equiv -1 \pmod{641}$, deduce that $2^{32} + 1 \equiv 0 \pmod{641}$; hence, $641 \mid F_5$.

2. Gauss (1796) discovered that a regular polygon with p sides, where p is a prime, can be constructed with ruler and compass if and only if $p - 1$ is a power of 2. Show that this condition is equivalent to requiring that p be a Fermat prime.

3. For $n > 0$, prove that
   (a) there are infinitely many composite numbers of the form $2^{2n} + 3$; [*Hint:* Use the fact that $2^{2n} = 3k + 1$ for some k to establish that $7 \mid 2^{2^{2n+1}} + 3$.]
   (b) each of the numbers $2^{2^n} + 5$ is composite.

4. Composite integers n for which $n \mid 2^n - 2$ are called *pseudoprimes*. Show that:
   (a) If n is odd pseudoprime, then the Mersenne number M_n is also pseudoprime; hence, there are infinitely many pseudoprimes. [*Hint:* The relation $2n \mid 2^n - 2$ produces $n \mid 2^{n-1} - 1$, whence $2^{n-1} - 1 = kn$ for some k. Then $2^{M_n - 1} - 1 = 2^{2^n - 1} - 1 = (2^n)^{2k} - 1$, which implies that $2^n - 1 \mid 2^{M_n - 1} - 1$.]
   (b) Every Fermat number F_n is either a prime or a pseudoprime. [*Hint:* Raise the congruence $2^{2^n} \equiv -1 \pmod{F_n}$ to the $2^{2^n - n}$ power.]

5. For $n \geq 2$, show that the last digit of the Fermat number $F_n = 2^{2^n} + 1$ is 7. [*Hint:* By induction on n, verify that $2^{2^n} \equiv 6 \pmod{10}$ for $n \geq 2$.]

6. Establish that $2^{2^n} - 1$ has at least n distinct prime divisors. [*Hint:* Use induction on n and the fact that

$$2^{2^n} - 1 = (2^{2^{n-1}} + 1)(2^{2^{n-1}} - 1).]$$

7. In 1869, Landry wrote: "No one of our numerous factorizations of the numbers $2^n \pm 1$ gave us as much trouble and labor as that of $2^{58} + 1$." Verify that $2^{58} + 1$ can be factored rather easily using the identity

$$4x^4 + 1 = (2x^2 - 2x + 1)(2x^2 + 2x + 1).$$

8. From Problem 5, conclude that
   (a) the Fermat number F_n is never a perfect square;
   (b) for $n > 0$, F_n is never a triangular number.

9. (a) For any odd integer n, show that $3 \mid 2^n + 1$.
   (b) Prove that if p and q are both odd primes and $q \mid 2^p + 1$, then either $q = 3$ or $q = 2kp + 1$ for some integer k. [*Hint:* Since $2^{2p} \equiv 1 \pmod{q}$, the order of 2 modulo q is either 2 or $2p$; in the latter case, $2p \mid \phi(q)$.]
   (c) Find the smallest prime divisor $q > 3$ of each of the integers $2^{29} + 1$ and $2^{41} + 1$.

10. Determine the smallest odd integer $n > 1$ such that $2^n - 1$ is divisible by a pair of twin primes p and q, where $p < q$. [*Hint:* Being the first member of a pair of twin primes, $p \equiv -1 \pmod 6$. Since $(2/p) = (2/q) = 1$, Theorem 9-6 gives $p \equiv q \equiv \pm 1 \pmod 8$; hence, $p \equiv -1 \pmod{24}$ and $q \equiv 1 \pmod{24}$. Now use the fact that the orders of 2 modulo p and q must divide n.]

11. Find all prime numbers p such that p divides $2^p + 1$; do the same for $2^p - 1$.

12. Let $p = 3 \cdot 2^n + 1$ be a prime, where $n \geq 1$. (Twenty-five primes of this form are currently known, the smallest occurring when $n = 1$ and the largest when $n = 3912$.) Prove each of the following assertions:
    (a) The order of 2 modulo p is either 3, 2^k or $3 \cdot 2^k$ for some $0 \leq k \leq n$.
    (b) Except when $p = 13$, 2 is not a primitive root of p. [*Hint:* If 2 is a primitive root of p, then $(2/p) = -1$.]
    (c) The order of 2 modulo p is not divisible by 3 if and only if p divides a Fermat number F_k with $0 \leq k \leq n - 1$. [*Hint:* Use the identity $2^{2^k} - 1 = F_0 F_1 F_2 \ldots F_{k-1}$.]

    (d) There is no Fermat number which is divisible by 7, 13, or 97.

13. For any Fermat number $F_n = 2^{2^n} + 1$, establish that $F_n \equiv 5$ or 8 (mod 9) according as n is odd or even. [*Hint:* Use induction to show, first, that $2^{2^n} \equiv 2^{2^{n-2}}$ (mod 9) for $n \geq 3$.]

14. Use the fact that the prime divisors of F_5 are of the form $2^7 k + 1 = 128k + 1$ to confirm that $641 \mid F_5$.

15. For any prime $p > 3$, prove the following:
    (a) $\frac{1}{3}(2^p + 1)$ is not divisible by 3. [*Hint:* Consider the identity

$$\frac{2^p + 1}{2 + 1} = 2^{p-1} - 2^{p-2} + \cdots - 2 + 1.]$$

    (b) $\frac{1}{3}(2^p + 1)$ has a prime divisor greater than p. [*Hint:* Problem 9(b).]

    (c) The integers $\frac{1}{3}(2^{19} + 1)$ and $\frac{1}{3}(2^{23} + 1)$ are both prime.

16. From the previous problem, deduce that there are infinitely many prime numbers.

17. (a) Prove that 3, 5, and 7 are quadratic nonresidues of any Fermat prime F_n, where $n \geq 2$. [*Hint:* Pepin's Test and Problem 15, Section 9.3.]
    (b) Show that every quadratic nonresidue of a Fermat prime F_n is a primitive root of F_n.

18. Establish that any Fermat prime F_n can be written as the difference of two squares, but not of two cubes. [*Hint:*

$$F_n = 2^{2^n} + 1 = (2^{2^n - 1} + 1)^2 - (2^{2^n - 1})^2.]$$

19. For $n \geq 1$, show that $\gcd(F_n, n) = 1$. [*Hint:* Theorem 10-11.]

20. Use Theorems 10–9 and 10–11 to deduce that there are infinitely many primes of the form $4k + 1$.

11

The Fermat Conjecture

"He who seeks for methods without having a definite problem in mind seeks for the most part in vain."

D. HILBERT

11.1 PYTHAGOREAN TRIPLES

Fermat, whom many regard as a father of modern number theory, nevertheless had a custom peculiarly ill-suited to this role. He published very little personally, preferring to communicate his discoveries in letters to friends (usually with no more than the terse statement that he possessed a proof) or to keep them to himself in notes. A number of such notes were jotted down in the margin of his copy of Bachet's translation of Diophantus' *Arithmetica*. By far the most famous of these marginal comments is the one—presumably written about 1637—which states:

> It is impossible to write a cube as a sum of two cubes, a fourth power as a sum of two fourth powers, and, in general, any power beyond the second as a sum of two similar powers. For this, I have discovered a truly wonderful proof, but the margin is too small to contain it.

In this tantalizing aside, Fermat was simply asserting that, if $n > 2$, then the Diophantine equation

$$x^n + y^n = z^n$$

has no solution in the integers, other than the trivial solutions in which at least one of the variables is zero.

The quotation just cited has come to be known as Fermat's Last Theorem or, more accurately, Fermat's Conjecture. All the results he enunciated in the margin of his *Arithmetica* were later found to be true with the one exception of the Last Theorem, which still awaits proof or disproof. If Fermat had a "truly wonderful proof," it has never come to light. To date, the conjecture has only been established for specific values of the exponent n (electronic computers have shown that there are no nontrivial solutions in the range $3 \leq n < 10^6$), but no general proof has been forthcoming.

Any explicit counterexample to the Fermat Conjecture would involve integers of immense size. It has been shown that if p is an odd prime, $xyz \neq 0$, and $x^p + y^p = z^p$, then x must have at least 1.8 million digits and x^p at least $2 \cdot 10^{11}$ digits. In decimal notation, x would require over one hundred pages to write down and x^p would fill up at least ten thousand volumes of one thousand pages each.

Fermat did however, leave a proof of his Last Theorem for the case $n = 4$. To carry through the argument, we first undertake the task of identifying all solutions in the positive integers of the equation

(1) $$x^2 + y^2 = z^2.$$

Since the length z of the hypotenuse of a right triangle is related to the lengths x and y of the sides by the famous Pythagorean equation $x^2 + y^2 = z^2$, the search for all positive integers which satisfy equation (1) is equivalent to the problem of finding all right triangles with sides of integral length. The latter problem was raised in the days of the Babylonians and was a favorite with the ancient Greek geometers. Pythagoras himself has been credited with a formula for infinitely many such triangles, namely

$$x = 2n + 1, \, y = 2n^2 + 2n, \, z = 2n^2 + 2n + 1,$$

where n is an arbitrary positive integer. This formula does not account for all right triangles with integral sides and it was not until Euclid wrote his *Elements* that a complete solution to the problem appeared.

The following definition gives us a concise way of referring to the solutions of equation (1):

DEFINITION 11-1.
A *Pythagorean triple* is a set of three integers x, y, z such that $x^2 + y^2 = z^2$; the triple is said to be *primitive* if $\gcd(x, y, z) = 1$.

Perhaps the best known examples of primitive Pythagorean triples are 3, 4, 5 and 5, 12, 13, while a less obvious one is 12, 35, 37.

There are several points that need to be noted. Suppose that x, y, z is any Pythagorean triple and $d = \gcd(x, y, z)$. If we write $x = dx_1, y = dy_1, z = dz_1$, then it is easily seen that

$$x_1^2 + y_1^2 = \frac{x^2 + y^2}{d^2} = \frac{z^2}{d^2} = z_1^2$$

with $\gcd(x_1, y_1, z_1) = 1$. In short, x_1, y_1, z_1 form a primitive Pythagorean triple. Thus, it is enough to occupy ourselves with finding all primitive Pythagorean triples; any Pythagorean triple can be obtained from a primitive one upon multiplying by a suitable nonzero integer. The search may be confined to those primitive Pythagorean triples x, y, z in which $x > 0, y > 0, z > 0$, inasmuch as all others arise from the positive ones through a simple change of sign.

Our development requires two preparatory lemmas, the first of which sets forth a basic fact regarding primitive Pythagorean triples.

LEMMA 1.

If x, y, z is a primitive Pythagorean triple, then one of the integers x and y is even, while the other is odd.

Proof: If x and y are both even, then $2 \mid (x^2 + y^2)$ or $2 \mid z^2$, so that $2 \mid z$. The inference is that $\gcd(x, y, z) \geq 2$, which we know to be false. If, on the other hand, x and y should both be odd, then $x^2 \equiv 1 \pmod 4$ and $y^2 \equiv 1 \pmod 4$, leading to

$$z^2 = x^2 + y^2 \equiv 2 \pmod 4.$$

But this is equally impossible, since the square of any integer must be congruent either to 0 or to 1 modulo 4.

Given a primitive Pythagorean triple x, y, z, exactly one of these integers is even, the other two being odd (if x, y, z were all odd, then $x^2 + y^2$ would be even, while z^2 is odd). The foregoing lemma indicates that the even integer is either x or y; to be definite, we shall hereafter write our Pythagorean triples so that x is even and y is odd; then, of course, z is odd.

It is worth noticing (and we will use this fact) that each pair of the integers x, y, and z must be relatively prime. Were it the case that $\gcd(x, y) = d > 1$, then there would exist a prime p with $p \mid d$. Since $d \mid x$ and $d \mid y$, we would have $p \mid x$ and $p \mid y$, whence $p \mid x^2$ and $p \mid y^2$. But then $p \mid (x^2 + y^2)$, or $p \mid z^2$, giving $p \mid z$. This would conflict with the assumption that $\gcd(x, y, z) = 1$, and so $d = 1$. In like manner, one can verify that $\gcd(y, z) = \gcd(x, z) = 1$.

By virtue of Lemma 1, there exists no primitive Pythagorean triple x, y, z all of whose values are prime numbers. There are primitive Pythagorean triples in which z and one of x or y is a prime; for instance, 3, 4, 5; 11, 60, 61; and 19, 180, 181. It is unknown whether there exist infinitely many such triples.

The next hurdle that stands in our way is to establish that if a and b are relatively prime positive integers having a square as their product, then a and b are themselves squares. With an assist from the Fundamental Theorem of Arithmetic, we can prove considerably more, to wit,

LEMMA 2.

If $ab = c^n$, where $\gcd(a, b) = 1$, then a and b are nth powers; that is, there exist positive integers a_1, b_1 for which $a = a_1^n$, $b = b_1^n$.

Proof: There is no harm in assuming that $a > 1$ and $b > 1$. If

$$a = p_1^{k_1} p_2^{k_2} \cdots p_r^{k_r}, \quad b = q_1^{j_1} q_2^{j_2} \cdots q_s^{j_s}$$

are the prime factorizations of a and b, then, bearing in mind that $\gcd(a, b) = 1$, no p_i can occur among the q_i. As a result, the prime factorization of ab is given by

$$ab = p_1^{k_1} \cdots p_r^{k_r} q_1^{j_1} \cdots q_s^{j_s}.$$

Let us suppose that c can be factored into primes as $c = u_1^{l_1} u_2^{l_2} \cdots u_t^{l_t}$. Then the condition $ab = c^n$ becomes

$$p_1^{k_1} \cdots p_r^{k_r} q_1^{j_1} \cdots q_s^{j_s} = u_1^{n l_1} \cdots u_t^{n l_t}.$$

From this, one sees that the primes $u_1, \ldots, u_t$ are $p_1, \ldots, p_r, q_1, \ldots, q_s$ (in some order) and $nl_1, \ldots, nl_t$ are the corresponding exponents $k_1, \ldots, k_r, j_1, \ldots, j_s$. The conclusion: each of the integers k_i and j_i must be divisible by n. If we now put

$$a_1 = p_1^{k_1/n} p_2^{k_2/n} \cdots p_r^{k_r/n}$$

$$b_1 = q_1^{j_1/n} q_2^{j_2/n} \cdots q_s^{j_s/n},$$

then $a_1{}^n = a$, $b_1{}^n = b$, as desired.

With the routine work now out of the way, the characterization of all primitive Pythagorean triples is fairly straightforward.

THEOREM 11-1.

All the solutions of the Pythagorean equation

$$x^2 + y^2 = z^2$$

satisfying the conditions

$$\gcd(x, y, z) = 1, \quad 2 \,|\, x, \quad x > 0, y > 0, z > 0$$

are given by the formulas

$$x = 2st, \, y = s^2 - t^2, z = s^2 + t^2$$

for integers $s > t > 0$ such that $\gcd(s, t) = 1$ and $s \not\equiv t \pmod 2$.

Proof: To start, let x, y, z be a (positive) primitive Pythagorean triple. Since we have agreed to take x even, and y and z both odd, $z - y$ and $z + y$ are even integers; say, $z - y = 2u$ and $z + y = 2v$. Now the equation $x^2 + y^2 = z^2$ may be rewritten as

$$x^2 = z^2 - y^2 = (z - y)(z + y),$$

whence

$$(x/2)^2 = \left(\frac{z - y}{2}\right)\left(\frac{z + y}{2}\right) = uv.$$

Notice that u and v are relatively prime; indeed, if $\gcd(u, v) = d > 1$, then $d \,|\, (u - v)$ and $d \,|\, (u + v)$, or equivalently, $d \,|\, y$ and $d \,|\, z$, which violates the fact that $\gcd(y, z) = 1$. Taking Lemma 2 into consideration, we may conclude that u and v are each perfect squares; to be specific, let

$$u = t^2, v = s^2$$

where s and t are positive integers. The result of substituting these values of u and v reads:

$$z = u + v = s^2 + t^2,$$

$$y = v - u = s^2 - t^2,$$

$$x^2 = 4uv = 4s^2t^2,$$

or, in the last case $x = 2st$. Since a common factor of s and t divides both y and z, the condition $\gcd(y, z) = 1$ forces $\gcd(s, t) = 1$. It remains for us to observe that if s and t were both even, or both odd, then this would make each of y and z even, which is an impossibility. Hence, exactly one of the pair s, t is even, while the other is odd; in symbols, $s \not\equiv t \pmod 2$.

Conversely, let s and t be two integers subject to the conditions described above. That $x = 2st$, $y = s^2 - t^2$, $z = s^2 + t^2$ form a Pythagorean triple follows from the easily verified identity

$$x^2 + y^2 = (2st)^2 + (s^2 - t^2)^2 = (s^2 + t^2)^2 = z^2.$$

To see that this triple is primitive, we assume that $\gcd(x, y, z) = d > 1$ and take p to be any prime divisor of d. Observe that $p \neq 2$, since p divides the odd integer z (one of s and t is odd, while the other is even, hence $s^2 + t^2 = z$ must be odd). From $p \mid y$ and $p \mid z$, we obtain $p \mid (z + y)$ and $p \mid (z - y)$, or put otherwise, $p \mid 2s^2$ and $p \mid 2t^2$. But then $p \mid s$ and $p \mid t$, which is incompatible with $\gcd(s, t) = 1$. The implication of all this is that $d = 1$ and so x, y, z constitutes a primitive Pythagorean triple. Theorem 11-1 is thus proven.

The table below lists some primitive Pythagorean triples arising from small values of s and t. For each value of $s = 1, 2, 3, \ldots, 7$, we have taken those values of t which are relatively prime to s, less than s, and even whenever s is odd.

		x	y	z
s	t	$(2st)$	$(s^2 - t^2)$	$(s^2 + t^2)$
2	1	4	3	5
3	2	12	5	13
4	1	8	15	17
4	3	24	7	25
5	2	20	21	29
5	4	40	9	41
6	1	12	35	37
6	5	60	11	61
7	2	28	45	53
7	4	56	33	65
7	6	84	13	85

From this, or from a more extensive table, the reader might be led to suspect that if x, y, z is a primitive Pythagorean triple, then exactly one of the integers x or y is divisible by 3. This is, in fact, the case. For, by Theorem 11-1, we have

$$x = 2st, \ y = s^2 - t^2, \ z = s^2 + t^2,$$

where $\gcd(s, t) = 1$. If either $3 \mid s$ or $3 \mid t$, then evidently $3 \mid x$, and we need go no farther. Suppose that $3 \nmid s$ and $3 \nmid t$. Fermat's Theorem asserts that

$$s^2 \equiv 1 \pmod 3, \ t^2 \equiv 1 \pmod 3$$

and so

$$y = s^2 - t^2 \equiv 0 \text{ (mod 3)}.$$

In other words, y is divisible by 3, which is what we were required to show.

Let us define a *Pythagorean triangle* to be a right triangle whose sides are of integral length. Our findings lead to an interesting geometric fact concerning Pythagorean triangles, recorded as

THEOREM 11-2.

The radius of the inscribed circle of a Pythagorean triangle is always an integer.

Proof: Let r denote the radius of the circle inscribed in a right triangle with hypotenuse of length z and sides of lengths x and y. The area of the triangle is equal to the sum of the areas of the three triangles having common vertex at the center of the circle, hence

$$\tfrac{1}{2}xy = \tfrac{1}{2}rx + \tfrac{1}{2}ry + \tfrac{1}{2}rz = \tfrac{1}{2}r(x + y + z).$$

The situation is illustrated below:

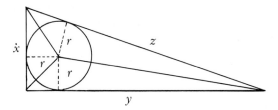

Now $x^2 + y^2 = z^2$. But we know that the positive integral solutions of this equation are given by

$$x = 2kst, \ y = k(s^2 - t^2), \ z = k(s^2 + t^2)$$

for an appropriate choice of positive integers k, s, t. Replacing x, y, z in the equation $xy = r(x + y + z)$ by these values and solving for r, it will be found that

$$r = \frac{2k^2st(s^2 - t^2)}{k(2st + s^2 - t^2 + s^2 + t^2)}$$

$$= \frac{kt(s^2 - t^2)}{s + t} = kt(s - t),$$

which is an integer.

We take the opportunity to mention another result relating to Pythagorean triangles. Notice that it is possible for different Pythagorean triangles to have the same area; for instance, the right triangles associated with the primitive Pythagorean triples 20, 21, 29 and 12, 35, 57 each have an area equal to 210. Fermat proved: for any integer $n > 1$, there exist n Pythagorean triangles with different hypotenuses and the same area. The details of this are omitted.

PROBLEMS 11.1

1. (a) Find three different Pythagorean triples, not necessarily primitive, of the form 16, y, z.
    (b) Obtain all primitive Pythagorean triples x, y, z in which $x = 40$; do the same for $x = 60$.

2. If x, y, z is a primitive Pythagorean triple, prove that $x + y$ and $x - y$ are congruent modulo 8 to either 1 or 7.

3. (a) Prove that if $n \not\equiv 2 \pmod 4$, then there is a primitive Pythagorean triple x, y, z in which x or y equals n.
    (b) If $n \geq 3$ is arbitrary, find a Pythagorean triple (not necessarily primitive) having n as one of its members. [*Hint:* Assuming n is odd, consider the triple n, $\frac{1}{2}(n^2 - 1)$, $\frac{1}{2}(n^2 + 1)$; for n even, consider the triple n, $(n^2/4) - 1$, $(n^2/4) + 1$.]

4. Prove that in a primitive Pythagorean triple x, y, z, the product xy is divisible by 12, hence $60 \mid xyz$.

5. For a given positive integer n, show that there are at least n Pythagorean triples having the same first member. [*Hint:* Let $y_k = 2^k(2^{2n-2k} - 1)$ and $z_k = 2^k(2^{2n-2k} + 1)$ for $k = 0$, $1, 2, \ldots , n - 1$. Then 2^{n+1}, y_k, z_k are all Pythagorean triples.]

6. Verify that 3, 4, 5 is the only primitive Pythagorean triple involving consecutive positive integers.

7. Show that $3n$, $4n$, $5n$ where $n = 1, 2, \ldots$ are the only Pythagorean triples whose terms are in arithmetic progression. [*Hint:* Call the triple in question $x - d$, x, $x + d$, and solve for x in terms of d.]

8. Find all Pythagorean triangles whose areas are equal to their perimeters. [*Hint:* The equations $x^2 + y^2 = z^2$ and $x + y + z = \frac{1}{2}xy$ imply that $(x - 4)(y - 4) = 8$.]

9. (a) Prove that if x, y, z is a primitive Pythagorean triple in which x and z are consecutive positive integers, then

$$x = 2t(t + 1), y = 2t + 1, z = 2t(t + 1) + 1$$

for some $t > 0$. [*Hint:* The equation

$$1 = z - x = s^2 + t^2 - 2st$$

implies that $s - t = 1$.]
    (b) Prove that if x, y, z is a primitive Pythagorean triple in which the difference $z - y = 2$, then

$$x = 2t, y = t^2 - 1, z = t^2 + 1$$

for some $t > 1$.

10. Show that there exist infinitely many primitive Pythagorean triples x, y, z whose even member x is a perfect square. [*Hint:* Consider the triple $4n^2$, $n^4 - 4$, $n^4 + 4$, where n is an arbitrary odd integer.]

11. For an arbitrary positive integer n, show that there exists a Pythagorean triangle the radius of whose inscribed circle is n. [*Hint:* If r denotes the radius of the circle inscribed in the Pythagorean triangle having sides a and b and hypotenuse c, then $r = \frac{1}{2}(a + b - c)$. Now consider the triple $2n + 1$, $2n^2 + 2n$, $2n^2 + 2n + 1$.]

12.   (a) Establish that there exist infinitely many primitive Pythagorean triples x, y, z in which x and y are consecutive positive integers. Exhibit five of these. [*Hint:* If x, $x + 1$, z forms a Pythagorean triple, then so does the triple $3x + 2z + 1$, $3x + 2z + 2$, $4x + 3z + 2$.]

   (b) Show that there exist infinitely many Pythagorean triples x, y, z in which x and y are consecutive triangular numbers. Exhibit three of these. [*Hint:* If x, $x + 1$, z forms a Pythagorean triple, then so does t_{2x}, $t_{2x + 1}$, $(2x + 1)z$.]

13. Use Problem 12 to prove that there exist infinitely many triangular numbers which are perfect squares. Exhibit five such triangular numbers. [*Hint:* If x, $x + 1$, z forms a Pythagorean triple, then upon setting $u = z - x - 1$, $v = x + \frac{1}{2}(1 - z)$, one obtains $u(u + 1)/2 = v^2$.]

11.2 THE FAMOUS "LAST THEOREM"

With our knowledge of Pythagorean triples, we are now prepared to take up the one case in which Fermat himself had a proof of his conjecture, the case $n = 4$. The technique used in the proof is a form of induction sometimes called "Fermat's method of infinite descent." In brief, the method may be described as follows: It is assumed that a solution of the problem in question is possible in the positive integers. From this solution, one constructs a new solution in smaller positive integers, which then leads to a still smaller solution and so on. Since the positive integers cannot be decreased in magnitude indefinitely, it follows that the initial assumption must be false and therefore no solution is possible.

Instead of giving a proof of the Fermat Conjecture for $n = 4$, it turns out to be easier to establish a fact which is slightly stronger; namely, the impossibility of solving the equation $x^4 + y^4 = z^2$ in the positive integers.

THEOREM 11-3 (Fermat).

   The Diophantine equation $x^4 + y^4 = z^2$ has no solution in positive integers x, y, z.

   Proof: With the idea of deriving a contradiction, let us assume that there exists a positive solution x_0, y_0, z_0 of $x^4 + y^4 = z^2$. Nothing is lost in supposing also that $\gcd(x_0, y_0) = 1$; otherwise, put $\gcd(x_0, y_0) = d$, $x_0 = dx_1$, $y_0 = dy_1$, $z_0 = d^2z_1$ to get $x_1^4 + y_1^4 = z_1^2$ with $\gcd(x_1, y_1) = 1$.

   Expressing the supposed equation $x_0^4 + y_0^4 = z_0^2$ in the form

$$(x_0^2)^2 + (y_0^2)^2 = z_0^2$$

we see that x_0^2, y_0^2, z_0 meet all the requirements of a primitive Pythagorean triple, and so Theorem 11-1 can be brought into play. In such triples, one of the integers x_0^2 or y_0^2 is necessarily even, while the other is odd. Taking x_0^2 (and hence x_0) to be even, there exist relatively prime integers $s > t > 0$ satisfying

$$x_0^2 = 2st,$$
$$y_0^2 = s^2 - t^2,$$
$$z_0 = s^2 + t^2,$$

where exactly one of s and t is even. If it happens that s is even, then we have

$$1 \equiv y_0^2 = s^2 - t^2 \equiv 0 - 1 \equiv 3 \;(\text{mod } 4),$$

which is an impossibility. Therefore, s must be the odd integer and, in consequence, t is the even one. Let us put $t = 2r$. Then the equation $x_0^2 = 2st$ becomes $x_0^2 = 4sr$, which says that

$$(x_0/2)^2 = sr.$$

But Lemma 2 asserts that the product of two relatively prime integers [note that $\gcd(s, t) = 1$ implies that $\gcd(s, r) = 1$] is a square only if each of the integers is itself a square; hence, $s = z_1^2$, $r = w_1^2$ for positive integers z_1, w_1.

We wish to apply Theorem 11-1 again, this time to the equation

$$t^2 + y_0^2 = s^2.$$

Since $\gcd(s, t) = 1$, it follows that $\gcd(t, y_0, s) = 1$, making t, y_0, s a primitive Pythagorean triple. With t even, we obtain

$$t = 2uv,$$
$$y_0 = u^2 - v^2,$$
$$s = u^2 + v^2,$$

for relatively prime integers $u > v > 0$. Now the relation

$$uv = t/2 = r = w_1^2$$

signifies that u and v are both squares (Lemma 2 serves its purpose once more); say, $u = x_1^2$ and $v = y_1^2$. When these values are substituted into the equation for s, the result is

$$z_1^2 = s = u^2 + v^2 = x_1^4 + y_1^4.$$

A crucial point is that, z_1 and t being positive, we also have the inequality

$$0 < z_1 \le z_1^2 = s \le s^2 < s^2 + t^2 = z_0.$$

What has happened is this: starting with one solution x_0, y_0, z_0 of $x^4 + y^4 = z^2$, we have constructed another solution x_1, y_1, z_1 such that $0 < z_1 < z_0$. Repeating the whole argument, our second solution would lead to a third solution x_2, y_2, z_2 with $0 < z_2 < z_1$, which in its turn gives rise to a fourth. This process can be carried out as many times as desired to produce an infinite decreasing sequence of positive integers

$$z_0 > z_1 > z_2 > \cdots .$$

Since there is only a finite supply of positive integers less than z_0, a contradiction occurs. We are forced to conclude that $x^4 + y^4 = z^2$ is not solvable in the positive integers.

As an immediate corollary, one gets the following.

COROLLARY.

The equation $x^4 + y^4 = z^4$ has no solution in the positive integers.

Proof: If x_0, y_0, z_0 were a positive solution of $x^4 + y^4 = z^4$, then x_0, y_0, z_0^2 would satisfy the equation $x^4 + y^4 = z^2$, in conflict with Theorem 11-3.

If $n > 2$, then n is either a power of 2 or divisible by an odd prime p. In the first case, $n = 4k$ for some $k \geq 1$ and the Fermat equation $x^n + y^n = z^n$ can be written as

$$(x^k)^4 + (y^k)^4 = (z^k)^4.$$

We have just seen that this equation is impossible in the positive integers. When $n = pk$, the Fermat equation is the same as

$$(x^k)^p + (y^k)^p = (z^k)^p.$$

If it could be shown that the equation $u^p + v^p = w^p$ has no solution, then, in particular, there would be no solution of the form $u = x^k$, $v = y^k$, $w = z^k$ and hence $x^n + y^n = z^n$ would not be solvable. Fermat's Conjecture therefore reduces to this: for no odd prime p does the equation

$$x^p + y^p = z^p$$

admit a solution in the positive integers.

Although the problem has challenged the foremost mathematicians of the last 300 years, their efforts have only produced partial results and proofs of individual cases. Euler gave the first proof of the Fermat Conjecture for the prime $p = 3$ in the year 1770; the reasoning was incomplete at one stage, but Legendre later supplied the missing steps. Using the method of infinite descent, Dirichlet and Legendre independently settled the case $p = 5$ around 1825. Not long thereafter, in 1839, Lamé proved the conjecture for seventh powers. With the increasing complexity of the arguments came the realization that a successful resolution of the general case called for different techniques. The best hope seemed to lie in extending the meaning of "integer" to include a wider class of numbers and, by attacking the problem within this enlarged system, obtaining more information than was possible by using ordinary integers only.

The German mathematician Kummer made the major breakthrough. In 1843, he submitted to Dirichlet a purported proof of the Fermat Conjecture based upon an extension of the integers to include the so-called "algebraic numbers" (that is, complex numbers satisfying polynomials with rational coefficients). Having spent considerable time on the problem himself, Dirichlet was immediately able to detect the flaw in the reasoning: Kummer had taken for granted that algebraic numbers admit a unique factorization similar to that of the ordinary integers, which is not always true.

But Kummer was undeterred by this perplexing situation and returned to his investigations with redoubled effort. To restore unique factorization to the algebraic numbers, he was led to invent the concept of *ideal numbers*. By adjoining these new entities to the algebraic numbers, Kummer successfully proved the Fermat Conjecture for a large class of primes which he termed "regular primes" (that this represented an enormous achievement is reflected in the fact that the only irregular primes less than 100 are 37, 59, and 67). Unfortunately, it is still not known whether there are an infinite number of regular primes, while, in the other direction, Jensen (1915) established that there exist infinitely many irregular ones. Almost all the subsequent progress on the problem has been within the framework suggested by Kummer.

In 1983, a 29-year-old West German mathematician, Gerd Faltings, proved that for each exponent $n > 2$, the Fermat equation $x^n + y^n = z^n$ can have at most a finite number (as opposed to an infinite number) of integral solutions. At first glance, this may not seem like much of an advance; but if it can now be shown that the finite number of solutions is zero in each case, then the Fermat Conjecture can be laid to rest once and for all.

For one brief moment in 1988, it appeared that the final breakthrough had been made. A flurry of reports in the press announced that Yoichi Miyaoka of Tokyo Metropolitan University had favorably resolved the conjecture. But closer scrutiny of the immensely complicated proof revealed subtle yet fundamental missteps. The failure of Miyaoka's initial attempt is not really surprising or unusual in mathematical research. Normally, proposed proofs are privately circulated and examined for possible flaws months in advance of any formal announcement. In Miyaoka's case, the notoriety of one of number theory's most elusive conjectures brought premature publicity and subsequent disappointment to the mathematical community.

To round out our historical digression, we might mention that in 1908 a prize of 100,000 marks was bequeathed to the Academy of Science at Göttingen to be paid for the first complete proof of Fermat's Conjecture. The immediate result was a deluge of incorrect demonstrations by amateur mathematicians. Since only printed solutions were eligible, Fermat's Conjecture is reputed to be the mathematical problem for which the greatest number of false proofs have been published; indeed, between 1908 and 1912 over 1000 alleged proofs appeared, mostly printed as private pamphlets. Suffice it to say, interest declined as the German inflation of the 1920s wiped out the monetary value of the prize.

From $x^4 + y^4 = z^2$, we move on to a closely related Diophantine equation, namely, $x^4 - y^4 = z^2$. The proof of its insolubility parallels that of Theorem 11-3, but we give a slight variation in the method of infinite descent.

THEOREM 11-4 (Fermat).

The Diophantine equation $x^4 - y^4 = z^2$ has no solution in positive integers x, y, z.

Proof: The proof proceeds by contradiction. Let us assume that the equation admits a solution in the positive integers and among these solutions x_0, y_0, z_0 is one with a least value of $x;$ in particular, this supposition forces x_0 to be odd (Why?). Were $\gcd(x_0, y_0) = d > 1$, then putting $x_0 = dx_1, y_0 = dy_1$, we would have $d^4(x_1^4 - y_1^4) = z_0^2$, whence $d^2 | z_0$ or $z_0 = d^2 z_1$ for some $z_1 > 0$. It follows that x_1, y_1, z_1 provides a solution to the equation under consideration with

$0 < x_1 < x_0$, which is an impossible situation. Thus, we are free to assume a solution x_0, y_0, z_0 in which $\gcd(x_0, y_0) = 1$. The ensuing argument falls into two stages, depending on whether y_0 is odd or even.

First, consider the case of an odd integer y_0. If the equation $x_0^4 - y_0^4 = z_0^2$ is written in the form $z_0^2 + (y_0^2)^2 = (x_0^2)^2$, one sees that z_0, y_0^2, x_0^2 constitute a primitive Pythagorean triple. Theorem 11-1 asserts the existence of relatively prime integers $s > t > 0$ for which

$$z_0 = 2st,$$

$$y_0^2 = s^2 - t^2,$$

$$x_0^2 = s^2 + t^2.$$

Thus it appears that

$$s^4 - t^4 = (s^2 + t^2)(s^2 - t^2) = x_0^2 y_0^2 = (x_0 y_0)^2,$$

making $s, t, x_0 y_0$ a (positive) solution to the equation $x^4 - y^4 = z^2$. Since

$$0 < s < \sqrt{s^2 + t^2} = x_0,$$

we arrive at a contradiction to the minimal nature of x_0.

For the second part of the proof, assume that y_0 is an even integer. Using the formulas for primitive Pythagorean triples, we now write

$$y_0^2 = 2st,$$

$$z_0 = s^2 - t^2,$$

$$x_0^2 = s^2 + t^2,$$

where s may be taken to be even and t to be odd. Then, in the relation $y_0^2 = 2st$, we have $\gcd(2s, t) = 1$. The by-now-customary application of Lemma 2 tells us that $2s$ and t are each squares of positive integers; say, $2s = w^2$, $t = v^2$. Since w must of necessity be an even integer, set $w = 2u$ to get $s = 2u^2$. Therefore,

$$x_0^2 = s^2 + t^2 = 4u^4 + v^4$$

and so $2u^2, v^2, x_0$ forms a primitive Pythagorean triple. Falling back on Theorem 11-1 again, there exist integers $a > b > 0$ for which

$$2u^2 = 2ab,$$

$$v^2 = a^2 - b^2,$$

$$x_0 = a^2 + b^2,$$

where $\gcd(a, b) = 1$. The equality $u^2 = ab$ ensures that a and b are perfect squares, so that $a = c^2$ and $b = d^2$. Knowing this, the rest of the proof is easy; for, upon substituting,

$$v^2 = a^2 - b^2 = c^4 - d^4.$$

The result is a new solution c, d, v of the given equation $x^4 - y^4 = z^2$ and what's more, a solution in which

$$0 < c = \sqrt{a} < a^2 + b^2 = x_0,$$

contrary to our assumption regarding x_0.

The only resolution of these contradictions is that the equation $x^4 - y^4 = z^2$ cannot be satisfied in the positive integers.

In the margin of his copy of Diophantus' *Arithmetica*, Fermat states and proves: the area of a right triangle with rational sides cannot be the square of a rational number. Clearing of fractions, this reduces to a theorem about Pythagorean triangles; to wit,

THEOREM 11-5.

The area of a Pythagorean triangle can never be equal to a perfect (integral) square.

Proof: Consider a Pythagorean triangle whose hypotenuse has length z and other two sides have lengths x and y, so that $x^2 + y^2 = z^2$. The area of the triangle in question is $\frac{1}{2} xy$ and if this were a square, say u^2, it would follow that $2xy = 4u^2$. By adding and subtracting the last-written equation from $x^2 + y^2 = z^2$, we are led to

$$(x + y)^2 = z^2 + 4u^2 \quad \text{and} \quad (x - y)^2 = z^2 - 4u^2.$$

When these last two equations are multiplied together, the outcome is that two fourth powers have as their difference a square:

$$(x^2 - y^2)^2 = z^4 - 16u^4 = z^4 - (2u)^4.$$

Since this amounts to an infringement of Theorem 11-4, there can be no Pythagorean triangle whose area is a square.

There are a number of simple problems pertaining to Pythagorean triangles that still await solution. The Corollary to Theorem 11-3 may be expressed by saying that there exists no Pythagorean triangle all the sides of which are squares. However, it is not difficult to produce Pythagorean triangles whose sides, if increased by 1, are squares; for instance, the triangles associated with the triples $13^2 - 1$, $10^2 - 1$, $14^2 - 1$, and $287^2 - 1$, $265^2 - 1$, $329^2 - 1$. An obvious—and as yet unanswered—question is whether there are an infinite number of such triangles. One can find Pythagorean triangles each side of which is a triangular number. [By a triangular number, we mean an integer of the form $t_n = n(n + 1)/2$.] An example of such is the triangle corresponding to $t_{132}, t_{143}, t_{164}$. It is not known if infinitely many Pythagorean triangles of this type exist.

As a closing comment, we should observe that all the effort expended on attempting to prove Fermat's Conjecture has been far from wasted. The new mathematics that was developed as a by-product laid the foundations for algebraic number theory, as well as the ideal theory of modern abstract algebra. It seems fair to say that the value of these far exceeds that of the conjecture itself.

PROBLEMS 11.2

1. Show that the equation $x^2 + y^2 = z^3$ has infinitely many solutions for x, y, z positive integers. [*Hint:* For any $n > 3$, let $x = n(n^2 - 3)$ and $y = 3n^2 - 1$.]

2. Prove the theorem: The only solutions in nonnegative integers of the equation $x^2 + 2y^2 = z^2$, with $\gcd(x, y, z) = 1$, are given by

$$x = \pm(2s^2 - t^2), \, y = 2st, \, z = 2s^2 + t^2$$

where s, t are arbitrary nonnegative integers. [*Hint:* If u, v, w are such that $y = 2w$, $z + x = 2u$, $z - x = 2v$, then the equation becomes $2w^2 = uv$.]

3. In a Pythagorean triple x, y, z, prove that not more than one of x, y, or z can be a perfect square.

4. Prove each of the following assertions:
   (a) The system of simultaneous equations

   $$x^2 + y^2 = z^2 - 1 \quad \text{and} \quad x^2 - y^2 = w^2 - 1$$

   has infinitely many solutions in positive integers x, y, z, w. [*Hint:* For any integer $n \geq 1$, take $x = 2n^2$ and $y = 2n$.]
   (b) The system of simultaneous equations

   $$x^2 + y^2 = z^2 \quad \text{and} \quad x^2 - y^2 = w^2$$

   admits no solution in positive integers x, y, z, w.
   (c) The system of simultaneous equations

   $$x^2 + y^2 = z^2 + 1 \quad \text{and} \quad x^2 - y^2 = w^2 + 1$$

   has infinitely many solutions in positive integers x, y, z, w. [*Hint:* For any integer $n \geq 1$, take $x = 8n^4 + 1$ and $y = 8n^3$.]

5. Use Problem 4 to establish that there is no solution in positive integers of the simultaneous equations

   $$x^2 + y^2 = z^2 \quad \text{and} \quad x^2 + 2y^2 = w^2.$$

   [*Hint:* Any solution of the given system also satisfies $z^2 + y^2 = w^2$ and $z^2 - y^2 = x^2$.]

6. Show that there is no solution in positive integers of the simultaneous equations

   $$x^2 + y^2 = z^2 \quad \text{and} \quad x^2 + z^2 = w^2;$$

   hence, there exists no Pythagorean triangle whose hypotenuse and one of whose sides form the sides of another Pythagorean triangle. [*Hint:* Any solution of the given system also satisfies $x^2 + (wy)^2 = z^4$.]

7. Prove that the equation $x^4 - y^4 = 2z^2$ has no solutions in positive integers x, y, z. [*Hint:* Since x, y must be both odd or both even, $x^2 + y^2 = 2a^2$, $x + y = 2b^2$, $x - y = 2c^2$ for some a, b, c; hence, $a^2 = b^4 + c^4$.]

8. Verify that the only solution in relatively prime positive integers of the equation $x^4 + y^4 = 2z^2$ is $x = y = z = 1$. [*Hint:* Any solution of the given equation also satisfies the equation

   $$z^4 - (xy)^4 = [(x^4 - y^4)/2]^2.]$$

9. Prove that the Diophantine equation $x^4 - 4y^4 = z^2$ has no solution in positive integers x, y, z. [*Hint:* Rewrite the given equation as $(2y^2)^2 + z^2 = (x^2)^2$ and appeal to Theorem 11-1.]

10. Use Problem 9 to prove that there exists no Pythagorean triangle whose area is twice a perfect square. [*Hint:* Assume to the contrary that $x^2 + y^2 = z^2$ and $\frac{1}{2} xy = 2w^2$. Then $(x + y)^2 = z^2 + 8w^2$, while $(x - y)^2 = z^2 - 8w^2$. This leads to $z^4 - 4(2w)^4 = (x^2 - y^2)^2$.]

11. Prove the theorem: The only solutions in positive integers of the equation

$$1/x^2 + 1/y^2 = 1/z^2, \quad \gcd(x, y, z) = 1$$

are given by

$$x = 2st(s^2 + t^2), y = s^4 - t^4, z = 2st(s^2 - t^2),$$

where s, t are relatively prime positive integers, one of which is even, with $s > t$.

12. Show that the equation $1/x^4 + 1/y^4 = 1/z^2$ has no solution in positive integers.

12

Representation of Integers as Sums of Squares

"The object of pure Physic is the unfolding of the laws of the intelligible world; the object of pure Mathematic that of unfolding the laws of human intelligence."

J. J. SYLVESTER

12.1 JOSEPH LOUIS LAGRANGE

After the deaths of Descartes, Pascal, and Fermat, no French mathematician of comparable stature appeared for over a century. In England meanwhile, mathematics was being pursued with restless zeal, first by Newton, then by Taylor, Sterling, and Maclaurin, while Leibniz came upon the scene in Germany. Mathematical activity in Switzerland was marked by the work of the Bernoullis and Euler. Towards the end of the 18th century, Paris did again become the center of mathematical studies, as Lagrange, Laplace, and Legendre brought fresh glory to France.

An Italian by birth, German by adoption, and Frenchman by choice, Joseph Louis Lagrange (1736–1813) was, next to Euler, the foremost mathematician of the 18th century. When he entered the University of Turin, his great interest was in physics, but, after chancing to read a tract by Halley on the merits of Newtonian calculus, he became excited about the new mathematics that was transforming celestial mechanics. He applied himself with such energy to mathematical studies that he was appointed, at the age of eighteen, Professor of Geometry at the Royal Artillery School in Turin. The French Academy of Sciences soon became accustomed to including Lagrange among the competitors for its biennial prizes: between 1764 and 1788, he won five of the coveted prizes for his applications of mathematics to problems in astronomy.

In 1766, when Euler left Berlin for St. Petersburg, Frederick the Great arranged for Lagrange to fill the vacated post, accompanying his invitation with a modest message which said, "It is necessary that the greatest geometer of Europe should live near the greatest of Kings." (To D'Alembert, who had suggested Lagrange's name, the King wrote, "To your care and recommendation am I indebted for having replaced a half-blind mathematician with a mathematician with both eyes, which will especially please

Joseph Louis Lagrange
(1736–1813)

(From A Concise History of Mathematics *by Dirk Struik, 1967, Dover Publications, Inc., N.Y.)*

the anatomical members of my academy.") For the next twenty years, Lagrange served as director of the mathematics section of the Berlin Academy, producing work of high distinction which culminated in his monumental treatise, the *Mécanique Analytique* (published in 1788 in four volumes). In this work he unified general mechanics and made of it, as the mathematician Hamilton was later to say, "a kind of scientific poem." Holding that mechanics was really a branch of pure mathematics, Lagrange so completely banished geometric ideas from the *Mécanique Analytique* that he could boast in the preface that not a single diagram appeared in its pages.

Frederick the Great died in 1787 and Lagrange, no longer finding a sympathetic atmosphere at the Prussian court, decided to accept the invitation of Louis XVI to settle in Paris, where he took French citizenship. But the years of constant activity had taken their toll: Lagrange fell into a deep mental depression which destroyed his interest in mathematics. So profound was his loathing for the subject that the first printed copy of the *Mécanique Analytique*—the work of a quarter century—lay unexamined on his desk for more than two years. Strange to say, it was the turmoil of the French Revolution that helped to awaken him from his lethargy. Following the abolition of all the old French universities (the Academy of Sciences was also suppressed) in 1793, the revolutionists created two new schools, with the humble titles of École Normale and École Polytechnique, and Lagrange was invited to lecture on analysis. Although he had not lectured since his early days in Turin, having been under royal patronage in the interim, he seemed to welcome the appointment. Subject to constant surveillance, the instructors were pledged "neither to read nor repeat from memory" and transcripts of their lectures as delivered were inspected by the authorities. Despite the petty harassments, Lagrange gained a reputation as an inspiring teacher. His lecture notes on differential calculus formed the basis of another classic in mathematics, the *Théorie des Fonctions Analytique* (1797).

While Lagrange's research covered an extraordinarily wide spectrum, he possessed, much like Diophantus and Fermat before him, a special talent for the theory of numbers. His work here included: the first proof of Wilson's Theorem that if n is a prime, then $(n - 1)! \equiv -1 \pmod{n}$; the investigation of the conditions under which ± 2 and ± 5 are quadratic residues or nonresidues of an odd prime (-1 and ± 3 having been discussed by Euler); finding all integral solutions of the equation $x^2 - ay^2 = 1$; and the solution of a number of problems posed by Fermat to the effect that certain primes can be represented in particular ways (typical of these is the result which asserts that every prime $p \equiv 3 \pmod 8$ is of the form $p = a^2 + 2b^2$). The present chapter focuses on the discovery for which Lagrange has acquired his greatest renown in number theory, the proof that every positive integer can be expressed as the sum of four squares.

12.2 SUMS OF TWO SQUARES

Historically, a problem which has received a good deal of attention has been that of representing numbers as sums of squares. In the present chapter, we develop enough material to settle completely the following question: What is the smallest value n such that every positive integer can be written as the sum of not more than n squares? Upon examining the first few positive integers, one finds that

$$1 = 1^2$$
$$2 = 1^2 + 1^2$$
$$3 = 1^2 + 1^2 + 1^2$$
$$4 = 2^2$$
$$5 = 2^2 + 1^2$$
$$6 = 2^2 + 1^2 + 1^2$$
$$7 = 2^2 + 1^2 + 1^2 + 1^2.$$

Since four squares are needed in the representation of 7, a partial answer to our question is that $n \geq 4$. Needless to say, there remains the possibility that some integers might require more than four squares. A justly famous theorem of Lagrange, proved in 1770, asserts that four squares are sufficient; that is, every positive integer is realizable as the sum of four squared integers, some of which may be $0 = 0^2$. This is our Theorem 12-7.

To begin with simpler things, we first find necessary and sufficient conditions that a positive integer be representable as the sum of two squares. The problem may be reduced to the consideration of primes by the lemma below.

LEMMA.

If m and n are each the sum of two squares, then so is their product mn.

Proof: If $m = a^2 + b^2$ and $n = c^2 + d^2$ for integers a, b, c, d, then

$$mn = (a^2 + b^2)(c^2 + d^2) = (ac + bd)^2 + (ad - bc)^2.$$

It is clear that not every prime can be written as the sum of two squares; for instance, $3 = a^2 + b^2$ has no solution for integral a and b. More generally, one can prove

THEOREM 12-1.

No prime p of the form $4k + 3$ is a sum of two squares.

Proof: Modulo 4, we have $a \equiv 0$, 1, 2, or 3 for any integer a; hence, $a^2 \equiv 0$ or 1 (mod 4). It follows that, for arbitrary integers a and b,

$$a^2 + b^2 \equiv 0, 1, \text{ or } 2 \pmod 4.$$

Since $p \equiv 3 \pmod 4$, the equation $p = a^2 + b^2$ is impossible.

On the other hand, any prime which is congruent to 1 modulo 4 is expressible as the sum of two squared integers. The proof, in the form we shall give it, employs a theorem on congruences due to the Norwegian mathematician Axel Thue. This, in its turn, relies on Dirichlet's famous "pigeon-hole principle":

PIGEON-HOLE PRINCIPLE.

If n objects are placed in m boxes (or pigeon-holes) and if $n > m$, then some box will contain at least two objects.

Phrased in more mathematical terms, this simple principle asserts that if a set with n elements is the union of m of its subsets and if $n > m$, then some subset has more than one element.

LEMMA (Thue's Lemma).

Let p be a prime and $\gcd(a , p) = 1$. Then the congruence

$$ax \equiv y \pmod p$$

admits a solution x_0, y_0, where

$$0 < |x_0| < \sqrt{p} \quad \text{and} \quad 0 < |y_0| < \sqrt{p}.$$

Proof: Let $k = [\sqrt{p}] + 1$ and consider the set of integers

$$S = \{ax - y \mid 0 \le x \le k - 1, 0 \le y \le k - 1\}.$$

Since $ax - y$ takes on $k^2 > p$ possible values, the Pigeon-hole Principle guarantees that at least two members of S must be congruent modulo p; call them $ax_1 - y_1$ and $ax_2 - y_2$, where $x_1 \ne x_2$ or $y_1 \ne y_2$. Then we can write

$$a(x_1 - x_2) \equiv y_1 - y_2 \pmod p.$$

Setting $x_0 = x_1 - x_2$ and $y_0 = y_1 - y_2$, it follows that x_0 and y_0 provide a solution to the congruence $ax \equiv y \pmod p$. If either x_0 or y_0 is equal to zero, then the fact that $\gcd(a , p) = 1$ can be used to show that the other must also be zero, contrary to assumption. Hence, $0 < |x_0| \le k - 1 < \sqrt{p}$ and $0 < |y_0| \le k - 1 < \sqrt{p}$.

We are now ready to derive the theorem of Fermat that every prime of the form $4k + 1$ can be expressed as the sum of squares of two integers. (In terms of priority, Albert Girard recognized this fact several years earlier and the result is sometimes referred to as Girard's Theorem.) Fermat communicated his theorem in a letter to Mersenne, dated December 25, 1640, stating that he possessed an irrefutable proof. However, the first published proof was given by Euler in 1754, who in addition succeeded in showing that the representation is unique.

THEOREM 12-2 (Fermat).

An odd prime p is expressible as a sum of two squares if and only if $p \equiv 1 \pmod 4$.

Proof: While the "only if" part is covered by Theorem 12-1, let us give a different proof here. Suppose that p can be written as the sum of two squares, let us say $p = a^2 + b^2$. Because p is a prime, we have $p \nmid a$ and $p \nmid b$. (If $p \mid a$, then $p \mid b^2$ and so $p \mid b$, leading to the contradiction that $p^2 \mid p$.) Thus, by the theory of linear congruences, there exists an integer c for which $bc \equiv 1 \pmod p$. Modulo p, the relation $(ac)^2 + (bc)^2 = pc^2$ becomes

$$(ac)^2 \equiv -1 \pmod p,$$

making -1 a quadratic residue of p. At this point, the corollary to Theorem 9-2 comes to our aid, for $(-1/p) = 1$ only when $p \equiv 1 \pmod 4$.

For the converse, assume that $p \equiv 1 \pmod 4$. Since -1 is a quadratic residue of p, we can find an integer a satisfying $a^2 \equiv -1 \pmod p$; in fact, by Theorem 5-3, $a = [(p - 1)/2]!$ is one such integer. Now $\gcd(a , p) = 1$, so that the congruence

$$ax \equiv y \pmod p$$

admits a solution x_0, y_0 for which the conclusion of Thue's Lemma holds. As a result,

$$-x_0^2 \equiv a^2 x_0^2 \equiv (ax_0)^2 \equiv y_0^2 \pmod p$$

or $x_0^2 + y_0^2 \equiv 0 \pmod p$. This says that

$$x_0^2 + y_0^2 = kp$$

for some integer $k \geq 1$. Inasmuch as $0 < |x_0| < \sqrt{p}$ and $0 < |y_0| < \sqrt{p}$, we obtain $0 < x_0^2 + y_0^2 < 2p$, the implication of which is that $k = 1$. Consequently, $x_0^2 + y_0^2 = p$, and we are finished.

Counting a^2 and $(-a)^2$ as the same, we have

COROLLARY.

Any prime p of the form $4k + 1$ can be represented uniquely (aside from the order of the summands) as a sum of two squares.

Proof: To establish the uniqueness assertion, suppose that

$$p = a^2 + b^2 = c^2 + d^2,$$

where a, b, c, d are all positive integers. Then

$$a^2 d^2 - b^2 c^2 = p(d^2 - b^2) \equiv 0 \pmod{p}$$

whence $ad \equiv bc \pmod{p}$ or $ad \equiv -bc \pmod{p}$. Since a, b, c, d are all less than $\sqrt{p}$, these relations imply that

$$ad - bc = 0 \quad \text{or} \quad ad + bc = p.$$

If the second equality holds, then we would have $ac = bd$; for,

$$p^2 = (a^2 + b^2)(c^2 + d^2) = (ad + bc)^2 + (ac - bd)^2$$
$$= p^2 + (ac - bd)^2$$

and so $ac - bd = 0$. It follows that either

$$ad = bc \quad \text{or} \quad ac = bd.$$

Suppose, for instance, that $ad = bc$. Then $a | bc$, with $\gcd(a, b) = 1$, which forces $a | c$; say, $c = ka$. The condition $ad = bc = b(ka)$ then reduces to $d = bk$. But

$$p = c^2 + d^2 = k^2(a^2 + b^2)$$

implies that $k = 1$. In this case, we get $a = c$ and $b = d$. By a similar argument, the condition $ac = bd$ leads to $a = d$ and $b = c$. What is important is that, in either event, our two representations of the prime p turn out to be identical.

Let us follow the steps in Theorem 12-2, using the prime $p = 13$. One choice for the integer a is $6! = 720$. A solution of the congruence $720x \equiv y \pmod{13}$, or rather,

$$5x \equiv y \pmod{13}$$

is obtained by considering the set

$$S = \{5x - y \mid 0 \le x, y < 4\}.$$

The elements of S are just the integers

0	5	10	15
-1	4	9	14
-2	3	8	13
-3	2	7	12

which, modulo 13, become

0	5	10	2
12	4	9	1
11	3	8	0
10	2	7	12.

Among the various possibilities, we have

$$5 \cdot 1 - 3 \equiv 2 \equiv 5 \cdot 3 - 0 \pmod{13}$$

or

$$5(1 - 3) \equiv 3 \pmod{13}.$$

Thus, we may take $x_0 = -2$ and $y_0 = 3$ to obtain

$$13 = x_0^2 + y_0^2 = 2^2 + 3^2.$$

REMARK: Some authors would claim that any prime $p \equiv 1 \pmod 4$ can be written as a sum of squares in eight ways. For with $p = 13$, we have

$$13 = 2^2 + 3^3 = 2^2 + (-3)^2 = (-2)^2 + 3^2 = (-2)^2 + (-3)^2$$
$$= 3^2 + 2^2 = 3^2 + (-2)^2 = (-3)^2 + 2^2 = (-3)^2 + (-2)^2.$$

Since these eight representations can all be obtained from any one of them by interchanging the signs of 2 and 3 or by interchanging the summands, there is "essentially" only one way of doing this. Thus, from our point of view, 13 is uniquely representable as the sum of two squares.

We have shown that every prime p such that $p \equiv 1 \pmod 4$ is expressible as the sum of two squares. But other integers also enjoy this property; for instance,

$$10 = 1^2 + 3^2.$$

The next step in our program is to characterize explicitly those positive integers which can be realized as the sum of two squares.

THEOREM 12-3.

Let the positive integer n be written as $n = N^2 m$, where m is square-free. Then n can be represented as the sum of two squares if and only if m contains no prime factor of the form $4k + 3$.

Proof: To start, suppose that m has no prime factor of the form $4k + 3$. If $m = 1$, then $n = N^2 + 0^2$ and we are through. In the case in which $m > 1$, let $m = p_1 p_2 \cdots p_r$ be the factorization of m into a product of distinct primes. Each of these primes p_i, being equal to 2 or of the form $4k + 1$, can be written as the sum of two squares. Now, the identity

$$(a^2 + b^2)(c^2 + d^2) = (ac + bd)^2 + (ad - bc)^2$$

shows the product of two (and, by induction, any finite number) integers each of which is representable as a sum of two squares is likewise so representable. Thus there exist integers x and y satisfying $m = x^2 + y^2$. We end up with

$$n = N^2 m = N^2(x^2 + y^2) = (Nx)^2 + (Ny)^2,$$

a sum of two squares.

Now for the opposite direction. Assume that n can be represented as the sum of two squares,

$$n = a^2 + b^2 = N^2 m$$

and let p be any odd prime divisor of m (without loss of generality, it may be assumed that $m > 1$). If $d = \gcd(a, b)$, then $a = rd$, $b = sd$, where $\gcd(r, s) = 1$. We get

$$d^2(r^2 + s^2) = N^2 m$$

and so, m being square-free, $d^2 \mid N^2$. But then

$$r^2 + s^2 = (N^2/d^2)m = tp$$

for some integer t, which leads to

$$r^2 + s^2 \equiv 0 \pmod{p}.$$

Now the condition $\gcd(r, s) = 1$ implies that one of r or s, say r, is relatively prime to p. Let r' satisfy the congruence

$$rr' \equiv 1 \pmod{p}.$$

When the equation $r^2 + s^2 \equiv 0 \pmod{p}$ is multiplied by $(r')^2$, we obtain

$$(sr')^2 + 1 \equiv 0 \pmod{p}$$

or, to put it differently, $(-1/p) = 1$. Since -1 is a quadratic residue of p, Theorem 9-2 ensures that $p \equiv 1 \pmod 4$. The implication of our reasoning is that there is no prime of the form $4k + 3$ which divides m.

As a corollary to the preceding analysis, we have

COROLLARY.

A positive integer n is representable as the sum of two squares if and only if each of its prime factors of the form $4k + 3$ occurs to an even power.

Example 12-1

The integer 459 cannot be written as the sum of two squares, since $459 = 3^3 \cdot 17$, with the prime 3 occurring to an odd exponent. On the other hand, $153 = 3^2 \cdot 17$ admits the representation

$$153 = 3^2(4^2 + 1^2) = 12^2 + 3^2.$$

Somewhat more complicated is the example $n = 5 \cdot 7^2 \cdot 13 \cdot 17$. In this case, we have

$$n = 7^2 \cdot 5 \cdot 13 \cdot 17 = 7^2(2^2 + 1^2)(3^2 + 2^2)(4^2 + 1^2).$$

Two applications of the identity appearing in Theorem 12-3 give

$$(3^2 + 2^2)(4^2 + 1^2) = (12 + 2)^2 + (3 - 8)^2 = 14^2 + 5^2$$

and

$$(2^2 + 1^2)(14^2 + 5^2) = (28 + 5)^2 + (10 - 14)^2 = 33^2 + 4^2.$$

When these are combined, we end up with

$$n = 7^2(33^2 + 4^2) = 231^2 + 28^2.$$

There exist certain positive integers (obviously, not primes of the form $4k + 1$) that can be represented in more than one way as the sum of two squares. The smallest is

$$25 = 4^2 + 3^2 = 5^2 + 0^2.$$

If $a \equiv b \pmod 2$, then the relation

$$ab = \left(\frac{a+b}{2}\right)^2 - \left(\frac{a-b}{2}\right)^2$$

allows us to manufacture a variety of such examples. Take $n = 153$ as an illustration; here,

$$153 = 17 \cdot 9 = \left(\frac{17+9}{2}\right)^2 - \left(\frac{17-9}{2}\right)^2 = 13^2 - 4^2$$

and

$$153 = 51 \cdot 3 = \left(\frac{51+3}{2}\right)^2 - \left(\frac{51-3}{2}\right)^2 = 27^2 - 24^2$$

so that

$$13^2 - 4^2 = 27^2 - 24^2.$$

This yields the two distinct representations

$$27^2 + 4^2 = 24^2 + 13^2 = 745.$$

At this stage, a natural question should suggest itself: What positive integers admit a representation as the difference of two squares? We answer this below.

THEOREM 12-4.

A positive integer n can be represented as the difference of two squares if and only if n is not of the form $4k + 2$.

Proof: Since $a^2 \equiv 0$ or $1 \pmod 4$ for all integers a, it follows that

$$a^2 - b^2 \equiv 0, 1, \text{ or } 3 \pmod 4.$$

Thus, if $n \equiv 2 \pmod 4$, we cannot have $n = a^2 - b^2$ for any choice of a and b.

Turning affairs around, suppose that the integer n is not of the form $4k + 2$; that is to say, $n \equiv 0, 1,$ or $3 \pmod 4$. If $n \equiv 1$ or $3 \pmod 4$, then $n + 1$ and $n - 1$ are both even integers; hence, n can be written as

$$n = \left(\frac{n+1}{2}\right)^2 - \left(\frac{n-1}{2}\right)^2,$$

a difference of squares. If $n \equiv 0 \pmod 4$, then we have

$$n = \left(\frac{n}{4} + 1\right)^2 - \left(\frac{n}{4} - 1\right)^2.$$

COROLLARY.

An odd prime is the difference of two successive squares.

Examples of this last corollary are afforded by

$$11 = 6^2 - 5^2, \ 17 = 9^2 - 8^2 \text{ and } 29 = 15^2 - 14^2.$$

Another point worth mentioning is that the representation of a given prime p as the difference of two squares is unique. To see this, suppose that

$$p = a^2 - b^2 = (a - b)(a + b),$$

where $a > b > 0$. Since 1 and p are the only factors of p, necessarily we have

$$a - b = 1 \quad \text{and} \quad a + b = p,$$

from which it may be inferred that

$$a = \frac{p + 1}{2} \quad \text{and} \quad b = \frac{p - 1}{2}.$$

Thus, any odd prime p can be written as the difference of the squares of two integers in precisely one way; namely, as

$$p = \left(\frac{p + 1}{2}\right)^2 - \left(\frac{p - 1}{2}\right)^2.$$

A different situation occurs when we pass from primes to arbitrary integers. Suppose that n is a positive integer which is neither prime nor of the form $4k + 2$. Starting with a divisor d of n, put $d' = n/d$ (it is harmless to assume that $d \geq d'$). Now if d and d' are both even, or both odd, then $(d + d')/2$ and $(d - d')/2$ are integers. Furthermore, we may write

$$n = dd' = \left(\frac{d + d'}{2}\right)^2 - \left(\frac{d - d'}{2}\right)^2.$$

By way of illustration, consider the integer $n = 24$. Here,

$$24 = 12 \cdot 2 = \left(\frac{12 + 2}{2}\right)^2 - \left(\frac{12 - 2}{2}\right)^2 = 7^2 - 5^2$$

and

$$24 = 6 \cdot 4 = \left(\frac{6 + 4}{2}\right)^2 - \left(\frac{6 - 4}{2}\right)^2 = 5^2 - 1^2,$$

giving us two representations for 24 as the difference of squares.

PROBLEMS 12.2

1. Represent each of the primes 113, 229, and 373 as a sum of two squares.

2. (a) It has been conjectured that there exist infinitely many prime numbers p such that $p = n^2 + (n + 1)^2$ for some positive integer n; for example, $5 = 1^2 + 2^2$ while $13 = 2^2 + 3^2$. Find five more of these primes.

   (b) Another conjecture is that there are infinitely many prime numbers p of the form $p = 2^2 + p_1^2$, where p_1 is a prime. Find five such primes.

3. Establish each of the following assertions:
    (a) each of the integers 2^n, where $n = 1, 2, 3, \ldots$, is a sum of two squares;
    (b) if $n \equiv 3$ or $6 \pmod 9$, then n cannot be represented as a sum of two squares;
    (c) if n is the sum of two triangular numbers, then $4n + 1$ is the sum of two squares;
    (d) every Fermat number $F_n = 2^{2^n} + 1$, where $n \geq 1$, can be expressed as the sum of two squares;
    (e) every odd perfect number (if one exists) is the sum of two squares. [*Hint:* See the Corollary to Theorem 10-7.]

4. Prove that a prime p can be written as a sum of two squares if and only if the congruence $x^2 + 1 \equiv 0 \pmod p$ admits a solution.

5. (a) Show that a positive integer n is a sum of two squares if and only if $n = 2^m a^2 b$, where $m \geq 0$, a is an odd integer, and every prime divisor of b is of the form $4k + 1$.
    (b) Write the integers $3185 = 5 \cdot 7^2 \cdot 13$; $39690 = 2 \cdot 3^4 \cdot 5 \cdot 7^2$; and $62920 = 2^3 \cdot 5 \cdot 11^2 \cdot 13$ as a sum of two squares.

6. Find a positive integer having at least three different representations as the sum of two squares, disregarding signs and the order of the summands. [*Hint:* Choose an integer which has three distinct prime factors, each of the form $4k + 1$.]

7. If the positive integer n is not the sum of squares of two integers, show that n cannot be represented as the sum of two squares of rational numbers. [*Hint:* By Theorem 12-3, there is a prime $p \equiv 3 \pmod 4$ and an odd integer k such that $p^k \mid n$, while $p^{k+1} \nmid n$. If $n = (a/b)^2 + (c/d)^2$, then p will occur to an odd power on the left-hand side of the equation $n(bd)^2 = (ad)^2 + (bc)^2$, but not on the right-hand side.]

8. Prove that the positive integer n has as many representations as the sum of two squares as does the integer $2n$. [*Hint:* Starting with a representation of n as a sum of two squares, obtain a similar representation for $2n$, and conversely.]

9. (a) If n is a triangular number, show that each of the three successive integers $8n^2$, $8n^2 + 1$, $8n^2 + 2$ can be written as a sum of two squares.
    (b) Prove that of any four consecutive integers, at least one is not representable as a sum of two squares.

10. Prove that:
    (a) if a prime number is the sum of two or four squares of different primes, then one of these primes must be equal to 2;
    (b) if a prime number is the sum of three squares of different primes, then one of these primes must be equal to 3.

11. (a) Let p be an odd prime. If $p \mid a^2 + b^2$, where $\gcd(a, b) = 1$, prove that the prime $p \equiv 1 \pmod 4$. [*Hint:* Raise the congruence $a^2 \equiv -b^2 \pmod p$ to the power $(p - 1)/2$ and apply Fermat's Theorem to conclude that $(-1)^{(p-1)/2} = 1$.]
    (b) Use part (a) to show that any positive divisor of a sum of two relatively prime squares is itself a sum of two squares.

12. Establish that every prime number p of the form $8k + 1$ or $8k + 3$ can be written as $p = a^2 + 2b^2$ for some choice of integers a and b. [*Hint:* Mimic the proof of Theorem 12-2.]

13. Prove that:
    (a) A positive integer is representable as the difference of two squares if and only if it is the product of two factors which are both even or both odd.
    (b) A positive even integer can be written as the difference of two squares if and only if it is divisible by 4.

14. Verify that 45 is the smallest positive integer admitting three distinct representations as the difference of two squares. [*Hint:* See part (a) of the previous problem.]

15. For any $n > 0$, show that there exists a positive integer which can be expressed in n distinct ways as the difference of two squares. [*Hint:* Note that

$$2^{2n+1} = (2^{2n-k} + 2^{k-1})^2 - (2^{2n-k} - 2^{k-1})^2$$

for $k = 1, 2, \ldots, n$.]

16. Prove that every prime $p \equiv 1 \pmod 4$ divides the sum of two relatively prime squares, where each square exceeds 3. [*Hint:* Given an odd primitive root r of p, $r^k \equiv 2 \pmod p$; hence $r^{2[k+(p-1)/4]} \equiv -4 \pmod p$.]

17. For a prime $p \equiv 1$ or $3 \pmod 8$, show that the equation $x^2 + 2y^2 = p$ has a solution.

18. The English number theorist G. H. Hardy relates the following story about his young protégé Ramanujan: "I remember going to see him once when he was lying ill in Putney. I had ridden in taxi-cab No. 1729, and remarked that the number seemed to me rather a dull one, and that I hoped it was not an unfavorable omen. 'No,' he reflected, 'it is a very interesting number; it is the smallest number expressible as the sum of two cubes in two different ways.' " Verify Ramanujan's assertion.

12.3 SUMS OF MORE THAN TWO SQUARES

While not every positive integer can be written as the sum of two squares, what about their representation in terms of three squares (0^2 still permitted)? With an extra square to add, it seems reasonable that there should be fewer exceptions. For instance, when only two squares are allowed, we have no representation for such integers as 14, 33, and 67, but

$$14 = 3^2 + 2^2 + 1^2, 33 = 5^2 + 2^2 + 2^2, 67 = 7^2 + 3^2 + 3^2.$$

It is still possible to find integers which are not expressible as the sum of three squares. A theorem which speaks to this point is

THEOREM 12-5.

No positive integer of the form $4^n(8m + 7)$ *can be represented as the sum of three squares.*

Proof: To start, let us show that the integer $8m + 7$ is not expressible as the sum of three squares. For any integer a, we have $a^2 \equiv 0, 1,$ or $4 \pmod 8$. It follows that

$$a^2 + b^2 + c^2 \equiv 0, 1, 2, 3, 4, 5, \text{ or } 6 \pmod 8$$

for any choice of integers a, b, c. Since we have $8m + 7 \equiv 7 \pmod 8$, the equation $a^2 + b^2 + c^2 = 8m + 7$ is impossible.

Next, let us suppose that $4^n(8m + 7)$, where $n \geq 1$, can be written as

$$4^n(8m + 7) = a^2 + b^2 + c^2.$$

Then each of the integers a, b, c must be even. Putting $a = 2a_1$, $b = 2b_1$, $c = 2c_1$, we get

$$4^{n-1}(8m + 7) = a_1^2 + b_1^2 + c_1^2.$$

If $n - 1 \geq 1$, the argument may be repeated until $8m + 7$ is eventually represented as the sum of three squared integers; this, of course, contradicts the result of the first paragraph.

One can prove that the condition of Theorem 12-5 is also sufficient in order that a positive integer be realizable as the sum of three squares; however, the argument is much too difficult for inclusion here. Part of the trouble is that, unlike the case of two (or even four) squares, there is no algebraic identity which expresses the product of sums of three squares as a sum of three squares.

With this trace of ignorance left showing, let us make a few historical remarks. Diophantus conjectured, in effect, that no number of the form $8m + 7$ is the sum of three squares, a fact easily verified by Descartes in 1638. It seems fair to credit Fermat with being the first to state in full the criterion that a number can be written as a sum of three squared integers if and only if it is not of the form $4^n(8m + 7)$, where m and n are nonnegative integers. This was proved in a complicated manner by Legendre in 1798 and more clearly (but by no means easily) by Gauss in 1801.

As just indicated, there exist positive integers which are not representable as the sum of either two or three squares (take 7 and 15, for simple examples). Things change dramatically when we turn to four squares: there are no exceptions at all!

The first explicit reference to the fact that every positive integer can be written as the sum of four squares, counting 0^2, was made by Bachet (in 1621) and he checked this conjecture for all integers up to 325. Fifteen years later, Fermat claimed that he had a proof using his favorite method of infinite descent, but, as usual, he gave no details. Both Bachet and Fermat felt that Diophantus must have known the result; the evidence is entirely conjectural: Diophantus gave necessary conditions in order that a number be the sum of two or three squares, while making no mention of a condition for a representation as a sum of four squares.

One measure of the difficulty of the problem is the fact that Euler, in spite of his brilliant achievements, wrestled with it for more than forty years without success. Nonetheless his contribution towards the eventual solution was substantial; Euler discovered the fundamental identity which allows one to express the product of two sums of four squares as such a sum, as well as the crucial result that the congruence $x^2 + y^2 + 1 \equiv 0 \pmod{p}$ is solvable for any prime p. A complete proof of the four-square conjecture was published by Lagrange in 1772, who acknowledged his indebtedness to the ideas of Euler. The next year, Euler offered a much simpler demonstration, which is essentially the version to be presented here.

It is convenient to establish two preparatory lemmas, so as not to interrupt the main argument at an awkward stage. The proof of the first contains the algebraic identity (Euler's Identity) which allows us to reduce the four-square problem to the consideration of prime numbers only.

LEMMA 1 (Euler).

If the integers m and n are each the sum of four squares, then mn is likewise so representable.

Proof: If $m = a_1{}^2 + a_2{}^2 + a_3{}^2 + a_4{}^2$ and $n = b_1{}^2 + b_2{}^2 + b_3{}^2 + b_4{}^2$ for integers a_i, b_i, then

$$mn = (a_1{}^2 + a_2{}^2 + a_3{}^2 + a_4{}^2)(b_1{}^2 + b_2{}^2 + b_3{}^2 + b_4{}^2)$$
$$= (a_1b_1 + a_2b_2 + a_3b_3 + a_4b_4)^2$$
$$+ (a_1b_2 - a_2b_1 + a_3b_4 - a_4b_3)^2$$
$$+ (a_1b_3 - a_2b_4 - a_3b_1 + a_4b_2)^2$$
$$+ (a_1b_4 + a_2b_3 - a_3b_2 - a_4b_1)^2.$$

One confirms this cumbersome identity by brute force: just multiply everything out and compare terms. The details are not suitable for the printed page.

Another basic ingredient in our development is

LEMMA 2.

If p is an odd prime, then the congruence

$$x^2 + y^2 + 1 \equiv 0 \pmod{p}$$

has a solution x_0, y_0 where $0 \le x_0 \le (p - 1)/2$ and $0 \le y_0 \le (p - 1)/2$.

Proof: The idea of the proof is to consider the following two sets:

$$S_1 = \left\{ 1 + 0^2, 1 + 1^2, 1 + 2^2, \ldots, 1 + \left(\frac{p-1}{2}\right)^2 \right\},$$

$$S_2 = \left\{ -0^2, -1^2, -2^2, \ldots, -\left(\frac{p-1}{2}\right)^2 \right\}.$$

Evidently, no two elements of the set S_1 are congruent modulo p. For if $1 + x_1{}^2 \equiv 1 + x_2{}^2 \pmod{p}$, then either $x_1 \equiv x_2 \pmod{p}$ or $x_1 \equiv -x_2 \pmod{p}$. But the latter consequence is impossible, since $0 < x_1 + x_2 < p$ (unless $x_1 = x_2 = 0$), whence $x_1 \equiv x_2 \pmod{p}$, which implies that $x_1 = x_2$. In the same vein, no two elements of S_2 are congruent modulo p.

Together S_1 and S_2 contain $2[1 + \frac{1}{2}(p - 1)] = p + 1$ integers. By the Pigeon-hole Principle, some integer in S_1 must be congruent modulo p to some integer in S_2; that is, there exist x_0, y_0 such that

$$1 + x_0{}^2 \equiv -y_0{}^2 \pmod{p},$$

where $0 \le x_0 \le (p - 1)/2$ and $0 \le y_0 \le (p - 1)/2$.

COROLLARY.

Given an odd prime p, there exists an integer $k < p$ such that kp is the sum of four squares.

Proof: According to the theorem, we can find integers x_0 and y_0,

$$0 \le x_0 < p/2, \quad 0 \le y_0 < p/2$$

such that

$$x_0^2 + y_0^2 + 1^2 + 0^2 = kp$$

for a suitable choice of k. The restrictions on the size of x_0 and y_0 imply that

$$kp = x_0^2 + y_0^2 + 1 < p^2/4 + p^2/4 + 1 < p^2$$

and so $k < p$, as asserted in the corollary.

Example 12–2

We digress for a moment to look at an example. If one takes $p = 17$, then the sets S_1 and S_2 become

$$S_1 = \{1, 2, 5, 10, 17, 26, 37, 50, 65\}$$

and

$$S_2 = \{0, -1, -4, -9, -16, -25, -36, -49, -64\}.$$

Modulo 17, the set S_1 consists of the integers 1, 2, 5, 10, 0, 9, 3, 16, 14, while those in S_2 are 0, 16, 13, 8, 1, 9, 15, 2, 4. Lemma 2 tells us that some member $1 + x^2$ of the first set is congruent to some member $-y^2$ of the second set. We have, among the various possibilities,

$$1 + 5^2 \equiv 9 \equiv -5^2 \text{ (mod 17)}$$

or $1 + 5^2 + 5^2 \equiv 0$ (mod 17). It follows that

$$3 \cdot 17 = 1^2 + 5^2 + 5^2 + 0^2$$

is a multiple of 17 written as a sum of four squares.

The last lemma is so essential to our work that it is worth pointing out another approach, this one involving the theory of quadratic residues. If $p \equiv 1$ (mod 4), we may choose x_0 to be a solution of $x^2 \equiv -1$ (mod p) (this is permissible by the corollary to Theorem 9-2) and $y_0 = 0$ to get

$$x_0^2 + y_0^2 + 1 \equiv 0 \text{ (mod } p).$$

Thus, it suffices to concentrate on the case $p \equiv 3$ (mod 4). We first pick the integer a to be the smallest positive quadratic nonresidue of p (keep in mind that $a \geq 2$, since 1 is a quadratic residue). Then

$$(-a/p) = (-1/p)(a/p) = (-1)(-1) = 1,$$

so that $-a$ is a quadratic residue of p. Hence, the congruence

$$x^2 \equiv -a \text{ (mod } p)$$

admits a solution x_0, with $0 < x_0 \leq (p - 1)/2$. Now $a - 1$, being positive and smaller than a, must itself be a quadratic residue of p. Thus, there exists an integer y_0, where $0 < y_0 \leq (p - 1)/2$, satisfying

$$y^2 \equiv a - 1 \text{ (mod } p).$$

The conclusion:

$$x_0^2 + y_0^2 + 1 \equiv -a + (a - 1) + 1 \equiv 0 \ (\text{mod } p).$$

With these two lemmas among our tools, we now have the necessary information to carry out a proof of the fact that any prime can be realized as the sum of four squared integers.

THEOREM 12-6.

Any prime p can be written as the sum of four squares.

Proof: The theorem is certainly true for $p = 2$, since $2 = 1^2 + 1^2 + 0^2 + 0^2$. Thus, we may hereafter restrict our attention to odd primes. Let k be the smallest positive integer such that kp is the sum of four squares; say,

$$kp = x^2 + y^2 + z^2 + w^2.$$

By virtue of the foregoing corollary, $k < p$. The crux of our argument is that $k = 1$.

We make a start by showing that k is an odd integer. For a proof by contradiction, assume that k is even. Then x, y, z, w are all even; or all are odd; or two are even and two are odd. In any event, we may rearrange them, so that

$$x \equiv y \ (\text{mod } 2) \quad \text{and} \quad z \equiv w \ (\text{mod } 2).$$

It follows that

$$\tfrac{1}{2}(x - y), \tfrac{1}{2}(x + y), \tfrac{1}{2}(z - w), \tfrac{1}{2}(z + w)$$

are all integers and

$$\tfrac{1}{2}(kp) = \left(\frac{x - y}{2}\right)^2 + \left(\frac{x + y}{2}\right)^2 + \left(\frac{z - w}{2}\right)^2 + \left(\frac{z + w}{2}\right)^2$$

is a representation of $(k/2)p$ as a sum of four squares. This violates the minimal nature of k, giving us our contradiction.

There still remains the problem of showing that $k = 1$. Assume $k \neq 1$; then k, being an odd integer, is at least 3. It is therefore possible to choose integers a, b, c, d such that

$$a \equiv x \ (\text{mod } k), \ b \equiv y \ (\text{mod } k), \ c \equiv z \ (\text{mod } k), \ d \equiv w \ (\text{mod } k)$$

and

$$|a| < k/2, |b| < k/2, |c| < k/2, |d| < k/2.$$

(To obtain the integer a, for instance, find the remainder r when x is divided by k; put $a = r$ or $a = r - k$ according as $r < k/2$ or $r > k/2$.) Then

$$a^2 + b^2 + c^2 + d^2 \equiv x^2 + y^2 + z^2 + w^2 \equiv 0 \ (\text{mod } k)$$

and so

$$a^2 + b^2 + c^2 + d^2 = nk$$

for some nonnegative integer n. Because of the restrictions on the size of $a, b, c, d,$

$$0 \le nk = a^2 + b^2 + c^2 + d^2 < 4(k/2)^2 = k^2.$$

We cannot have $n = 0$, since this would signify that $a = b = c = d = 0$ and, in consequence, that k divides each of the integers x, y, z, w. Then $k^2 \mid kp$, or $k \mid p$, which is impossible in light of the inequality $1 < k < p$. The relation $nk < k^2$ also allows us to conclude that $n < k$. In sum: $0 < n < k$. Combining the various pieces, we get

$$k^2 np = (kp)(kn) = (x^2 + y^2 + z^2 + w^2)(a^2 + b^2 + c^2 + d^2)$$
$$= r^2 + s^2 + t^2 + u^2,$$

where $r = xa + yb + zc + wd,$

$\quad\quad s = xb - ya + zd - wc,$

$\quad\quad t = xc - yd - za + wb,$

$\quad\quad u = xd + yc - zb - wa.$

It is important to observe that all four of r, s, t, u are divisible by k. In the case of the integer r, for example, one has

$$r = xa + yb + zc + wd \equiv a^2 + b^2 + c^2 + d^2 \equiv 0 \pmod{k}.$$

Similarly, $s \equiv t \equiv u \equiv 0 \pmod{k}$. This leads to the representation

$$np = (r/k)^2 + (s/k)^2 + (t/k)^2 + (u/k)^2,$$

where $r/k, s/k, t/k, u/k$ are all integers. Since $0 < n < k$, we therefore arrive at a contradiction to the choice of k as the smallest positive integer for which kp is the sum of four squares. With this contradiction, $k = 1$, and the proof is finally complete.

This brings us to our ultimate objective, the classical result of Lagrange:

THEOREM 12-7 (Lagrange).
> *Any positive integer n can be written as the sum of four squares, some of which may be zero.*

Proof: Clearly, the integer 1 is expressible as $1 = 1^2 + 0^2 + 0^2 + 0^2$, a sum of four squares. Assume that $n > 1$ and let $n = p_1 p_2 \cdots p_r$ be the factorization of n into (not necessarily distinct) primes. Since each p_i is realizable as a sum of four squares, Euler's Identity permits us to express the product of any two primes as a sum of four squares. This, by induction, extends to any finite number of prime factors, so that applying the identity r times, we obtain the desired representation for n.

Example 12-3

To write the integer $459 = 3^3 \cdot 17$ as the sum of four squares, we use Euler's Identity as follows:

$$459 = 3^2 \cdot 3 \cdot 17$$
$$= 3^2(1^2 + 1^2 + 1^2 + 0^2)\,(4^2 + 1^2 + 0^2 + 0^2)$$
$$= 3^2[(4 + 1 + 0 + 0)^2 + (1 - 4 + 0 - 0)^2$$
$$+ (0 - 0 - 4 + 0)^2 + (0 + 0 - 1 - 0)^2]$$
$$= 3^2[5^2 + 3^2 + 4^2 + 1^2]$$
$$= 15^2 + 9^2 + 12^2 + 3^2.$$

While squares have received all our attention so far, many of the ideas involved generalize to higher powers.

In his book, *Meditationes Algebraicae* (1770), Edward Waring stated that each positive integer is expressible as a sum of at most 9 cubes, also a sum of at most 19 fourth powers, and so on. This assertion has been interpreted to mean: Can each positive integer be written as the sum of no more than a fixed number $g(k)$ of kth powers, where $g(k)$ depends only on k, not the integer being represented? In other words, for a given k, a number $g(k)$ is sought such that every $n > 0$ can be represented in at least one way as

$$n = a_1{}^k + a_2{}^k + \cdots + a_{g(k)}{}^k$$

where the a_i are nonnegative integers, not necessarily distinct. The resulting problem was the starting point of a large body of research in number theory on what has become known as "Waring's Problem." There seems little doubt that Waring had limited numerical grounds in favor of his assertion and no shadow of a proof.

As we have reported in Theorem 12-7, $g(2) = 4$. Except for squares, the first case of a Waring-type theorem actually proved is attributed to Liouville (1859): every positive integer is a sum of at most 53 fourth powers. This bound for $g(4)$ is somewhat inflated, and through the years was progressively reduced. The existence of $g(k)$ for each value of k was resolved in the affirmative by Hilbert in 1909; unfortunately, his proof relies on heavy machinery (including a 25-fold integral at one stage) and is in no way constructive.

Once it is known that Waring's Problem admits a solution, a natural question to pose is "How big is $g(k)$?" There is an extensive literature on this aspect of the problem, but the question itself is still open. A sample result, due to Leonard Dickson, is that $g(3) = 9$, while

$$23 = 2^3 + 2^3 + 1^3 + 1^3 + 1^3 + 1^3 + 1^3 + 1^3 + 1^3$$

and

$$239 = 4^3 + 4^3 + 3^3 + 3^3 + 3^3 + 3^3 + 1^3 + 1^3 + 1^3$$

are the only integers that actually require so many as 9 cubes in their representation; each integer greater than 239 can be realized as the sum of at most 8 cubes. In 1942,

Linnik proved that only a finite number of integers need 8 cubes; from some point onwards 7 will suffice. Whether 6 cubes are also sufficient to obtain all but finitely many positive integers is still unsettled.

The cases $k = 4$ and $k = 5$ have turned out to be the most subtle. For many years, the best known result was that $g(4)$ lay somewhere in the range $19 \leq g(4) \leq 35$, while $g(5)$ satisfied $37 \leq g(5) \leq 54$. Subsequent work (1964) has shown that $g(5) = 37$. The upper bound on $g(4)$ has decreased dramatically in the past decade, the sharpest estimate being $g(4) \leq 20$. It has also been proved that every integer less than 10^{140} or greater than 10^{367} can be written as a sum of at most 19 fourth powers; thus, in principle, $g(4)$ can be calculated. The recent (1986) announcement that in fact 19 fourth powers suffice to represent all integers appears to settle this case completely. As far as $k \geq 6$ is concerned, it has been established that the formula

$$g(k) = [(3/2)^k] + 2^k - 2$$

holds, except possibly for a finite number of values of k. There is considerable evidence to suggest that this expression is correct for all k.

For $k \geq 3$, all sufficiently large integers require fewer than $g(k)$ kth powers in their representations. This suggests a general definition: let $G(k)$ denote the smallest integer r with the property that every sufficiently large integer is the sum of at most r kth powers. Clearly, $G(k) \leq g(k)$. Exact values of $G(k)$ are known only in two cases; namely, $G(2) = 4$ and $G(4) = 16$. Linnik's result on cubes indicates that $G(3) \leq 7$, while as far back as 1851 Jacobi conjectured that $G(3) \leq 5$. In recent years the bounds $G(5) \leq 18$ and $G(6) \leq 28$ have been established.

Another problem that has attracted considerable attention is whether an nth power can be written as a sum of n nth powers, with $n > 3$. The first progress was made in 1911 with the discovery of the smallest solution in 4th powers,

$$353^4 = 30^4 + 120^4 + 272^4 + 315^4.$$

In the 5th powers, the smallest solution is

$$72^5 = 19^5 + 43^5 + 46^5 + 47^5 + 67^5.$$

However, for 6th or higher powers no solution is yet known.

There is a related question; it may be asked, "Can an nth power ever be the sum of less than n nth powers?" Euler conjectured that this is impossible, but in 1968 Lander and Parkin came across the representation

$$144^5 = 27^5 + 84^5 + 110^5 + 133^5.$$

With the subsequent increase in computer power and sophistication, N. Elkies was able to show (1987) that for fourth powers there are infinitely many counterexamples to Euler's conjecture. The one with the smallest values is

$$422481^4 = 95800^4 + 217519^4 + 414560^4.$$

PROBLEMS 12.3

1. Without actually adding the squares, confirm that the following relations hold:
   (a) $1^2 + 2^2 + 3^2 + \cdots + 23^2 + 24^2 = 70^2$;
   (b) $18^2 + 19^2 + 20^2 + \cdots + 27^2 + 28^2 = 77^2$;
   (c) $2^2 + 5^2 + 8^2 + \cdots + 23^2 + 26^2 = 48^2$;
   (d) $6^2 + 12^2 + 18^2 + \cdots + 42^2 + 48^2 = 95^2 - 41^2$.

2. Regiomontanus proposed the problem of finding twenty squares whose sum is a square greater than 300,000. Furnish two solutions. [*Hint:* Consider the identity

$$(a_1^2 + a_2^2 + \cdots + a_n^2)^2 = (a_1^2 + a_2^2 + \cdots + a_{n-1}^2 - a_n^2)^2$$
$$+ (2a_1a_n)^2 + (2a_2a_n)^2 + \cdots + (2a_{n-1}a_n)^2.]$$

3. Show that $n^2 + (n + 1)^2 + (n + 2)^2 + \cdots + (n + k)^2$ is not equal to a square whenever $1^2 + 2^2 + 3^2 + \cdots + k^2$ is a quadratic nonresidue of $k + 1$.

4. Establish that the equation $a^2 + b^2 + c^2 + a + b + c = 1$ has no solution in the integers. [*Hint:* The equation in question is equivalent to the equation $(2a + 1)^2 + (2b + 1)^2 + (2c + 1)^2 = 7$.]

5. For a given positive integer n, show that either n or $2n$ is a sum of three squares.

6. An unanswered question is whether there exist infinitely many prime numbers p such that $p = n^2 + (n + 1)^2 + (n + 2)^2$, for some $n > 0$. Find three of these primes.

7. In our examination of $n = 459$, no representation as a sum of two squares was found. Express 459 as a sum of three squares.

8. Verify each of the statements below:
   (a) Every positive odd integer is of the form $a^2 + b^2 + 2c^2$, where a, b, c are integers. [*Hint:* Given $n > 0$, $4n + 2$ can be written as $4n + 2 = x^2 + y^2 + z^2$, with x and y odd and z even. Then

$$2n + 1 = \left(\frac{x + y}{2}\right)^2 + \left(\frac{x - y}{2}\right)^2 + 2(z/2)^2.]$$

   (b) Every positive integer is either of the form $a^2 + b^2 + c^2$ or $a^2 + b^2 + 2c^2$, where a, b, c are integers. [*Hint:* If $n > 0$ cannot be written as a sum $a^2 + b^2 + c^2$, then it is of the form $4^m(8k + 7)$. Apply part (a) to the odd integer $8k + 7$.]
   (c) Every positive integer is of the form $a^2 + b^2 - c^2$, where a, b, c are integers. [*Hint:* Given $n > 0$, choose a such that $n - a^2$ is a positive odd integer and use Theorem 12-4.]

9. Establish the following:
   (a) No integer of the form $9k + 4$ or $9k + 5$ can be the sum of three or fewer cubes. [*Hint:* Notice that $a^3 \equiv 0, 1,$ or $8 \pmod 9$ for any integer a.]
   (b) The only prime p which is representable as the sum of two cubes is $p = 2$. [*Hint:* Use the identity

$$a^3 + b^3 = (a + b)((a - b)^2 + ab).]$$

   (c) A prime p can be represented as the difference of two cubes if and only if it is of the form $p = 3k(k + 1) + 1$, for some k.

10. Express each of the primes 7, 19, 37, 61, and 127 as the difference of two cubes.

11. Prove that every positive integer can be represented as a sum of three or fewer triangular numbers. [*Hint:* Given $n > 0$, express $8n + 3$ as a sum of three odd squares and then solve for n.]

12. Show that there are infinitely many primes p of the form $p = a^2 + b^2 + c^2 + 1$, where a, b, c are integers. [*Hint:* By Theorem 9-8, there are infinitely many primes of the form $p = 8k + 7$. Write $p - 1 = 8k + 6 = a^2 + b^2 + c^2$ for some a, b, c.]

13. Express the integers $231 = 3 \cdot 7 \cdot 11$, $391 = 17 \cdot 23$, and $2109 = 37 \cdot 57$ as sums of four squares.

14. (a) Prove that every integer $n \geq 170$ is a sum of five squares, none of which are equal to zero. [*Hint:* Write $n - 169 = a^2 + b^2 + c^2 + d^2$ for some integers a, b, c, d and consider the cases in which one or more of a, b, c is zero.]

    (b) Prove that any positive multiple of 8 is a sum of eight odd squares. [*Hint:* Assuming $n = a^2 + b^2 + c^2 + d^2$, then $8n + 8$ is the sum of the squares of $2a \pm 1$, $2b \pm 1$, $2c \pm 1$, and $2d \pm 1$.]

15. From the fact that $n^3 \equiv n \pmod 6$ conclude that every integer n can be represented as the sum of the cubes of five integers, allowing negative cubes. [*Hint:* Utilize the identity

$$n^3 - 6k = n^3 - (k + 1)^3 - (k - 1)^3 + k^3 + k^3.]$$

16. Prove that every odd integer is the sum of four squares, two of which are consecutive. [*Hint:* For $n > 0$, $4n + 1$ is a sum of three squares, only one being odd; notice that $4n + 1 = (2a)^2 + (2b)^2 + (2c + 1)^2$ gives $2n + 1 = (a + b)^2 + (a - b)^2 + c^2 + (c + 1)^2.]$

17. Prove that there are infinitely many triangular numbers which are simultaneously expressible as the sum of two cubes and the difference of two cubes. Exhibit the representations for one such triangular number. [*Hint:* In the identity

$$(27k^6)^2 - 1 = (9k^4 - 3k)^3 + (9k^3 - 1)^3$$
$$= (9k^4 + 3k)^3 - (9k^3 + 1)^3,$$

take k to be an odd integer to get

$$(2n + 1)^2 - 1 = (2a)^3 + (2b)^3 = (2c)^3 - (2d)^3,$$

or equivalently, $t_n = a^3 + b^3 = c^3 - d^3.]$

18. (a) If $n - 1$ and $n + 1$ are both primes, establish that the integer $2n^2 + 2$ can be represented as the sum of 2, 3, 4, and 5 squares.

    (b) Illustrate the result of part (a) in the cases in which $n = 4, 6,$ and 12.

13

Fibonacci Numbers

". . . what is physical is subject to the laws of mathematics, and what is spiritual to the laws of God, and the laws of mathematics are but the expression of the thoughts of God."
THOMAS HILL

13.1 THE FIBONACCI SEQUENCE

Perhaps the greatest mathematician of the Middle Ages was Leonardo of Pisa, who wrote under the name of Fibonacci—a contraction of "filius Bonacci," that is, son of Bonacci. The Hindu-Arabic numeral system became known to Western Europe through his work *Liber Abaci* which was written in 1202, but survives only in a revised 1228 edition (the word "Abaci" in the title does not refer to the abacus; rather it means computation in general). It is ironic that despite his many achievements Fibonacci is remembered today mainly because the 19th century number theorist Edouard Lucas attached his name to a sequence that appears in a trivial problem in the *Liber Abaci*. Specifically, Fibonacci posed the following problem dealing with the number of offspring generated by a pair of rabbits conjured up in the imagination:

> A man put one pair of rabbits in a certain place entirely surrounded by a wall. How many pairs of rabbits can be produced from that pair in a year, if the nature of these rabbits is such that every month each pair bears a new pair which from the second month on becomes productive?

Assuming that none of the rabbits dies, then a pair is born during the first month, so that there are two pairs present. During the second month, the original pair has produced another pair. One month later, both the original pair and the firstborn pair have produced new pairs, so that three adult and two young pairs are present, and so on. (The figures are tabulated in the chart on page 264.) The point to bear in mind is that each month the young pairs grow up and become adult pairs, making the new "adult" entry the previous one plus the previous "young" entry. Each of the pairs that was adult last month produces one young pair, so that the new "young" entry is equal to the previous "adult" entry.

When continued indefinitely, the sequence encountered in the rabbit problem

$$1, 1, 2, 3, 5, 8, 13, 21, 34, 55, 89, 144, 233, 377, \ldots$$

is called the *Fibonacci sequence* and its terms the *Fibonacci numbers*. The position of each number in this sequence is traditionally indicated by a subscript, so that $u_1 = 1$, $u_2 = 1$, $u_3 = 2$, and so forth, with u_n denoting the nth Fibonacci number.

Growth of Rabbit Colony

Months	Adult Pairs	Young Pairs	Total
1	1	1	2
2	2	1	3
3	3	2	5
4	5	3	8
5	8	5	13
6	13	8	21
7	21	13	34
8	34	21	55
9	55	34	89
10	89	55	144
11	144	89	233
12	233	144	377

The Fibonacci sequence exhibits an intriguing property, namely,

$$
\begin{array}{llll}
2 = 1 + 1 & \text{or} & u_3 = u_2 + u_1, \\
3 = 2 + 1 & \text{or} & u_4 = u_3 + u_2, \\
5 = 3 + 2 & \text{or} & u_5 = u_4 + u_3, \\
8 = 5 + 3 & \text{or} & u_6 = u_5 + u_4.
\end{array}
$$

By this time, the general rule of formulation should be discernible:

$$u_1 = u_2 = 1, \quad u_n = u_{n-1} + u_{n-2} \qquad \text{for } n \geq 3.$$

That is, each term in the sequence (after the second) is the sum of the two that immediately precede it. Such sequences, in which from a certain point on every term can be represented as a linear combination of preceding terms, are said to be *recursive sequences*. The Fibonacci sequence is the first known recursive sequence in mathematical work. Fibonacci himself was probably aware of the recursive nature of his sequence, but it was not until 1634—by which time mathematical notation had made sufficient progress—that Albert Girard wrote down the formula.

It may not have escaped attention that in the portion of the Fibonacci sequence which we have written down, successive terms are relatively prime. This is no accident, as is now proved.

THEOREM 13-1.

For the Fibonacci sequence, $\gcd(u_n, u_{n+1}) = 1$ for every $n \geq 1$.

Proof: Let us suppose that the integer $d > 1$ divides both u_n and u_{n+1}. Then their difference $u_{n+1} - u_n = u_{n-1}$ will also be divisible by d. From this and from the relation $u_n - u_{n-1} = u_{n-2}$, it may be concluded that $d \mid u_{n-2}$. Working backward, the same argument shows that $d \mid u_{n-3}$, $d \mid u_{n-4}$, . . . , and finally that $d \mid u_1$. But $u_1 = 1$, which is certainly not divisible by any $d > 1$. This contradiction ends our proof.

Since $u_3 = 2$, $u_5 = 5$, $u_7 = 13$, and $u_{11} = 89$ are all prime numbers, one might be tempted to guess that u_n is prime whenever the subscript $n > 2$ is a prime. This conjecture fails at an early stage, for a little figuring indicates that

$$u_{19} = 4181 = 37 \cdot 113.$$

Not only is there no known device for predicting which u_n are prime, but it is not even certain whether the number of prime Fibonacci numbers is infinite. There is nonetheless a useful positive result whose cumbersome proof is omitted: For any prime p, there are infinitely many Fibonacci numbers which are divisible by p and these are all equally spaced in the Fibonacci sequence. To illustrate, 3 divides every fourth term of the Fibonacci sequence, 5 divides every fifth term, while 7 divides every eighth term.

With the exception of u_1, u_2, u_6 and u_{12}, each Fibonacci number has a "new" prime factor; that is, a prime factor which does not occur in any Fibonacci number with a smaller subscript. For example, 19 divides $u_{14} = 377 = 13 \cdot 19$, but divides no earlier Fibonacci number. We might also add that, to date, the largest Fibonacci number recognized to be prime is u_{2971}.

As we know, the greatest common divisor of two positive integers can be found from the Euclidean Algorithm after finitely many divisions. By suitably choosing the integers, the number of divisions required can be made arbitrarily large. The precise statement is this: Given $n > 0$, there exist positive integers a and b such that in order to calculate $\gcd(a, b)$ by means of the Euclidean Algorithm exactly n divisions are needed. To verify the contention, it is enough to let $a = u_{n+2}$ and $b = u_{n+1}$. The Euclidean Algorithm for obtaining $\gcd(u_{n+2}, u_{n+1})$ leads to the system of equations

$$u_{n+2} = 1 \cdot u_{n+1} + u_n,$$

$$u_{n+1} = 1 \cdot u_n + u_{n-1},$$

$$\vdots$$

$$u_4 = 1 \cdot u_3 + u_2,$$

$$u_3 = 2 \cdot u_2 + 0.$$

Evidently, the number of divisions necessary here is n. The reader will no doubt recall that the last nonzero remainder appearing in the algorithm furnishes the value of $\gcd(u_{n+2}, u_{n+1})$. Hence,

$$\gcd(u_{n+2}, u_{n+1}) = u_2 = 1,$$

which confirms anew that successive Fibonacci numbers are relatively prime.

Suppose, for instance, that $n = 6$. The following calculations show that one needs 6 divisions in order to find the greatest common divisor of the integers $u_8 = 21$ and $u_7 = 13$:

$$21 = 1 \cdot 13 + 8,$$
$$13 = 1 \cdot 8 + 5,$$
$$8 = 1 \cdot 5 + 3,$$
$$5 = 1 \cdot 3 + 2,$$
$$3 = 1 \cdot 2 + 1,$$
$$2 = 2 \cdot 1 + 0.$$

One of the striking features of the Fibonacci sequence is that the greatest common divisor of two Fibonacci numbers is itself a Fibonacci number. The identity

(1) $$u_{m+n} = u_{m-1} u_n + u_m u_{n+1}$$

is central to bringing out this fact. For fixed m, this identity is established by induction on n. When $n = 1$, (1) takes the form

$$u_{m+1} = u_{m-1} u_1 + u_m u_2 = u_{m-1} + u_m$$

which is obviously true. Let us therefore assume that the formula in question holds when n is one of the integers $1, 2, \ldots, k$ and try to verify it when $n = k + 1$. By the induction assumption,

$$u_{m+k} = u_{m-1} u_k + u_m u_{k+1},$$
$$u_{m+(k-1)} = u_{m-1} u_{k-1} + u_m u_k.$$

Addition of these two equations gives us

$$u_{m+k} + u_{m+(k-1)} = u_{m-1}(u_k + u_{k-1}) + u_m(u_{k+1} + u_k).$$

By the way in which the Fibonacci numbers are defined, this expression is the same as

$$u_{m+(k+1)} = u_{m-1} u_{k+1} + u_m u_{k+2},$$

which is precisely formula (1) with n replaced by $k + 1$. The induction step is thus complete and (1) holds for all m and n.

One example of formula (1) should suffice:

$$u_9 = u_{6+3} = u_5 u_3 + u_6 u_4 = 5 \cdot 2 + 8 \cdot 3 = 34.$$

The next theorem, aside from its importance to the ultimate result which we seek, has an interest all its own.

THEOREM 13-2.

For $m \geq 1$, $n \geq 1$, u_{mn} is divisible by u_m.

Proof: We again argue by induction on n, the result being certainly true when $n = 1$. For our induction hypothesis, let us assume that u_{mn} is divisible by u_m for $n = 1, 2, \ldots, k$. The transition to the case $u_{m(k+1)} = u_{mk+m}$ is realized using formula (1); indeed,

$$u_{m(k+1)} = u_{mk-1} u_m + u_{mk} u_{m+1}.$$

Since u_m divides u_{mk} by supposition, the right-hand side of this expression (and hence, the left-hand side) must be divisible by u_m. Accordingly, $u_m \mid u_{m(k+1)}$, which was to be proved.

Preparatory to evaluating $\gcd(u_m, u_n)$, we dispose of a technical lemma.

LEMMA.

If $m = qn + r$, then $\gcd(u_m, u_n) = \gcd(u_r, u_n)$.

Proof: To begin with, formula (1) allows us to write

$$\gcd(u_m, u_n) = \gcd(u_{qn+r}, u_n)$$
$$= \gcd(u_{qn-1} u_r + u_{qn} u_{r+1}, u_n).$$

An appeal to Theorem 13-2 and the fact that $\gcd(a + c, b) = \gcd(a, b)$, whenever $b \mid c$, gives

$$\gcd(u_{qn-1} u_r + u_{qn} u_{r+1}, u_n) = \gcd(u_{qn-1} u_r, u_n).$$

Our claim is that $\gcd(u_{qn-1}, u_n) = 1$. To see this, set $d = \gcd(u_{qn-1}, u_n)$. The relations $d \mid u_n$ and $u_n \mid u_{qn}$ imply that $d \mid u_{qn}$, and so d is a (positive) common divisor of the successive Fibonacci numbers u_{qn-1} and u_{qn}. Since successive Fibonacci numbers are relatively prime, the effect of this is that $d = 1$.

To finish the proof, the reader is left the task of showing that whenever $\gcd(a, c) = 1$, then $\gcd(a, bc) = \gcd(a, b)$. Knowing this, we can immediately pass on to

$$\gcd(u_m, u_n) = \gcd(u_{qn-1} u_r, u_n) = \gcd(u_r, u_n),$$

the desired equality.

This lemma leaves us in the happy position in which all that is required is to put the pieces together.

THEOREM 13-3.

The greatest common divisor of two Fibonacci numbers is again a Fibonacci number; specifically,

$$\gcd(u_m, u_n) = u_d, \text{ where } d = \gcd(m, n).$$

Proof: Assume that $m \geq n$. Applying the Euclidean Algorithm to m and n, we get the following system of equations:

$$m = q_1 n + r_1, \qquad\qquad 0 < r_1 < n$$

$$n = q_2 r_1 + r_2, \qquad\qquad 0 < r_2 < r_1$$

$$r_1 = q_3 r_2 + r_3, \qquad\qquad 0 < r_3 < r_2$$

$$\vdots$$

$$r_{n-2} = q_n r_{n-1} + r_n, \qquad\qquad 0 < r_n < r_{n-1}$$

$$r_{n-1} = q_{n+1} r_n + 0.$$

In accordance with the previous lemma,

$$\gcd(u_m, u_n) = \gcd(u_{r_1}, u_n) = \gcd(u_{r_1}, u_{r_2}) = \cdots = \gcd(u_{r_{n-1}}, u_{r_n}).$$

Since $r_n \mid r_{n-1}$, Theorem 13-2 tells us that $u_{r_n} \mid u_{r_{n-1}}$, whence $\gcd(u_{r_{n-1}}, u_{r_n}) = u_{r_n}$. But r_n, being the last nonzero remainder in the Euclidean Algorithm for m and n, is equal to $\gcd(m, n)$. Tying up the loose ends, we get

$$\gcd(u_m, u_n) = u_{\gcd(m, n)}$$

and in this way the theorem is established.

It is interesting to note that the converse of Theorem 13-2 can be obtained from the theorem just proved; in other words, if u_n is divisible by u_m, then we can conclude that n is divisible by m. Indeed, if $u_m \mid u_n$, then $\gcd(u_m, u_n) = u_m$. But according to Theorem 13-3, the value of $\gcd(u_m, u_n)$ must be equal to $u_{\gcd(m, n)}$. The implication of all this is that $\gcd(m, n) = m$, from which it follows that $m \mid n$. We summarize these remarks in:

COROLLARY.

In the Fibonacci sequence, $u_m \mid u_n$ if and only if $m \mid n$ for $n \geq m \geq 2$.

A good illustration of Theorem 13-3 is provided by calculating $\gcd(u_{16}, u_{12}) = \gcd(987, 144)$. From the Euclidean Algorithm,

$$987 = 6 \cdot 144 + 123,$$

$$144 = 1 \cdot 123 + 21,$$

$$123 = 5 \cdot 21 + 18,$$

$$21 = 1 \cdot 18 + 3,$$

$$18 = 6 \cdot 3 + 0,$$

and so $\gcd(987, 144) = 3$. The net result is that

$$\gcd(u_{16}, u_{12}) = 3 = u_4 = u_{\gcd(16, 12)},$$

as asserted by Theorem 13-3.

PROBLEMS 13.1

1. Given any prime $p \neq 5$, it is known that either u_{p-1} or u_{p+1} is divisible by p. Confirm this in the cases of the primes 7, 11, 13, and 17.

2. For $n = 1, 2, \ldots, 10$, show that $5u_n^2 + 4(-1)^n$ is always a perfect square.

3. Prove that if $2 \mid u_n$, then $4 \mid (u_{n+1}^2 - u_{n-1}^2)$; and correspondingly, if $3 \mid u_n$, then $9 \mid (u_{n+1}^3 - u_{n-1}^3)$.

4. For the Fibonacci sequence, establish that
    (a) $u_{n+3} \equiv u_n \pmod 2$, hence $u_3, u_6, u_9, \ldots$ are all even integers;
    (b) $u_{n+5} \equiv 3u_n \pmod 5$, hence $u_5, u_{10}, u_{15}, \ldots$ are all divisible by 5.

5. Show that the sum of the squares of the first n Fibonacci numbers is given by the formula

$$u_1^2 + u_2^2 + u_3^2 + \cdots + u_n^2 = u_n u_{n+1}.$$

[*Hint:* For $n \geq 2$, $u_n^2 = u_n u_{n+1} - u_n u_{n-1}$.]

6. Utilize the identity in Problem 5 to prove that for $n \geq 3$

$$u_{n+1}^2 = u_n^2 + 3u_{n-1}^2 + 2(u_{n-2}^2 + u_{n-3}^2 + \cdots + u_2^2 + u_1^2).$$

7. Evaluate $\gcd(u_9, u_{12})$, $\gcd(u_{15}, u_{20})$, and $\gcd(u_{24}, u_{36})$.

8. Find the Fibonacci numbers which divide both u_{24} and u_{36}.

9. Use the fact that $u_m \mid u_n$ if and only if $m \mid n$ to verify each of the assertions below:
    (a) $2 \mid u_n$ if and only if $3 \mid n$;
    (b) $3 \mid u_n$ if and only if $4 \mid n$;
    (c) $4 \mid u_n$ if and only if $6 \mid n$;
    (d) $5 \mid u_n$ if and only if $5 \mid n$.

10. If $\gcd(m, n) = 1$, prove that $u_m u_n$ divides u_{mn} for all $m, n \geq 1$.

11. It can be shown that if u_n is divided by u_m $(n > m)$, then either the remainder r is a Fibonacci number or else $u_m - r$ is a Fibonacci number. Give examples illustrating both cases.

12. It is conjectured that there are only five Fibonacci numbers which are also triangular numbers. Find them.

13. For $n \geq 1$, prove that $2^{n-1} u_n \equiv n \pmod 5$. [*Hint:* Use induction and the fact that $2^n u_{n+1} = 2(2^{n-1}u_n) + 4(2^{n-2}u_{n-1})$.]

14. If $u_n < a < u_{n+1} < b < u_{n+2}$ for some $n \geq 3$, establish that the sum $a + b$ cannot be a Fibonacci number.

15. Prove that there is no positive integer n for which

$$u_1 + u_2 + u_3 + \cdots + u_{3n} = 16!.$$

[*Hint:* By Wilson's Theorem, the equation is equivalent to $u_{3n+2} \equiv 0 \pmod{17}$. Since $17 \mid u_9$, $17 \mid u_m$ if and only if $9 \mid m$.]

16. If 3 divides $n + m$, show that $u_{n-m-1}u_n + u_{n-m}u_{n+1}$ is an even integer.

17. For $n \geq 1$, verify that there exist n consecutive composite Fibonacci numbers.

18. Prove that $9 \mid u_{n+24}$ if and only if $9 \mid u_n$. [*Hint:* Use formula (1) to establish that $u_{n+24} \equiv u_n \pmod 9$.]

19. Use induction to show that $u_{2n} \equiv n(-1)^{n+1} \pmod 5$ for $n \geq 1$.

20. Derive the identity

$$u_{n+3} = 3u_{n+1} - u_{n-1}. \qquad\qquad n \geq 2.$$

[*Hint:* Apply formula (1).]

13.2 CERTAIN IDENTITIES INVOLVING FIBONACCI NUMBERS

We move on and develop several of the basic identities involving Fibonacci numbers; these should be useful in doing the problems at the end of the section. One of the simplest asserts that the sum of the first n Fibonacci numbers is equal to $u_{n+2} - 1$. For instance, when the first eight Fibonacci numbers are added together, we obtain

$$1 + 1 + 2 + 3 + 5 + 8 + 13 + 21 = 54 = 55 - 1 = u_{10} - 1.$$

That this is typical of the general situation follows by adding the relations

$$u_1 = u_3 - u_2,$$

$$u_2 = u_4 - u_3,$$

$$u_3 = u_5 - u_4,$$

$$\vdots$$

$$u_{n-1} = u_{n+1} - u_n,$$

$$u_n = u_{n+2} - u_{n+1}.$$

On doing so, the left-hand side yields the sum of the first n Fibonacci numbers, while on the right-hand side the terms cancel in pairs leaving only $u_{n+2} - u_2$. But $u_2 = 1$. The consequence is that

$$(2) \qquad\qquad u_1 + u_2 + u_3 + \cdots + u_n = u_{n+2} - 1.$$

Another Fibonacci property worth recording is the identity

$$(3) \qquad\qquad u_n^2 = u_{n+1}u_{n-1} + (-1)^{n-1}, \qquad\qquad n \geq 2.$$

This may be illustrated by taking, say, $n = 6$ and $n = 7$; then

$$u_6^2 = 8^2 = 13 \cdot 5 - 1 = u_7 u_5 - 1,$$

$$u_7^2 = 13^2 = 21 \cdot 8 + 1 = u_8 u_6 + 1.$$

The plan for establishing formula (3) is to start with the equation

$$u_n^2 - u_{n+1}u_{n-1} = u_n(u_{n-1} + u_{n-2}) - u_{n+1}u_{n-1}$$

$$= (u_n - u_{n+1})u_{n-1} + u_n u_{n-2}.$$

From the rule of formation of the Fibonacci sequence, we have $u_{n+1} = u_n + u_{n-1}$, and so the expression in parentheses may be replaced by the term $-u_{n-1}$ to produce

$$u_n^2 - u_{n+1}u_{n-1} = (-1)(u_{n-1}^2 - u_n u_{n-2}).$$

The important point is that except for the initial sign the right-hand side of this equation is the same as the left-hand side, but with all the subscripts decreased by 1. By repeating the argument $u_{n-1}^2 - u_n u_{n-2}$ can be shown to be equal to the expression $(-1)(u_{n-2}^2 - u_{n-1} u_{n-3})$, whence

$$u_n^2 - u_{n+1} u_{n-1} = (-1)^2(u_{n-2}^2 - u_{n-1} u_{n-3}).$$

Continue in this pattern. After $n - 2$ such steps, we arrive at

$$u_n^2 - u_{n+1} u_{n-1} = (-1)^{n-2}(u_2^2 - u_3 u_1)$$
$$= (-1)^{n-2}(1^2 - 2 \cdot 1) = (-1)^{n-1},$$

which we sought to prove.

For $n = 2k$, formula (3) becomes

(4) $$u_{2k}^2 = u_{2k+1} u_{2k-1} - 1.$$

While we are on the subject, we might observe that this last identity is the basis of a well-known geometric deception whereby a square 8 units by 8 can be broken up into pieces which seemingly fit together to form a rectangle 5 by 13. To accomplish this, divide the square into four parts as shown below on the left and rearrange them as indicated on the right.

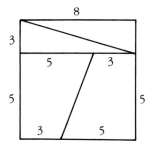
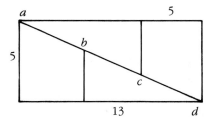

The area of the square is $8^2 = 64$, while that of the rectangle which seems to have the same constituent parts is $5 \cdot 13 = 65$, and so the area has apparently been increased by 1 square unit. The puzzle is easy to explain: the points a, b, c, d do not all lie on the diagonal of the rectangle, but instead are the vertices of a parallelogram whose area is of course exactly equal to the extra unit of area.

The foregoing construction can be carried out with any square whose sides are equal to a Fibonacci number u_{2k}. When partitioned in the manner indicated

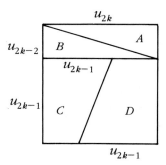

the pieces may be reformed to produce a rectangle having a slot in the shape of a slim parallelogram (our figure is somewhat exaggerated):

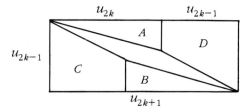

The identity $u_{2k-1} u_{2k+1} - 1 = u_{2k}^2$ may be interpreted as asserting that the area of the rectangle minus the area of the parallelogram is precisely equal to the area of the original square. It can be shown that the height of the parallelogram—that is, the width of the slot at its widest point—is

$$\frac{1}{\sqrt{u_{2k}^2 + u_{2k-2}^2}}.$$

When u_{2k} has a reasonably large value (say, $u_{2k} = 144$, so that $u_{2k-2} = 55$), the slot is so narrow it is almost imperceptible to the eye.

The First Fifty Fibonacci Numbers

u_1	1	u_{26}	121393
u_2	1	u_{27}	196418
u_3	2	u_{28}	317811
u_4	3	u_{29}	514229
u_5	5	u_{30}	832040
u_6	8	u_{31}	1346269
u_7	13	u_{32}	2178309
u_8	21	u_{33}	3524578
u_9	34	u_{34}	5702887
u_{10}	55	u_{35}	9227465
u_{11}	89	u_{36}	14930352
u_{12}	144	u_{37}	24157817
u_{13}	233	u_{38}	39088169
u_{14}	377	u_{39}	63245986
u_{15}	610	u_{40}	102334155
u_{16}	987	u_{41}	165580141
u_{17}	1597	u_{42}	267914296
u_{18}	2584	u_{43}	433494437
u_{19}	4181	u_{44}	701408733
u_{20}	6765	u_{45}	1134903170
u_{21}	10946	u_{46}	1836311903
u_{22}	17711	u_{47}	2971215073
u_{23}	28657	u_{48}	4807526976
u_{24}	46368	u_{49}	7778724049
u_{25}	75025	u_{50}	12586269025

The next result to be proved is that every positive integer can be written as a sum of distinct Fibonacci numbers. For instance, looking at the first few positive integers:

$$1 = u_1 \qquad\qquad 5 = u_5 = u_4 + u_3$$
$$2 = u_3 \qquad\qquad 6 = u_5 + u_1 = u_4 + u_3 + u_1$$
$$3 = u_4 \qquad\qquad 7 = u_5 + u_3 = u_4 + u_3 + u_2 + u_1$$
$$4 = u_4 + u_1 \qquad\quad 8 = u_6 = u_5 + u_4.$$

It will be enough to show by induction on $n > 2$ that each of the integers $1, 2, 3,$ $\ldots, u_n - 1$ is a sum of numbers from the set $\{u_1, u_2, \ldots, u_{n-2}\}$, none repeated. Assuming that this holds for $n = k$, choose N with $u_k - 1 < N < u_{k+1}$. Since $N - u_{k-1} < u_{k+1} - u_{k-1} = u_k$, we infer that the integer $N - u_{k-1}$ is representable as a sum of distinct numbers from $\{u_1, u_2, \ldots, u_{k-2}\}$. Then N and, in consequence, each of the integers $1, 2, 3, \ldots, u_{k+1} - 1$ can be expressed as a sum (without repetitions) of numbers from the set $\{u_1, u_2, \ldots, u_{k-2}, u_{k-1}\}$. This completes the induction step.

For the reader's convenience, we explicitly record this fact as

THEOREM 13-4.

Every positive integer can be represented as a finite sum of Fibonacci numbers, none used more than once.

In 1843, the French mathematician Jacques-Philippe-Marie Binet (1786–1856) discovered a formula for expressing u_n in terms of the integer n; namely,

$$u_n = \frac{1}{\sqrt{5}}\left[\left(\frac{1+\sqrt{5}}{2}\right)^n - \left(\frac{1-\sqrt{5}}{2}\right)^n\right].$$

This formula can be obtained by considering the two roots

$$\alpha = \frac{1+\sqrt{5}}{2} \text{ and } \beta = \frac{1-\sqrt{5}}{2}$$

of the quadratic equation $x^2 - x - 1 = 0$. As roots of this equation, they must satisfy

$$\alpha^2 = \alpha + 1 \text{ and } \beta^2 = \beta + 1.$$

When the first of these relations is multiplied by α^n, and the second by β^n, the result is

$$\alpha^{n+2} = \alpha^{n+1} + \alpha^n \text{ and } \beta^{n+2} = \beta^{n+1} + \beta^n.$$

Subtracting the second equation from the first, and dividing by $\alpha - \beta$, leads to

$$\frac{\alpha^{n+2} - \beta^{n+2}}{\alpha - \beta} = \frac{\alpha^{n+1} - \beta^{n+1}}{\alpha - \beta} + \frac{\alpha^n - \beta^n}{\alpha - \beta}.$$

If we put $H_n = (\alpha^n - \beta^n)/(\alpha - \beta)$, the equation above can be restated more concisely as

$$H_{n+2} = H_{n+1} + H_n \qquad\qquad n \geq 1.$$

Now notice a few things about α and β:

$$\alpha + \beta = 1, \quad \alpha - \beta = \sqrt{5}, \quad \alpha\beta = -1.$$

Hence,

$$H_1 = \frac{\alpha - \beta}{\alpha - \beta} = 1, \quad H_2 = \frac{\alpha^2 - \beta^2}{\alpha - \beta} = \alpha + \beta = 1.$$

What all this means is that the sequence $H_1, H_2, H_3, \cdots$ is precisely the Fibonacci sequence, which gives

$$u_n = \frac{\alpha^n - \beta^n}{\alpha - \beta}, \qquad n \geq 1.$$

With the help of this awkward-looking expression for u_n known as the *Binet formula,* it is possible conveniently to derive many results connected with the Fibonacci numbers. Let us, for example, show that

$$u_{n+2}^2 - u_n^2 = u_{2n+2}.$$

As we start, recall that $\alpha\beta = -1$ with the immediate consequence that $(\alpha\beta)^{2k} = 1$ for $k \geq 1$. Then

$$u_{n+2}^2 - u_n^2 = \left(\frac{\alpha^{n+2} - \beta^{n+2}}{\alpha - \beta}\right)^2 - \left(\frac{\alpha^n - \beta^n}{\alpha - \beta}\right)^2$$

$$= \frac{\alpha^{2(n+2)} - 2 + \beta^{2(n+2)}}{(\alpha - \beta)^2} - \frac{\alpha^{2n} - 2 + \beta^{2n}}{(\alpha - \beta)^2}$$

$$= \frac{\alpha^{2(n+2)} + \beta^{2(n+2)} - \alpha^{2n} - \beta^{2n}}{(\alpha - \beta)^2}.$$

Now the expression in the numerator may be rewritten as

$$\alpha^{2(n+2)} - (\alpha\beta)^2\alpha^{2n} - (\alpha\beta)^2\beta^{2n} + \beta^{2(n+2)} = (\alpha^2 - \beta^2)(\alpha^{2n+2} - \beta^{2n+2}).$$

On doing so, we get

$$u_{n+2}^2 - u_n^2 = \frac{(\alpha^2 - \beta^2)(\alpha^{2n+2} - \beta^{2n+2})}{(\alpha - \beta)^2}$$

$$= (\alpha + \beta)\left(\frac{\alpha^{2n+2} - \beta^{2n+2}}{\alpha - \beta}\right)$$

$$= 1 \cdot u_{2n+2} = u_{2n+2}.$$

For a second illustration of the usefulness of the Binet formula, let us once again derive the relation $u_{2n+1}u_{2n-1} - 1 = u_{2n}^2$. First we calculate

$$u_{2n+1}u_{2n-1} - 1 = \left(\frac{\alpha^{2n+1} - \beta^{2n+1}}{\sqrt{5}}\right)\left(\frac{\alpha^{2n-1} - \beta^{2n-1}}{\sqrt{5}}\right) - 1$$

$$= \tfrac{1}{5}\left(\alpha^{4n} - \beta^{4n} - (\alpha\beta)^{2n-1}\alpha^2 - (\alpha\beta)^{2n-1}\beta^2 - 5\right)$$

$$= \tfrac{1}{5}\left(\alpha^{4n} - \beta^{4n} + (\alpha^2 + \beta^2) - 5\right)$$

Since $\alpha^2 + \beta^2 = 3$, this last expression becomes

$$\tfrac{1}{5}(\alpha^{4n} - \beta^{4n} - 2) = \tfrac{1}{5}(\alpha^{4n} - \beta^{4n} - 2(\alpha\beta)^{2n})$$

$$= \left(\frac{\alpha^{2n} - \beta^{2n}}{\sqrt{5}}\right)^2 = u_{2n}^2,$$

which is the required identity.

Problems 13.2

1. Using induction on the positive integer n, establish the formulas
   (a) $u_1 + 2u_2 + 3u_3 + \cdots + nu_n = (n + 1)u_{n+2} - u_{n+4} + 2$;
   (b) $u_2 + 2u_4 + 3u_6 + \cdots + nu_{2n} = nu_{2n+1} - u_{2n}$.

2. (a) Show that the sum of the first n Fibonacci numbers with odd indices is given by the formula

$$u_1 + u_3 + u_5 + \cdots + u_{2n-1} = u_{2n}.$$

   [*Hint:* Add the equalities $u_1 = u_2$, $u_3 = u_4 - u_2$, $u_5 = u_6 - u_4$, $\cdots$.]
   (b) Show that the sum of the first n Fibonacci numbers with even indices is given by the formula

$$u_2 + u_4 + u_6 + \cdots + u_{2n} = u_{2n+1} - 1.$$

   [*Hint:* Apply part (a) in conjunction with identity (2).]
   (c) Derive the following expression for the alternating sum of the first n Fibonacci numbers:

$$u_1 - u_2 + u_3 - u_4 + \cdots + (-1)^{n+1}u_n = 1 + (-1)^{n+1}u_{n-1}.$$

3. From formula (1), deduce that

$$u_{2n-1} = u_n^2 + u_{n-1}^2, \quad u_{2n} = u_{n+1}^2 - u_{n-1}^2 \text{ with } n \geq 2.$$

4. Use the results of Problem 3 to obtain the following identities:
   (a) $u_{n+1}^2 + u_{n-2}^2 = 2u_{2n-1}$, $\qquad\qquad\qquad\qquad n \geq 3.$
   (b) $u_{n+2}^2 + u_{n-1}^2 = 2(u_n^2 + u_{n+1}^2)$, $\qquad\qquad\qquad n \geq 2.$

5. Establish that the formula

$$u_n u_{n-1} = u_n^2 - u_{n-1}^2 + (-1)^n$$

   holds for $n \geq 2$ and use this to conclude that consecutive Fibonacci numbers are relatively prime.

6. Without resorting to induction, derive the following identities:
   (a) $u_{n+1}^2 - 4u_n u_{n-1} = u_{n-2}^2$, $\qquad\qquad\qquad\qquad n \geq 3.$
   [*Hint:* Start by squaring both $u_{n-2} = u_n - u_{n-1}$ and $u_{n+1} = u_n + u_{n-1}$.]
   (b) $u_{n+1}u_{n-1} - u_{n+2}u_{n-2} = 2(-1)^n$, $\qquad\qquad n \geq 3.$
   [*Hint:* Put $u_{n+2} = u_{n+1} + u_n$, $u_{n-2} = u_n - u_{n-1}$ and use formula (3).]
   (c) $u_n^2 - u_{n+2}u_{n-2} = (-1)^n$, $\qquad\qquad\qquad\qquad n \geq 3.$
   [*Hint:* Mimic the proof of formula (3).]
   (d) $u_n^2 - u_{n+3}u_{n-3} = 4(-1)^{n+1}$, $\qquad\qquad\qquad n \geq 4.$
   (e) $u_n u_{n+1} u_{n+3} u_{n+4} = u_{n+2}^4 - 1$, $\qquad\qquad\qquad n \geq 1.$
   [*Hint:* By part (c), $u_n u_{n+4} = u_{n+2}^2 + (-1)^{n+1}$, while by formula (3), $u_{n+1}u_{n+3}$
   $= u_{n+2}^2 + (-1)^{n+2}$.]

7. Represent the integers 50, 75, 100, and 125 as sums of distinct Fibonacci numbers.

8. Prove that every positive integer can be written as a sum of distinct terms from the sequence $u_2, u_3, u_4, \ldots$ (that is, the Fibonacci sequence with u_1 deleted).

9. Establish the identity

$$(u_n u_{n+3})^2 + (2u_{n+1}u_{n+2})^2 = (u_{2n+3})^2, \qquad n \geq 1$$

   and use this to generate five primitive Pythagorean triples.

10. Prove that the product $u_n u_{n+1} u_{n+2} u_{n+3}$ of any four consecutive Fibonacci numbers is the area of a Pythagorean triangle. [*Hint:* See the previous problem.]

11. From the Binet formula for Fibonacci numbers, derive the relation

$$u_{2n+2}\, u_{2n-1} - u_{2n}\, u_{2n+1} = 1 \qquad n \geq 1.$$

12. For $n \geq 1$, show that the product $u_{2n-1}u_{2n+5}$ can be expressed as the sum of two squares. [*Hint:* Problem 6(e).]

13. (a) Prove that if $p = 4k + 3$ is prime, then p cannot divide a Fibonacci number with odd index; that is, $p \nmid u_{2n-1}$ for all $n \geq 1$. [*Hint:* In the contrary case, $u_n^2 + u_{n-1}^2 = u_{2n-1} \equiv 0 \pmod{p}$. See Problem 12, Section 5.4.]

    (b) From part (a) conclude that there are infinitely many primes of the form $4k + 1$. [*Hint:* Consider the sequence $\{u_p\}$, where $p > 5$ is prime.]

14. Verify that the product $u_{2n}u_{2n+2}u_{2n+4}$ of three consecutive Fibonacci numbers with even indices is the product of three consecutive integers; for instance, we have $u_4 u_6 u_8 = 504 = 7 \cdot 8 \cdot 9$. [*Hint:* First show that $u_{2n}u_{2n+4} = u_{2n+2}^2 - 1$.]

15. Use formulas (1) and (2) to show that the sum of any twenty consecutive Fibonacci numbers is divisible by u_{10}.

16. For $n \geq 4$, prove that $u_n + 1$ is not a prime. [*Hint:* It suffices to establish the identities

$$u_{4k} + 1 = u_{2k-1}(u_{2k} + u_{2k+2}),$$

$$u_{4k+1} + 1 = u_{2k+1}(u_{2k-1} + u_{2k+1}),$$

$$u_{4k+2} + 1 = u_{2k+2}(u_{2k+1} + u_{2k-1}),$$

$$u_{4k+3} + 1 = u_{2k+1}(u_{2k+1} + u_{2k+3}).]$$

17. The *Lucas numbers* are defined by the same recurrence formula as the Fibonacci numbers,

$$L_n = L_{n-1} + L_{n-2}, \qquad n \geq 3,$$

   but with $L_1 = 1$ and $L_2 = 3$; this gives the sequence 1, 3, 4, 7, 11, 18, 29, 47, 76, 123, 199, 322, ... For the Lucas numbers, derive each of the identities below:
    (a) $L_1 + L_2 + L_3 + \cdots + L_n = L_{n+2} - 3,$ $n \geq 1.$
    (b) $L_1 + L_3 + L_5 + \cdots + L_{2n-1} = L_{2n} - 2,$ $n \geq 1.$
    (c) $L_2 + L_4 + L_6 + \cdots + L_{2n} = L_{2n+1} - 1,$ $n \geq 1.$
    (d) $L_n^2 = L_{n+1}L_{n-1} + 5(-1)^n,$ $n \geq 2.$
    (e) $L_1^2 + L_2^2 + L_3^2 + \cdots + L_n^2 = L_n L_{n+1} - 2,$ $n \geq 1.$
    (f) $L_{n+1}^2 - L_n^2 = L_{n-1}L_{n+2},$ $n \geq 2.$

18. Establish the following relations between the Fibonacci and Lucas numbers:

(a) $L_n = u_{n+1} + u_{n-1} = u_n + 2u_{n-1}$, $n \geq 2$. [*Hint:* Argue by induction on n.]

(b) $L_n = u_{n+2} - u_{n-2}$, $n \geq 3$.

(c) $u_{2n} = u_n L_n$, $n \geq 1$.

(d) $L_{n+1} + L_{n-1} = 5u_n$, $n \geq 2$.

(e) $L_n^2 = u_n^2 + 4u_{n+1}u_{n-1}$, $n \geq 2$.

(f) $2u_{m+n} = u_m L_n + L_m u_n$, $m \geq 1, n \geq 1$.

(g) $\gcd(u_n, L_n) = 1$ or 2, $n \geq 1$.

19. If $\alpha = \dfrac{1 + \sqrt{5}}{2}$ and $\beta = \dfrac{1 - \sqrt{5}}{2}$, obtain the Binet formula for the Lucas numbers

$$L_n = \alpha^n + \beta^n$$

for $n \geq 1$.

20. For the Lucas sequence, establish the following results without resorting to induction:

(a) $L_n^2 = L_{2n} + 2(-1)^n$, $n \geq 1$.

(b) $L_n L_{n+1} - L_{2n+1} = (-1)^n$, $n \geq 1$.

(c) $L_n^2 - L_{n-1}L_{n+1} = 5(-1)^n$, $n \geq 2$.

(d) $L_{2n} + 7(-1)^n = L_{n-2}L_{n+2}$, $n \geq 3$.

21. Use the Binet formulas to obtain the relations below:

(a) $L_n^2 - 5u_n^2 = 4(-1)^n$, $n \geq 1$.

(b) $L_{2n-1} = 5u_n u_{n+1} + (-1)^{n+1}$, $n \geq 1$.

(c) $L_n^2 - u_n^2 = 4u_{n-1}u_{n+1}$, $n \geq 2$.

(d) $L_m L_n + 5u_m u_n = 2L_{m+n}$, $m \geq 1, n \geq 1$.

22. Show that the Lucas numbers $L_4, L_8, L_{16}, L_{32}, \ldots$ all have 7 as the final digit; that is, $L_{2n} \equiv 7 \pmod{10}$ for $n \geq 2$. [*Hint:* Induct on the integer n and appeal to the formula $L_n^2 = L_{2n} + 2(-1)^n$.]

23. In 1876, Lucas discovered the following formula for the Fibonacci numbers in terms of the binomial coefficients:

$$u_n = \binom{n-1}{0} + \binom{n-2}{1} + \binom{n-3}{2} + \cdots + \binom{n-j}{j-1} + \binom{n-j-1}{j}$$

where j is the largest integer less than or equal to $(n-1)/2$. Derive this result. [*Hint:* Argue by induction, using the relation $u_n = u_{n-1} + u_{n-2}$; note also that $\binom{m}{k} = \binom{m-1}{k} + \binom{m-1}{k-1}$.]

24. Establish that for $n \geq 1$,

(a) $\binom{n}{1}u_1 + \binom{n}{2}u_2 + \binom{n}{3}u_3 + \cdots + \binom{n}{n}u_n = u_{2n}$;

(b) $-\binom{n}{1}u_1 + \binom{n}{2}u_2 - \binom{n}{3}u_3 + \cdots + (-1)^n\binom{n}{n}u_n = -u_n$.

[*Hint:* Use the Binet formula for u_n, and then the Binomial Theorem.]

25. Prove that 24 divides the sum of any 24 consecutive Fibonacci numbers. [*Hint:* Consider the identity

$$u_n + u_{n+1} + \cdots + u_{n+k-1} = u_{n-1}(u_{k+1} - 1) + u_n(u_{k+2} - 1).]$$

26. Let n be a positive integer and $m = n^{13} - n$. Show that u_m is divisible by 30290. [*Hint:* See Problem 1(b) of Section 7.3.]

14

Continued Fractions

". . . A mathematician, like a painter or a poet, is a maker of patterns. If his patterns are more permanent than theirs, it is because they are made with ideas."

G. H. HARDY

14.1 SRINIVASA RAMANUJAN

India has from time to time produced mathematicians of remarkable power, but Srinivasa Ramanujan (1887–1920) is universally considered to have been its greatest genius. He was born in the southern Indian town of Erode, near Madras, the son of a bookkeeper in a cloth merchant's shop. He began his single-minded pursuit of mathematics when, at the age of 15 or 16, he borrowed a copy of Carr's *Synopsis of Pure Mathematics*. This unusual book contained the statements of over 6000 theorems, very few with proofs. Ramanujan undertook the task of establishing, without help, all the formulas in the book. In 1903, he won a scholarship to the University of Madras, only to lose it a year later for neglecting other subjects in favor of mathematics. He dropped out of college in disappointment and wandered the countryside for the next several years, impoverished and unemployed. Compelled to seek a regular livelihood after marrying, Ramanujan secured (1912) a clerical position with the Madras Port Trust Office, a job that left him enough time to continue his work in mathematics. After publishing his first paper in 1911, and two more the next year, he gradually gained recognition.

At the urging of influential friends, Ramanujan began a correspondence with the leading British pure mathematician of the day, G. H. Hardy. Appended to his letters to Hardy were lists of theorems, 120 in all, some definitely proved and others only conjectured. Examining these with bewilderment, Hardy concluded that "they could only be written down by a mathematician of the highest class; they must be true because if they were not true, no one would have the imagination to invent them." Hardy immediately invited Ramanujan to come to Cambridge University to develop his already great, but untrained, mathematical talent. Up to that time Ramanujan had worked almost totally isolated from modern European mathematics.

Srinivasa Ramanujan
(1887–1920)
(Master and Fellows of Trinity College, Cambridge)

Supported by a special scholarship, Ramanujan arrived in Cambridge in April 1914. There he had three years of uninterrupted activity, doing much of his best work in collaboration with Hardy. Hardy wrote to Madras University saying, "He will return to India with a scientific standing and reputation such as no Indian has enjoyed before." But in 1917 Ramanujan became incurably ill. His disease was diagnosed at that time as tuberculosis, but it is now thought to have been a severe vitamin deficiency. (A strict vegetarian who cooked all of his own food, Ramanujan had difficulty maintaining an adequate diet in war-rationed England.) Early in 1919 when the seas were finally considered safe for travel, he returned to India. In extreme pain, Ramanujan continued to do mathematics while lying in bed. He died the following April, at the age of 32.

The theory of partitions is one of the outstanding examples of the success of the Hardy-Ramanujan collaboration. A *partition* of a positive integer n is a way of writing n as a sum of positive integers, the order of the summands being irrelevant. The integer 5, for example, may be partitioned in seven ways: $5, 4 + 1, 3 + 2, 3 + 1 + 1, 2 + 2 + 1, 2 + 1 + 1 + 1, 1 + 1 + 1 + 1 + 1$. If $p(n)$ denotes the total number of partitions of n, then the values of $p(n)$ for the first six positive integers are $p(1) = 1, p(2) = 2, p(3) = 3, p(4) = 5, p(5) = 7$ and $p(6) = 11$. Actual computation shows that the partition function $p(n)$ increases very rapidly with n; for instance, $p(200)$ has the enormous value

$$p(200) = 3{,}972{,}999{,}029{,}388.$$

While no simple formula for $p(n)$ exists, one can look for an approximate formula giving its general order of magnitude. In 1918, Hardy and Ramanujan proved what is considered one of the masterpieces in number theory: namely, that for large n the partition function satisfies the relation

$$p(n) \approx \frac{e^{c\sqrt{n}}}{4n\sqrt{3}}$$

where the constant $c = \pi (2/3)^{1/2}$. For $n = 200$, the right-hand side of the above relation is approximately $4 \cdot 10^{12}$, which is remarkably close to the actual value of $p(200)$.

Hardy and Ramanujan proved considerably more. They obtained a fairly complicated infinite series for $p(n)$ which could be used to calculate $p(n)$ exactly, for any positive integer n. When $n = 200$, the initial term of this series produces the approximation 3,972,998,993,185,896, agreeing with the first six significant figures of $p(200)$; truncated at five terms, the series approximates the exact value with an error of 0.004.

Ramanujan was the first to discover (in 1919) several remarkable congruence properties involving the partition function $p(n)$; namely, he proved that

$$p(5k + 4) \equiv 0 \ (\text{mod } 5), \ p(7k + 5) \equiv 0 \ (\text{mod } 7), \ p(11k + 6) \equiv 0 \ (\text{mod } 11),$$

as well as similar divisibility relations for the moduli 5^2, 7^2 and 11^2, such as $p(25k + 24) \equiv 0 \ (\text{mod } 5^2)$. These results were embodied in his famous conjecture: For $q = 5, 7$, or 11, if $24n \equiv 1 \ (\text{mod } q^k)$, then $p(n) \equiv 0 \ (\text{mod } q^k)$ for all $k \geq 0$. From extensive tables of values of $p(n)$, it was later noticed that the conjectured congruence relating to powers of 7 is false when $k = 3$; that is, when $n = 243$, we have $24n = 5832 \equiv 1 \ (\text{mod } 7^3)$, but

$$p(243) = 133,978,259,344,888 \equiv 245 \not\equiv 0 \ (\text{mod } 7^3).$$

Yet Ramanujan's inspired guesses were illuminating even when incorrect, for it is now known that if $24n \equiv 1 \ (\text{mod } 7^{2k-2})$, then $p(n) \equiv 0 \ (\text{mod } 7^k)$ for $k \geq 2$.

In 1915, Ramanujan published an elaborate 63-page memoir on highly composite numbers. An integer $n > 1$ is termed *highly composite* if it has more divisors than any preceding integer; in other words, the divisor function τ satisfies $\tau(m) < \tau(n)$ for all $m < n$. The first ten highly composite numbers are 2, 4, 6, 12, 24, 36, 48, 60, 120, and 180. Ramanujan obtained some surprisingly accurate information concerning their structure. It was known that highly composite numbers could be expressed as

$$n = 2^{k_1} 3^{k_2} 5^{k_3} \cdots p_r^{k_r}, \text{ where } k_1 \geq k_2 \geq k_3 \geq \cdots \geq k_r.$$

What Ramanujan showed was that the beginning exponents form a strictly decreasing sequence $k_1 > k_2 > k_3 > \cdots$, but that later on groups of equal exponents occur; and that the final exponent $k_r = 1$, except when $n = 4$ or $n = 36$, in which case $k_r = 2$. As an example,

$$6,746,328,888,800 = 2^6 \cdot 3^4 \cdot 5^2 \cdot 7^2 \cdot 11 \cdot 13 \cdot 17 \cdot 19 \cdot 23.$$

As a final example of Ramanujan's creativity, we mention his unparalleled ability to come up with infinite series representations for π. Computer scientists have exploited his series

$$\frac{1}{\pi} = \frac{\sqrt{8}}{9801} \sum_{n=0}^{\infty} \frac{(4n)!}{(n!)^4} \frac{[1103 + 26390n]}{396^{4n}}$$

to calculate the value of π to millions of decimal digits; each successive term in the series adds roughly eight more correct digits. Ramanujan discovered fourteen other series for $1/\pi$, but he gave almost no explanation as to their origin. The most remarkable of these is

$$\frac{1}{\pi} = \sum_{n=0}^{\infty} \binom{2n}{n}^3 \frac{42n+3}{2^{12n+4}}.$$

This series has the property that it can be used to compute the second block of k (binary) digits in the decimal expansion of π without calculating the first k digits.

14.2 FINITE CONTINUED FRACTIONS

In that part of the *Liber Abaci* dealing with the resolution of fractions into unit fractions, Fibonacci introduced a kind of "continued fraction." For example, he employed the symbol $\frac{1\,1\,1}{3\,4\,5}$ as an abbreviation for

$$\frac{1 + \dfrac{1 + \dfrac{1}{5}}{4}}{3} = \frac{1}{3} + \frac{1}{3 \cdot 4} + \frac{1}{3 \cdot 4 \cdot 5}.$$

The modern practice is, however, to write continued fractions in a descending fashion, as with

$$2 + \cfrac{1}{4 + \cfrac{1}{1 + \cfrac{1}{3 + \frac{1}{2}}}}.$$

A multiple-decked expression of this type is said to be a finite simple continued fraction. To put the matter formally:

DEFINITION 14-1.
  By a *finite continued fraction* is meant a fraction of the form

$$a_0 + \cfrac{1}{a_1 + \cfrac{1}{a_2 + \cfrac{1}{a_3 + \cfrac{1}{\ddots \cfrac{1}{a_{n-1} + \cfrac{1}{a_n}}}}}},$$

where $a_0, a_1, \ldots, a_n$ are real numbers, all of which except possibly a_0 are positive. The numbers $a_1, a_2, \ldots, a_n$ are the *partial denominators* of this fraction. Such a fraction is called *simple* if all of the a_i are integers.

While giving due credit to Fibonacci, most authorities agree that the theory of continued fractions begins with Rafael Bombelli, the last of the great algebraists of Renaissance Italy. In his *L'Algebra Opera* (1572), Bombelli attempted to find square roots by means of infinite continued fractions—a method both ingenious and novel. He essentially proved that $\sqrt{13}$ could be expressed as the continued fraction

$$\sqrt{13} = 3 + \cfrac{4}{6 + \cfrac{4}{6 + \cfrac{4}{6 + \cfrac{4}{6 + \ddots}}}}.$$

It may be interesting to mention that Bombelli was the first to popularize the work of Diophantus in the Latin West. He set out initially to translate the Vatican Library's copy of Diophantus' *Arithmetica* (probably the same manuscript uncovered by Regiomontanus), but, carried away by other labors, never finished the project. Instead he took all the problems of the first four Books and embodied them in his *Algebra*, interspersing them with his own problems. Although Bombelli did not distinguish between the problems, he nonetheless acknowledged that he had borrowed freely from the *Arithmetica*.

Evidently, the value of any finite simple continued fraction will always be a rational number. For instance, the continued fraction

$$3 + \cfrac{1}{4 + \cfrac{1}{1 + \cfrac{1}{4 + \frac{1}{2}}}}$$

can be condensed to the value $170/53$:

$$3 + \cfrac{1}{4 + \cfrac{1}{1 + \cfrac{1}{4 + \frac{1}{2}}}} = 3 + \cfrac{1}{4 + \cfrac{1}{1 + \frac{2}{9}}}$$

$$= 3 + \cfrac{1}{4 + \frac{9}{11}}$$

$$= 3 + \frac{11}{53}$$

$$= \frac{170}{53}.$$

THEOREM 14-1.

Any rational number can be written as a finite simple continued fraction.

Proof: Let a/b, where $b > 0$, be any rational number. Euclid's algorithm for finding the greatest common divisor of a and b gives us the equations

$$a = ba_0 + r_1, \qquad\qquad 0 < r_1 < b$$
$$b = r_1 a_1 + r_2, \qquad\qquad 0 < r_2 < r_1$$
$$r_1 = r_2 a_2 + r_3, \qquad\qquad 0 < r_3 < r_2$$

$$\vdots$$

$$r_{n-2} = r_{n-1} a_{n-1} + r_n, \qquad 0 < r_n < r_{n-1}$$
$$r_{n-1} = r_n a_n + 0.$$

Notice that since each remainder r_k is a positive integer, $a_1, a_2, \ldots, a_n$ are all positive. Rewrite the equations of the algorithm in the following manner:

$$a/b = a_0 + r_1/b = a_0 + 1/(b/r_1),$$
$$b/r_1 = a_1 + r_2/r_1 = a_1 + 1/(r_1/r_2),$$
$$r_1/r_2 = a_2 + r_3/r_2 = a_2 + 1/(r_2/r_3),$$

$$\vdots$$

$$r_{n-1}/r_n = a_n.$$

If we use the second of these equations to eliminate b/r_1 from the first equation, then

$$a/b = a_0 + 1/(b/r_1) = a_0 + \cfrac{1}{a_1 + \cfrac{1}{(r_1/r_2)}}.$$

In this result, substitute the value of r_1/r_2 as given in the third equation:

$$a/b = a_0 + \cfrac{1}{a_1 + \cfrac{1}{a_2 + \cfrac{1}{(r_2/r_3)}}}.$$

Continuing in this way, we can go on to get

$$a/b = a_0 + \cfrac{1}{a_1 + \cfrac{1}{a_2 + \cfrac{1}{a_3 + \cfrac{}{\ddots \cfrac{1}{a_{n-1} + \cfrac{1}{a_n}}}}}},$$

thereby finishing the proof.

To illustrate the procedure involved in the proof of Theorem 14-1, let us represent 19/51 as a continued fraction. An application of Euclid's algorithm to the integers 19 and 51 gives the equations

$$
\begin{array}{llll}
51 = 2 \cdot 19 + 13 & \text{or} & 51/19 = 2 + 13/19, \\
19 = 1 \cdot 13 + 6 & \text{or} & 19/13 = 1 + 6/13, \\
13 = 2 \cdot 6 + 1 & \text{or} & 13/6 = 2 + 1/6, \\
6 = 6 \cdot 1 + 0 & \text{or} & 6/6 = 1.
\end{array}
$$

Making the appropriate substitutions, it is seen that

$$\frac{19}{51} = \frac{1}{(51/19)} = \cfrac{1}{2 + \frac{13}{19}}$$

$$= \cfrac{1}{2 + \cfrac{1}{\frac{19}{13}}}$$

$$= \cfrac{1}{2 + \cfrac{1}{1 + \frac{6}{13}}}$$

$$= \cfrac{1}{2 + \cfrac{1}{1 + \cfrac{1}{\frac{13}{6}}}}$$

$$= \cfrac{1}{2 + \cfrac{1}{1 + \cfrac{1}{2 + \frac{1}{6}}}}$$

which is the continued fraction expansion for 19/51.

Since continued fractions are unwieldy to print or write, we adopt the convention of denoting a continued fraction by a symbol which displays its partial quotients; say, by the symbol $[a_0; a_1, \ldots, a_n]$. In this notation, the expansion for $19/51$ is indicated by

$$[0; 2, 1, 2, 6]$$

and for $172/51 = 3 + 19/51$ by

$$[3; 2, 1, 2, 6].$$

The initial integer in the symbol $[a_0; a_1, \ldots, a_n]$ will be zero when the value of the fraction is positive but less than one.

The representation of a rational number as a finite simple continued fraction is not unique: once the representation has been obtained, we can always modify the last term. For, if $a_n > 1$, then

$$a_n = (a_n - 1) + 1 = (a_n - 1) + \frac{1}{1},$$

where $a_n - 1$ is a positive integer, hence

$$[a_0; a_1, \ldots, a_n] = [a_0; a_1, \ldots, a_n - 1, 1].$$

On the other hand, if $a_n = 1$, then

$$a_{n-1} + \frac{1}{a_n} = a_{n-1} + \frac{1}{1} = a_{n-1} + 1,$$

so that

$$[a_0; a_1, \ldots, a_{n-1}, a_n] = [a_0; a_1, \ldots, a_{n-2}, a_{n-1} + 1].$$

Every rational number has two representations as a simple continued fraction, one with an even number of partial denominators and one with an odd number (it turns out that these are the only two representations). In the case of $19/51$,

$$19/51 = [0; 2, 1, 2, 6] = [0; 2, 1, 2, 5, 1].$$

Example 14-1

We go back to the Fibonacci sequence and consider the quotient of two successive Fibonacci numbers (that is, the rational number u_{n+1}/u_n) written as a simple continued fraction. As pointed out earlier, the Euclidean Algorithm for the greatest common divisor of u_n and u_{n+1} produces the $n - 1$ equations

$$u_{n+1} = 1 \cdot u_n + u_{n-1},$$

$$u_n = 1 \cdot u_{n-1} + u_{n-2},$$

$$\vdots$$

$$u_4 = 1 \cdot u_3 + u_2,$$

$$u_3 = 2 \cdot u_2 + 0.$$

Since the quotients generated by the algorithm become the partial denominators of the continued fraction, we may write

$$u_{n+1}/u_n = [1; 1, 1, \ldots, 1, 2].$$

But u_{n+1}/u_n is also represented by a continued fraction having one more partial denominator than does $[1; 1, 1, \ldots, 1, 2]$; namely,

$$u_{n+1}/u_n = [1; 1, 1, \ldots, 1, 1, 1],$$

where the integer 1 appears $n + 1$ times. Thus, the fraction u_{n+1}/u_n has a continued fraction expansion which is very easy to describe: there are n partial denominators all equal to 1.

As a final item on this part of our program, we would like to indicate how the theory of continued fractions can be applied to the solution of linear Diophantine equations. This requires knowing a few pertinent facts about the "convergents" of a continued fraction, so let us begin proving them here.

DEFINITION 14-2.

The continued fraction made from $[a_0; a_1, \ldots, a_n]$ by cutting off the expansion after the kth partial denominator a_k is called the kth *convergent* of the given continued fraction and denoted by C_k; in symbols,

$$C_k = [a_0; a_1, \ldots, a_k], \qquad\qquad (1 \le k \le n).$$

We let the zero'th convergent C_0 be equal to the number a_0.

A point worth calling attention to is that for $k < n$ if a_k is replaced by the value $a_k + 1/a_{k+1}$, then the convergent C_k becomes the convergent C_{k+1};

$$[a_0; a_1, \ldots, a_{k-1}, a_k + 1/a_{k+1}]$$
$$= [a_0; a_1, \ldots, a_{k-1}, a_k, a_{k+1}] = C_{k+1}.$$

It hardly needs remarking that the last convergent C_n always equals the rational number represented by the original continued fraction.

Going back to our example $19/51 = [0; 2, 1, 2, 6]$, the successive convergents are

$$C_0 = 0,$$

$$C_1 = [0; 2] = 0 + \frac{1}{2} = \frac{1}{2},$$

$$C_2 = [0; 2, 1] = 0 + \cfrac{1}{2 + \frac{1}{1}} = \frac{1}{3},$$

$$C_3 = [0; 2, 1, 2] = 0 + \cfrac{1}{2 + \cfrac{1}{1 + \frac{1}{2}}} = \frac{3}{8},$$

$$C_4 = [0; 2, 1, 2, 6] = 19/51.$$

Except for the last convergent C_4, these are alternately less than or greater than $19/51$, each convergent being closer to $19/51$ than the previous one.

Much of the labor in calculating the convergents of a finite continued fraction $[a_0; a_1, \ldots, a_n]$ can be avoided by establishing formulas for their numerators and denominators. To this end, let us define numbers p_k and q_k $(k = 0, 1, \ldots, n)$ as follows:

$$p_0 = a_0 \qquad\qquad\qquad q_0 = 1$$

$$p_1 = a_1 a_0 + 1 \qquad\qquad q_1 = a_1$$

$$p_k = a_k p_{k-1} + p_{k-2} \qquad q_k = a_k q_{k-1} + q_{k-2}$$

for $k = 2, 3, \ldots, n$.

A direct computation shows that the first few convergents of $[a_0; a_1, \ldots, a_n]$ are

$$C_0 = a_0 = \frac{a_0}{1} = \frac{p_0}{q_0},$$

$$C_1 = a_0 + \frac{1}{a_1} = \frac{a_1 a_0 + 1}{a_1} = \frac{p_1}{q_1},$$

$$C_2 = a_0 + \cfrac{1}{a_1 + \cfrac{1}{a_2}} = \frac{a_2(a_1 a_0 + 1) + a_0}{a_2 a_1 + 1} = \frac{p_2}{q_2}.$$

Success hinges on being able to show that this relationship continues to hold. This is the content of

THEOREM 14-2.

The kth convergent of the simple continued fraction $[a_0; a_1, \ldots, a_n]$ has the value

$$C_k = p_k / q_k \qquad\qquad (0 \le k \le n).$$

Proof: The remarks above indicate that the theorem is true for $k = 0, 1, 2$. Let us assume that it is true for $k = m$, where $2 \le m < n$; that is, for this m,

(1) $$C_m = p_m / q_m = \frac{a_m p_{m-1} + p_{m-2}}{a_m q_{m-1} + q_{m-2}}.$$

Note that the integers $p_{m-1}, q_{m-1}, p_{m-2}, q_{m-2}$ depend on the first $m-1$ partial denominators $a_1, a_2, \ldots, a_{m-1}$, hence are independent of a_m. Thus formula (1) remains valid if a_m is replaced by the value $a_m + 1/a_{m+1}$:

$$\left[a_0; a_1, \ldots, a_{m-1}, a_m + \frac{1}{a_{m+1}} \right]$$

$$= \frac{\left(a_m + \dfrac{1}{a_{m+1}} \right) p_{m-1} + p_{m-2}}{\left(a_m + \dfrac{1}{a_{m+1}} \right) q_{m-1} + q_{m-2}}.$$

As we have explained earlier, the effect of this substitution is to change C_m into the convergent C_{m+1}, so that

$$C_{m+1} = \frac{\left(a_m + \dfrac{1}{a_{m+1}}\right)p_{m-1} + p_{m-2}}{\left(a_m + \dfrac{1}{a_{m+1}}\right)q_{m-1} + q_{m-2}}$$

$$= \frac{a_{m+1}(a_m p_{m-1} + p_{m-2}) + p_{m-1}}{a_{m+1}(a_m q_{m-1} + q_{m-2}) + q_{m-1}}$$

$$= \frac{a_{m+1}p_m + p_{m-1}}{a_{m+1}q_m + q_{m-1}}.$$

But this is precisely the form the theorem should take in the case $k = m + 1$. So, by induction, the stated result holds.

Let us see how this works in a specific instance, say, $19/51 = [0; 2, 1, 6, 2]$:

$p_0 = 0$	and	$q_0 = 1,$
$p_1 = 0 \cdot 2 + 1 = 1$		$q_1 = 2,$
$p_2 = 1 \cdot 1 + 0 = 1$		$q_2 = 1 \cdot 2 + 1 = 3,$
$p_3 = 2 \cdot 1 + 1 = 3$		$q_3 = 2 \cdot 3 + 2 = 8,$
$p_4 = 6 \cdot 3 + 1 = 19$		$q_4 = 6 \cdot 8 + 3 = 51.$

This says that the convergents of $[0; 2, 1, 2, 6]$ are

$$C_0 = p_0/q_0 = 0, \; C_1 = p_1/q_1 = 1/2, \; C_2 = p_2/q_2 = 1/3,$$

$$C_3 = p_3/q_3 = 3/8, \; C_4 = p_4/q_4 = 19/51,$$

as we know that they should be.

We continue our development of the properties of convergents by proving

THEOREM 14-3.

If $C_k = p_k/q_k$ is the kth convergent of the finite simple continued fraction $[a_0; a_1, \ldots, a_n]$, then

$$p_k q_{k-1} - q_k p_{k-1} = (-1)^{k-1}, \qquad\qquad 1 \le k \le n.$$

Proof: Induction on k works quite simply, with the relation

$$p_1 q_0 - q_1 p_0 = (a_1 a_0 + 1) \cdot 1 - a_1 \cdot a_0 = 1 = (-1)^{1-1},$$

disposing of the case $k = 1$. We assume that the formula in question is also true for $k = m$, where $1 \le m < n$. Then

$$p_{m+1}q_m - q_{m+1}p_m = (a_{m+1}p_m + p_{m-1})q_m$$

$$- (a_{m+1}q_m + q_{m-1})p_m$$

$$= -(p_m q_{m-1} - q_m p_{m-1})$$

$$= -(-1)^{m-1} = (-1)^m$$

and so the formula holds for $m + 1$, whenever it holds for m. It follows by induction that it is valid for all k with $1 \leq k \leq n$.

A notable consequence of this result is that the numerator and denominator of any convergent are relatively prime, so that the convergents are always given in lowest terms.

COROLLARY.

For $1 \leq k \leq n$, p_k and q_k are relatively prime.

Proof: If $d = \gcd(p_k, q_k)$, then from the theorem, $d \mid (-1)^{k-1}$; since $d > 0$, this forces us to conclude that $d = 1$.

Example 14-2

Consider the continued fraction $[0; 1, 1, \ldots, 1]$ in which the partial denominators are all equal to 1. Here, the first few convergents are

$$C_0 = 0/1, C_1 = 1/1, C_2 = 2/1, C_3 = 3/2, C_4 = 5/3, \ldots .$$

Since the numerator of the kth convergent C_k is

$$p_k = 1 \cdot p_{k-1} + p_{k-2} = p_{k-1} + p_{k-2}$$

and the denominator is

$$q_k = 1 \cdot q_{k-1} + q_{k-2} = q_{k-1} + q_{k-2},$$

it is apparent that

$$C_k = u_{k+1}/u_k \qquad (k \geq 2),$$

where u_k denotes the kth Fibonacci number. In the present context, the identity $p_k q_{k-1} - q_k p_{k-1} = (-1)^{k-1}$ of Theorem 14-3 assumes the form

$$u_{k+1} u_{k-1} - u_k^2 = (-1)^{k-1};$$

this is precisely formula (3) on page 270.

Let us now turn to the linear Diophantine equation

$$ax + by = c,$$

where a, b, c are given integers. Since no solution of this equation exists if $d \nmid c$, where $d = \gcd(a, b)$, there is no harm in assuming that $d \mid c$. In fact, we need only concern ourselves with the situation in which the coefficients are relatively prime. For if $\gcd(a, b) = d > 1$, then the equation may be divided by d to produce

$$(a/d)x + (b/d)y = c/d.$$

Both equations have the same solutions and, in the latter case, we know that $\gcd(a/d, b/d) = 1$.

Observe, too, that a solution of the equation

$$ax + by = c, \quad \gcd(a, b) = 1$$

may be obtained by first solving the Diophantine equation

$$ax + by = 1, \quad \gcd(a, b) = 1.$$

Indeed, if integers x_0 and y_0 can be found for which $ax_0 + by_0 = 1$, then multiplication of both sides by c gives

$$a(cx_0) + b(cy_0) = c.$$

Hence, $x = cx_0$ and $y = cy_0$ is the desired solution of $ax + by = c$.

To secure a pair of integers x and y satisfying the equation $ax + by = 1$, expand the rational number a/b as a simple continued fraction; say,

$$a/b = [a_0; a_1, \ldots, a_n].$$

Now the last two convergents of this continued fraction are

$$C_{n-1} = p_{n-1}/q_{n-1} \text{ and } C_n = p_n/q_n = a/b.$$

Since $\gcd(p_n, q_n) = 1 = \gcd(a, b)$, it may be concluded that

$$p_n = a \quad \text{and} \quad q_n = b.$$

By virtue of Theorem 14-3, we have

$$p_n q_{n-1} - q_n p_{n-1} = (-1)^{n-1}$$

or, with a change of notation,

$$a q_{n-1} - b p_{n-1} = (-1)^{n-1}.$$

Thus, with $x = q_{n-1}$ and $y = -p_{n-1}$, we have

$$ax + by = (-1)^{n-1}.$$

If n is odd, the equation $ax + by = 1$ has the particular solution $x_0 = q_{n-1}$, $y_0 = -p_{n-1}$, while if n is an even integer, then a solution is given by $x_0 = -q_{n-1}$, $y_0 = p_{n-1}$. Our earlier theory tells us that the general solution is

$$x = x_0 + bt, \, y = y_0 - at, \quad (t = 0, \pm 1, \pm 2, \ldots).$$

Example 14-3

Let us solve the linear Diophantine equation

$$172x + 20y = 1000$$

by means of simple continued fractions. Since $\gcd(172, 20) = 4$, this equation may be replaced by the equation

$$43x + 5y = 250.$$

The first step is to find a particular solution to

$$43x + 5y = 1.$$

To accomplish this, we begin by writing 43/5 (or if one prefers, 5/43) as a simple continued fraction. The sequence of equalities obtained by applying the Euclidean Algorithm to the numbers 43 and 5 is

$$43 = 8 \cdot 5 + 3,$$
$$5 = 1 \cdot 3 + 2,$$
$$3 = 1 \cdot 2 + 1,$$
$$2 = 2 \cdot 1,$$

so that $43/5 = [8; 1, 1, 2] = 8 + \cfrac{1}{1 + \cfrac{1}{1 + \frac{1}{2}}}$. The convergents of this continued

fraction are

$$C_0 = 8/1, \; C_1 = 9/1, \; C_2 = 17/2, \; C_3 = 43/5,$$

from which it follows that $p_2 = 17$, $q_2 = 2$, $p_3 = 43$ and $q_3 = 5$. Falling back on Theorem 14-3 again,

$$p_3 q_2 - q_3 p_2 = (-1)^{3-1},$$

or in equivalent terms,

$$43 \cdot 2 - 5 \cdot 17 = 1.$$

When this relation is multiplied by 250, we obtain

$$43 \cdot 500 + 5(-4250) = 250.$$

Thus a particular solution of the Diophantine equation $43x + 5y = 250$ is

$$x_0 = 500, \; y_0 = -4250.$$

The general solution is given by the equations

$$x = 500 + 5t, \; y = -4250 - 43t, \; (t = 0, \pm 1, \pm 2, \ldots).$$

Before proving a theorem concerning the behavior of the odd and even numbered convergents of a simple continued fraction, a preliminary lemma is required.

LEMMA.

If q_k is the denominator of the kth convergent C_k of the simple continued fraction $[a_0; a_1, \ldots, a_n]$, then $q_{k-1} \leq q_k$ for $1 \leq k \leq n$, with strict inequality when $k > 1$.

Proof: We establish the lemma by induction. In the first place, $q_0 = 1 \leq a_1 = q_1$, so that the asserted equality holds when $k = 1$. Assume, then, that it is true for $k = m$, where $1 \leq m < n$. Then

$$q_{m+1} = a_{m+1}q_m + q_{m-1} > a_{m+1}q_m \geq 1 \cdot q_m = q_m$$

so that the inequality is also true for $k = m + 1$.

With this information available, it is an easy matter to prove

THEOREM 14-4.

(1) *The convergents with even subscripts form a strictly increasing sequence; that is,*

$$C_0 < C_2 < C_4 < \cdots.$$

(2) *The convergents with odd subscripts form a strictly decreasing sequence; that is,*

$$C_1 > C_3 > C_5 > \cdots.$$

(3) *Every convergent with an odd subscript is greater than every convergent with an even subscript.*

Proof: With the aid of Theorem 14-3, we find that

$$C_{k+2} - C_k = (C_{k+2} - C_{k+1}) + (C_{k+1} - C_k)$$

$$= \left(\frac{p_{k+2}}{q_{k+2}} - \frac{p_{k+1}}{q_{k+1}}\right) + \left(\frac{p_{k+1}}{q_{k+1}} - \frac{p_k}{q_k}\right)$$

$$= \frac{(-1)^{k+1}}{q_{k+2}q_{k+1}} + \frac{(-1)^k}{q_{k+1}q_k}$$

$$= \frac{(-1)^k(q_{k+2} - q_k)}{q_k q_{k+1} q_{k+2}}.$$

Recalling that $q_i > 0$ for all $i \geq 0$ and that $q_{k+2} - q_k > 0$ by the lemma, it is evident that $C_{k+2} - C_k$ has the same algebraic sign as does $(-1)^k$. Thus, if k is an even integer, say $k = 2j$, then $C_{2j+2} > C_{2j}$; whence

$$C_0 < C_2 < C_4 < \cdots.$$

Similarly, if k is an odd integer, say $k = 2j - 1$, then $C_{2j+1} < C_{2j-1}$; whence

$$C_1 > C_3 > C_5 > \cdots.$$

It remains only to show that any odd-numbered convergent C_{2r-1} is greater than any even-numbered convergent C_{2s}. Since $p_k q_{k-1} - q_k p_{k-1} = (-1)^{k-1}$, upon dividing both sides of the equation by $q_k q_{k-1}$, we obtain

$$C_k - C_{k-1} = \frac{p_k}{q_k} - \frac{p_{k-1}}{q_{k-1}} = \frac{(-1)^{k-1}}{q_k q_{k-1}}.$$

This means that $C_{2j} < C_{2j-1}$. The effect of tying the various inequalities together is that

$$C_{2s} < C_{2s+2r} < C_{2s+2r-1} < C_{2r-1},$$

as desired.

To take an actual example, consider the continued fraction $[2; 3, 2, 5, 2, 4, 2]$. A little calculation gives the convergents

$$C_0 = 2/1, \ C_1 = 7/3, \ C_2 = 16/7, \ C_3 = 87/38,$$

$$C_4 = 190/83, \ C_5 = 847/370, \ C_6 = 1884/823.$$

According to Theorem 14-4, these convergents satisfy the chain of inequalities

$$2 < 16/7 < 190/83 < 1884/823 < 847/370 < 87/38 < 7/3.$$

This is readily visible when the numbers are expressed in decimal notation:

$$2 < 2.28571 \cdots < 2.28915 \cdots < 2.28918 \cdots < 2.28947 \cdots < 2.33333 \cdots.$$

PROBLEMS 14.2

1. Express each of the rational numbers below as finite simple continued fractions:
   (a) $-19/51$ (b) $187/57$ (c) $71/55$ (d) $118/303$

2. Determine the rational numbers represented by the following simple continued fractions:
   (a) $[-2; 2, 4, 6, 8]$ (b) $[4; 2, 1, 3, 1, 2, 4]$ (c) $[0; 1, 2, 3, 4, 3, 2, 1]$

3. If $r = [a_0; a_1, a_2, \ldots, a_n]$, where $r > 1$, show that

$$1/r = [0; a_0, a_1, \ldots, a_n].$$

4. Represent the following simple continued fractions in an equivalent form, but with an odd number of partial denominators:
   (a) $[0; 3, 1, 2, 3]$ (b) $[-1; 2, 1, 6, 1]$ (c) $[2; 3, 1, 2, 1, 1, 1]$

5. Compute the convergents of the following simple continued fractions:
   (a) $[1; 2, 3, 3, 2, 1]$ (b) $[-3; 1, 1, 1, 1, 3]$ (c) $[0; 2, 4, 1, 8, 2]$

6. (a) If $C_k = p_k/q_k$ denotes the kth convergent of the simple finite continued fraction $[1; 2, 3, 4, \ldots, n, n+1]$, show that

$$p_n = np_{n-1} + np_{n-2} + (n-1)p_{n-3} + \cdots + 3p_1 + 2p_0 + (p_0 + 1).$$

   [*Hint:* Add the relations $p_0 = 1$, $p_1 = 3$, $p_k = (k+1)p_{k-1} + p_{k-2}$ for $k = 2$, $\ldots, n$.]
   (b) Illustrate part (a) by calculating the numerator p_4 for the fraction $[1; 2, 3, 4, 5]$.

7. Evaluate p_k, q_k, and C_k ($k = 0, 1, \ldots, 8$) for the simple continued fractions below; notice that the convergents provide an approximation to the irrational numbers in parentheses:
   (a) $[1; 2, 2, 2, 2, 2, 2, 2, 2]$ ($\sqrt{2}$)
   (b) $[1; 1, 2, 1, 2, 1, 2, 1, 2]$ ($\sqrt{3}$)
   (c) $[2; 4, 4, 4, 4, 4, 4, 4, 4]$ ($\sqrt{5}$)
   (d) $[2; 2, 4, 2, 4, 2, 4, 2, 4]$ ($\sqrt{6}$)
   (e) $[2; 1, 1, 1, 4, 1, 1, 1, 4]$ ($\sqrt{7}$)

8. If $C_k = p_k/q_k$ is the kth convergent of the simple continued fraction $[a_0; a_1, \ldots, a_n]$, establish that

$$q_k \geq 2^{(k-1)/2}, \qquad\qquad (2 \leq k \leq n).$$

   [*Hint:* Observe that $q_k = a_k q_{k-1} + q_{k-2} \geq 2q_{k-2}$.]

9. Find the simple continued fraction representation of 3.1416, and that of 3.14159.

10. If $C_k = p_k/q_k$ is the kth convergent of the simple continued fraction $[a_0; a_1, \ldots, a_n]$ and $a_0 > 0$, show that

$$p_k/p_{k-1} = [a_k; a_{k-1}, \ldots, a_1, a_0],$$

and

$$q_k/q_{k-1} = [a_k; a_{k-1}, \ldots, a_2, a_1].$$

[*Hint:* In the first case, notice that

$$p_k/p_{k-1} = a_k + (p_{k-2}/p_{k-1})$$

$$= a_k + \cfrac{1}{(p_{k-1}/p_{k-2})} \,.]$$

11. By means of continued fractions determine the general solutions of each of the following Diophantine equations:
    (a) $19x + 51y = 1$; (b) $364x + 227y = 1$;
    (c) $18x + 5y = 24$; (d) $158x - 57y = 1$.

12. Verify Theorem 14-4 for the simple continued fraction $[1; 1, 1, 1, 1, 1, 1, 1]$.

14.3 INFINITE CONTINUED FRACTIONS

Up to this point, only finite continued fractions have been considered; and these, when simple, represent rational numbers. One of the main uses of the theory of continued fractions is finding approximate values of irrational numbers. For this, the notion of an infinite continued fraction is necessary.

An infinite continued fraction is an expression of the form

$$a_0 + \cfrac{b_1}{a_1 + \cfrac{b_2}{a_2 + \cfrac{b_3}{a_4 + \cdots}}}$$

where $a_0, a_1, a_2, \cdots$ and $b_1, b_2, b_3, \cdots$ are real numbers. An early example of a fraction of this type is found in the work of William Brouncker who converted (in 1655) Wallis's famous infinite product

$$4/\pi = \frac{3 \cdot 3 \cdot 5 \cdot 5 \cdot 7 \cdot 7 \cdots}{2 \cdot 4 \cdot 4 \cdot 6 \cdot 6 \cdot 8 \cdots}$$

into the identity

$$4/\pi = 1 + \cfrac{1^2}{2 + \cfrac{3^2}{2 + \cfrac{5^2}{2 + \cfrac{7^2}{2 + \cdots}}}} \,.$$

Both Wallis's and Brouncker's discoveries aroused considerable interest, but their direct use in calculating approximations to π is impractical.

In evaluating infinite continued fractions and in expanding functions in continued fractions, Srinivasa Ramanujan has no rival in the history of mathematics. He contributed many problems on continued fractions to the *Journal of the Indian Mathematical Society,* and his notebooks contain about two hundred results on such fractions. G. H. Hardy, commenting on Ramanujan's work, said, "On this side [of mathematics] most certainly I have never met his equal, and I can only compare him with Euler or Jacobi." Perhaps the most celebrated of Ramanujan's fraction expansions is his assertion that

$$e^{2\pi/5}\left(\frac{\sqrt{5+\sqrt{5}}}{2} - \frac{1+\sqrt{5}}{2}\right) = \cfrac{1}{1 + \cfrac{e^{-2\pi}}{1 + \cfrac{e^{-4\pi}}{1 + \cfrac{e^{-6\pi}}{1 + \ldots}}}}.$$

Part of its fame rests on its inclusion by Ramanujan in his first letter to Hardy in 1913. Hardy found the identity startling and was unable to derive it, confessing later that a proof "completely defeated" him. Although we now have proved most of Ramanujan's marvelous formulas, it is still not known what passage he took to discover them.

In this section, our discussion will be restricted to infinite simple continued fractions. These have the form

$$a_0 + \cfrac{1}{a_1 + \cfrac{1}{a_2 + \cfrac{1}{a_4 + \ldots}}}$$

where $a_0, a_1, a_2, \ldots$ is an infinite sequence of integers, all positive except possibly for a_0. We shall use the compact notation $[a_0; a_1, a_2, \ldots]$ to denote such a fraction. To attach a mathematical meaning to this expression, observe that each of the finite continued fractions

$$C_n = [a_0; a_1, a_2, \ldots, a_n] \qquad\qquad (n \geq 0)$$

is defined. It seems reasonable therefore to define the value of the infinite continued fraction $[a_0; a_1, a_2, \ldots]$ to be the limit of the sequence of rational numbers C_n, provided, of course, that this limit exists. In something of an abuse of notation, we shall use $[a_0; a_1, a_2, \ldots]$ to indicate not only the infinite continued fraction, but also its value.

The question of the existence of the above limit is easily settled. For, under our hypothesis, the limit not only exists but is always an irrational number. To see this, observe that formulas previously obtained for finite continued fractions remain valid for infinite continued fractions, since the derivation of these relations did not depend on the finiteness of the fraction. When the upper limits on the indices are removed, Theorem 14-4 tells us that the convergents C_n of $[a_0; a_1, a_2, \ldots]$ satisfy the infinite chain of inequalities.

$$C_0 < C_2 < C_4 < \cdots < C_{2n} < \cdots < C_{2n+1} < \cdots < C_5 < C_3 < C_1.$$

Since the even-numbered convergents C_{2n} form a monotonically increasing sequence, bounded above by C_1, they will converge to a limit α which is greater than each C_{2n}. Similarly, the monotonically decreasing sequence of odd-numbered convergents C_{2n+1} is bounded below by C_0 and so has a limit α' which is less than each C_{2n+1}. Let us show that these limits are equal. On the basis of the relation $p_{2n+1}q_{2n} - q_{2n+1}p_{2n} = (-1)^{2n}$ we see that

$$\alpha' - \alpha < C_{2n+1} - C_{2n} = \frac{p_{2n+1}}{q_{2n+1}} - \frac{p_{2n}}{q_{2n}} = \frac{1}{q_{2n}q_{2n+1}},$$

whence,

$$0 \leq |\alpha' - \alpha| < \frac{1}{q_{2n}q_{2n+1}} < \frac{1}{q_{2n}^2}.$$

Since the q_i increase without bound as i becomes large, the right-hand side of this inequality can be made arbitrarily small. If α' and α were not the same, then a contradiction would result (more precisely, $1/q_{2n}^2$ could be made less than the value of $|\alpha' - \alpha|$). Thus, the two sequences of odd- and even-numbered convergents have the same limiting value α, which means that the sequence of convergents C_n has the limit α.

Taking our cue from these remarks, we make the following definition:

DEFINITION 14-3.

If $a_0, a_1, a_2, \ldots$ is an infinite sequence of integers, all positive except possibly a_0, then the infinite simple continued fraction $[a_0; a_1, a_2, \ldots]$ has the value

$$\lim_{n \to \infty} [a_0; a_1, a_2, \ldots, a_n].$$

It should be emphasized again that the adjective "simple" indicates that the partial denominators a_k are all integers; since the only infinite continued fractions to be considered are simple, we shall often omit the term in what follows and call them infinite continued fractions.

Perhaps the most elementary example is afforded by the infinite continued fraction $[1; 1, 1, 1, \ldots]$. The argument of Example 14-1 showed that the nth convergent $C_n = [1; 1, 1, \ldots, 1]$, where the integer 1 appears $n + 1$ times, is equal to

$$C_n = \frac{u_{n+1}}{u_n} \qquad\qquad (n \geq 0),$$

a quotient of successive Fibonacci numbers. If x denotes the value of the continued fraction $[1; 1, 1, 1, \ldots]$, then

$$x = \lim_{n \to \infty} C_n = \lim_{n \to \infty} \frac{u_{n+1}}{u_n} = \lim_{n \to \infty} \frac{u_n + u_{n-1}}{u_n}$$

$$= \lim_{n \to \infty} 1 + \frac{1}{\dfrac{u_n}{u_{n-1}}} = 1 + \frac{1}{\lim\limits_{n \to \infty} \left(\dfrac{u_n}{u_{n-1}} \right)} = 1 + \frac{1}{x}.$$

This gives rise to the quadratic equation $x^2 - x - 1 = 0$, whose only positive root is $x = (1 + \sqrt{5})/2$. Hence,

$$\frac{1 + \sqrt{5}}{2} = [1; 1, 1, 1, \ldots].$$

There is one situation that occurs often enough to merit special terminology. If an infinite continued fraction, such as $[3; 1, 2, 1, 6, 1, 2, 1, 6, \ldots]$, contains a block of partial denominators $b_1, b_2, \ldots, b_n$ which repeats indefinitely, the fraction is called *periodic*. The custom is to write a periodic continued fraction

$$[a_0; a_1, \ldots, a_m, b_1, \ldots, b_n, b_1, \ldots, b_n, \ldots]$$

more compactly as

$$[a_0; a_1, \ldots, a_m, \overline{b_1, \ldots, b_n}],$$

where the bar over $b_1, b_2, \ldots, b_n$ indicates that this block of integers repeats over and over. If $b_1, b_2, \ldots, b_n$ is the smallest block of integers which constantly repeats, we say that $b_1, b_2, \ldots, b_n$ is the *period* of the expansion and that the *length* of the period is n. Thus, for example, $[3; \overline{1, 2, 1, 6}]$ would denote $[3; 1, 2, 1, 6, 1, 2, 1, 6, \ldots]$, a continued fraction whose period 1, 2, 1, 6 has length 4.

We saw earlier that every finite continued fraction is represented by a rational number. Let us now consider the value of an infinite continued fraction.

THEOREM 14-5.

The value of any infinite continued fraction is an irrational number.

Proof: Let us suppose that x denotes the value of the infinite continued fraction $[a_0; a_1, a_2, \ldots]$; that is, x is the limit of the sequence of convergents

$$C_n = [a_0; a_1, a_2, \ldots, a_n] = \frac{p_n}{q_n}.$$

Since x lies strictly between the successive convergents C_n and C_{n+1}, we have

$$0 < |x - C_n| < |C_{n+1} - C_n| = \left| \frac{p_{n+1}}{q_{n+1}} - \frac{p_n}{q_n} \right| = \frac{1}{q_n q_{n+1}}.$$

With the view to obtaining a contradiction, assume that x is a rational number; say, $x = a/b$, where a and $b > 0$ are integers. Then

$$0 < \left| \frac{a}{b} - \frac{p_n}{q_n} \right| < \frac{1}{q_n q_{n+1}}$$

and so, upon multiplication by the positive number bq_n,

$$0 < |aq_n - bp_n| < \frac{b}{q_{n+1}}.$$

We recall that the value of q_i increases without bound as i increases. If n is chosen so large that $b < q_{n+1}$, the result is

$$0 < |aq_n - bp_n| < 1.$$

This says that there is a positive integer, namely $|aq_n - bp_n|$, between 0 and 1—an obvious impossibility.

We now ask whether two different infinite continued fractions can represent the same irrational number. Before giving the pertinent result, let us observe that the properties of limits allow us to write an infinite continued fraction $[a_0; a_1, a_2, \ldots]$ as

$$[a_0; a_1, a_2, \ldots] = \lim_{n \to \infty} [a_0; a_1, \ldots, a_n]$$

$$= \lim_{n \to \infty} \left(a_0 + \frac{1}{[a_1; a_2, \ldots, a_n]} \right)$$

$$= a_0 + \frac{1}{\lim_{n \to \infty} [a_1; a_2, \ldots, a_n]}$$

$$= a_0 + \frac{1}{[a_1; a_2, a_3, \ldots]}.$$

Our theorem is stated as:

THEOREM 14-6.

If the infinite continued fractions $[a_0; a_1, a_2, \ldots]$ and $[b_0; b_1, b_2, \ldots]$ are equal, then $a_n = b_n$ for all $n \geq 0$.

Proof: If $x = [a_0; a_1, a_2, \ldots]$, then $C_0 < x < C_1$, which is the same as saying that $a_0 < x < a_0 + 1/a_1$. Knowing that the integer $a_1 \geq 1$, this produces the inequality $a_0 < x < a_0 + 1$. Hence, $[x] = a_0$, where $[x]$ is the traditional notation for the greatest integer or "bracket" function (page 117).

Now assume that $[a_0; a_1, a_2, \ldots] = x = [b_0; b_1, b_2, \ldots]$ or, to put it in a different form,

$$a_0 + \frac{1}{[a_1; a_2, \ldots]} = x = b_0 + \frac{1}{[b_1; b_2, \ldots]}.$$

By virtue of the conclusion of the first paragraph, we have $a_0 = [x] = b_0$, from which it may then be deduced that $[a_1; a_2, \ldots] = [b_1; b_2, \ldots]$. When the reasoning is repeated, we next conclude that $a_1 = b_1$ and that $[a_2; a_3, \ldots] = [b_2; b_3, \ldots]$. The process continues by mathematical induction, thereby giving $a_n = b_n$ for all $n \geq 0$.

COROLLARY.

Two distinct infinite continued fractions represent two distinct irrational numbers.

Example 14-4

To determine the unique irrational number represented by the infinite continued fraction $x = [3; 6, \overline{1,4}]$, let us write $x = [3; 6, y]$, where

$$y = [\overline{1; 4}] = [1; 4, y].$$

Then

$$y = 1 + \frac{1}{4 + 1/y} = 1 + \frac{y}{4y + 1} = \frac{5y + 1}{4y + 1},$$

which leads to the quadratic equation

$$4y^2 - 4y - 1 = 0.$$

Inasmuch as $y > 0$ and this equation has only one positive root, we may infer that

$$y = \frac{1 + \sqrt{2}}{2}.$$

From $x = [3; 6, y]$, we then find that

$$x = 3 + \cfrac{1}{6 + \cfrac{1}{\cfrac{1 + \sqrt{2}}{2}}} = \frac{25 + 19\sqrt{2}}{8 + 6\sqrt{2}}$$

$$= \frac{(25 + 19\sqrt{2})(8 - 6\sqrt{2})}{(8 + 6\sqrt{2})(8 - 6\sqrt{2})} = \frac{14 - \sqrt{2}}{4};$$

that is, $[3; 6, \overline{1, 4}] = \dfrac{14 - \sqrt{2}}{4}.$

Our last theorem shows that every infinite continued fraction represents a unique irrational number. Turning matters around, we next establish that an arbitrary irrational number x_0 can be expanded into an infinite continued fraction $[a_0; a_1, a_2, \ldots]$ which converges to the value x_0. The sequence of integers $a_0, a_1, a_2, \ldots$ is defined as follows: using the bracket function, we first let

$$x_1 = \frac{1}{x_0 - [x_0]}, \, x_2 = \frac{1}{x_1 - [x_1]}, \, x_3 = \frac{1}{x_2 - [x_2]}, \, \cdots$$

and then take

$$a_0 = [x_0], \, a_1 = [x_1], \, a_2 = [x_2], \, a_3 = [x_3], \, \cdots.$$

In general, the a_k are given inductively by

$$a_k = [x_k], \quad x_{k+1} = \frac{1}{x_k - a_k}, \qquad\qquad k \geq 0.$$

It is evident that x_{k+1} is irrational, whenever x_k is irrational; and because we are confining ourselves to the case in which x_0 is an irrational number, all x_k are irrational by induction. Thus,

$$0 < x_k - a_k = x_k - [x_k] < 1$$

and we see that

$$x_{k+1} = \frac{1}{x_k - a_k} > 1$$

so that the integer $a_{k+1} = [x_{k+1}] \geq 1$ for all $k \geq 0$. This process therefore leads to an infinite sequence of integers $a_0, a_1, a_2, \ldots$, all positive except perhaps for a_0.

Employing our inductive definition in the form

$$x_k = a_k + \frac{1}{x_{k+1}} \qquad\qquad (k \geq 0)$$

we obtain through successive substitution

$$x_0 = a_0 + \frac{1}{x_1}$$

$$= a_0 + \cfrac{1}{a_1 + \cfrac{1}{x_2}}$$

$$= a_0 + \cfrac{1}{a_1 + \cfrac{1}{a_2 + \cfrac{1}{x_3}}}$$

$$\vdots$$

$$= [a_0; a_1, a_2, \ldots, a_n, x_{n+1}]$$

for every positive integer n. This makes one suspect—and it is our task to show—that x_0 is the value of the infinite continued fraction $[a_0; a_1, a_2, \ldots]$.

For any fixed integer n, the first $n+1$ convergents $C_k = p_k/q_k$, where $\leq k \leq n$, of $[a_0; a_1, a_2, \ldots]$ are the same as the first $n+1$ convergents of the finite continued fraction $[a_0; a_1, a_2, \ldots, a_n, x_{n+1}]$. If we denote the $(n+2)$th convergent of the latter by C'_{n+1}, then the argument used in the proof of Theorem 14-2 to obtain C_{n+1} from C_n by replacing a_n by $a_n + 1/a_{n+1}$ works eqally well in the present setting; this enables us to obtain C'_{n+1} from C_n by replacing a_{n+1} by x_{n+1}:

$$x_0 = C'_{n+1} = [a_0; a_1, a_2, \ldots, a_n, x_{n+1}]$$

$$= \frac{x_{n+1}p_n + p_{n-1}}{x_{n+1}q_n + q_{n-1}}.$$

Because of this,

$$x_0 - C_n = \frac{x_{n+1}p_n + p_{n-1}}{x_{n+1}q_n + q_{n-1}} - \frac{p_n}{q_n}$$

$$= \frac{(-1)(p_n q_{n-1} - q_n p_{n-1})}{(x_{n+1}q_n + q_{n-1})q_n} = \frac{(-1)^n}{(x_{n+1}q_n + q_{n-1})q_n},$$

where the last equality relies on Theorem 14-3. Now $x_{n+1} > a_{n+1}$ and so

$$|x_0 - C_n| = \frac{1}{(x_{n+1}q_n + q_{n-1})q_n} < \frac{1}{(a_{n+1}q_n + q_{n-1})q_n} = \frac{1}{q_{n+1}q_n}.$$

Since the integers q_k are increasing, the implication is that

$$x_0 = \lim_{n \to \infty} C_n = [a_0; a_1, a_2, \ldots].$$

Let us sum up our conclusions in

THEOREM 14-7.

Every irrational number has a unique representation as an infinite continued fraction, the representation being obtained from the continued fraction algorithm described above.

Incidentally, our argument reveals a fact worth recording separately.

COROLLARY.

If p_n/q_n is the nth convergent to the irrational number x, then

$$\left| x - \frac{p_n}{q_n} \right| < \frac{1}{q_{n+1}q_n} < \frac{1}{q_n^2}.$$

We give two examples to illustrate the use of the continued fraction algorithm in finding the representation of a given irrational number as an infinite continued fraction.

Example 14-5

For our first example, consider $x = \sqrt{23} \approx 4.8$. The successive irrational numbers x_k (and therefore the integers $a_k = [x_k]$) can be computed rather easily, with the calculations exhibited below:

$$x_0 = \sqrt{23} = 4 + (\sqrt{23} - 4), \qquad\qquad a_0 = 4,$$

$$x_1 = \frac{1}{x_0 - [x_0]} = \frac{1}{\sqrt{23} - 4} = \frac{\sqrt{23} + 4}{7} = 1 + \frac{\sqrt{23} - 3}{7}, \qquad a_1 = 1,$$

$$x_2 = \frac{1}{x_1 - [x_1]} = \frac{7}{\sqrt{23} - 3} = \frac{\sqrt{23} + 3}{2} = 3 + \frac{\sqrt{23} - 3}{2}, \qquad a_2 = 3,$$

$$x_3 = \frac{1}{x_2 - [x_2]} = \frac{2}{\sqrt{23} - 3} = \frac{\sqrt{23} + 3}{7} = 1 + \frac{\sqrt{23} - 4}{7}, \qquad a_3 = 1,$$

$$x_4 = \frac{1}{x_3 - [x_3]} = \frac{7}{\sqrt{23} - 4} = \sqrt{23} + 4 = 8 + (\sqrt{23} - 4), \quad a_4 = 8.$$

Since $x_5 = x_1$, also $x_6 = x_2$, $x_7 = x_3$, $x_8 = x_4$; then we get $x_9 = x_5 = x_1$, and so on, which means that the block of integers 1, 3, 1, 8 repeats indefinitely. We find that the continued fraction expansion of $\sqrt{23}$ is periodic with the form

$$\sqrt{23} = [4; 1, 3, 1, 8, 1, 3, 1, 8, \ldots] = [4; \overline{1, 3, 1, 8}].$$

Example 14-6

To furnish a second illustration, let us obtain several of the convergents of the continued fraction of the number

$$\pi = 3.141592653 \cdots,$$

defined by the ancient Greeks as the ratio of the circumference of a circle to its diameter. The letter π, from the Greek word *perimetros*, was never employed in antiquity for this ratio; it was Euler's adoption of the symbol in his many popular textbooks that made it widely known and used.

By straightforward calculations, one sees that

$$x_0 = \pi = 3 + (\pi - 3), \qquad\qquad\qquad\qquad a_0 = 3,$$

$$x_1 = \frac{1}{x_0 - [x_0]} = \frac{1}{0.14159265 \ldots} = 7.06251330. \ldots, \quad a_1 = 7,$$

$$x_2 = \frac{1}{x_1 - [x_1]} = \frac{1}{0.06251330 \ldots} = 15.99659440. \ldots, \quad a_2 = 15,$$

$$x_3 = \frac{1}{x_2 - [x_2]} = \frac{1}{0.99659440 \ldots} = 1.00341723. \ldots, \quad a_3 = 1,$$

$$x_4 = \frac{1}{x_3 - [x_3]} = \frac{1}{0.00341723 \ldots} = 292.63724. \ldots, \quad a_4 = 292,$$

$$\vdots$$

Thus, the infinite continued fraction for π starts out as

$$\pi = [3; 7, 15, 1, 292, \ldots];$$

but, unlike the case of $\sqrt{23}$ in which all the partial denominators a_n are explicitly known, there is no pattern which gives the complete sequence of a_n. The first five convergents are

$$\frac{3}{1}, \frac{22}{7}, \frac{333}{106}, \frac{355}{113}, \frac{103993}{33102}.$$

As a check on the Corollary to Theorem 14-7, notice that we should have

$$\left| \pi - \frac{22}{7} \right| < \frac{1}{7^2}.$$

Now $314/100 < \pi < 22/7$ and therefore

$$\left| \pi - \frac{22}{7} \right| < \frac{22}{7} - \frac{314}{100} = \frac{1}{7 \cdot 50} < \frac{1}{7^2},$$

as expected.

Unless the irrational number x assumes some very special form, it may be impossible to give the complete continued fraction expansion of x. One can prove, for instance, that the expansion for x becomes ultimately periodic if and only if x is an irrational root of a quadratic equation with integral coefficients; that is, if x takes the form $r + s\sqrt{d}$, where r and $s \neq 0$ are rational numbers and d is a positive integer which is not a perfect square. But among other irrational numbers, there are very few whose representations seem to exhibit any regularity. An exception is another positive constant which has occupied the attention of mathematicians for many centuries, namely

$$e = 2.718281828 \ldots ,$$

the base of the system of natural logarithms. In 1737, Euler showed that

$$\frac{e - 1}{e + 1} = [0; 2, 6, 10, 14, 18, \ldots],$$

where the partial denominators form an arithmetic progression, and that,

$$\frac{e^2 - 1}{e^2 + 1} = [0; 1, 3, 5, 7, 9, \ldots].$$

The continued fraction representation of e itself (also found by Euler) is a bit more complicated, yet still has a pattern:

$$e = [2; 1, 2, 1, 1, 4, 1, 1, 6, 1, 1, 8, \ldots],$$

with the even integers subsequently occurring in order and separated by two 1's. With regard to the symbol e, its use is also original with Euler and it appeared in print for the first time in one of his textbooks.

In the introduction to analysis, it is usually demonstrated that e can be defined by the infinite series

$$e = 1 + \frac{1}{1!} + \frac{1}{2!} + \frac{1}{3!} + \frac{1}{4!} + \cdots .$$

If the reader is willing to accept this fact, then Euler's proof of the irrationality of e can be given very quickly: Suppose to the contrary that e is rational, say $e = a/b$, where a and b are positive integers. Then for $n > b$ and also $n > 1$, the number

$$N = n! \left(e - \left(1 + \frac{1}{1!} + \frac{1}{2!} + \cdots + \frac{1}{n!} \right) \right)$$

$$= n! \left(\frac{a}{b} - 1 - \frac{1}{1!} - \frac{1}{2!} - \cdots - \frac{1}{n!} \right)$$

is a positive integer because multiplication by $n!$ clears all the denominators. When e is replaced by its series expansion, this becomes

$$N = n! \left(\frac{1}{(n+1)!} + \frac{1}{(n+2)!} + \frac{1}{(n+3)!} + \cdots \right)$$

$$= \frac{1}{n+1} + \frac{1}{(n+1)(n+2)} + \frac{1}{(n+1)(n+2)(n+3)} + \cdots$$

$$< \frac{1}{n+1} + \frac{1}{(n+1)(n+2)} + \frac{1}{(n+2)(n+3)} + \cdots$$

$$= \frac{1}{n+1} + \left(\frac{1}{n+1} - \frac{1}{n+2} \right) + \left(\frac{1}{n+2} - \frac{1}{n+3} \right) + \cdots$$

$$= \frac{2}{n+1} < 1.$$

Since the inequality $0 < N < 1$ is impossible for an integer, e must be irrational. The exact nature of the number π offers greater difficulties; J. H. Lambert (1728–1777), in 1761, communicated to the Berlin Academy an essentially rigorous proof of the irrationality of π.

Given an irrational number x, a natural question is to ask how closely, or with what degree of accuracy, it can be approximated by rational numbers. One way of approaching the problem is to consider all rational numbers with a fixed denominator $b > 0$. Since x lies between two such rational numbers, say $c/b < x < (c+1)/b$, it follows that

$$\left| x - \frac{c}{b} \right| < \frac{1}{b}.$$

Better yet, we can write

$$\left| x - \frac{a}{b} \right| < \frac{1}{2b}.$$

where $a = c$ or $a = c + 1$, whichever choice may be appropriate. The continued fraction process permitted us to prove a result which considerably strengthens the last-written inequality, namely: given any irrational number x, there exist infinitely many rational numbers a/b in lowest terms which satisfy

$$\left| x - \frac{a}{b} \right| < \frac{1}{b^2}.$$

In fact, by the corollary to Theorem 14-7, any of the convergents p_n/q_n of the continued fraction expansion of x can play the role of the rational number a/b. The forthcoming theorem asserts that the convergents p_n/q_n have the property of being the best approximations, in the sense of giving the closest approximation to x among all rational numbers a/b with denominators q_n or less.

For clarity, the technical core of the theorem is placed in the following lemma.

LEMMA.

Let p_n/q_n be the nth convergent of the continued fraction representing the irrational number x. If a and b are integers, with $1 \le b < q_{n+1}$, then

$$|q_n x - p_n| \le |bx - a|.$$

Proof: Consider the system of equations

$$p_n \alpha + p_{n+1} \beta = a,$$

$$q_n \alpha + q_{n+1} \beta = b.$$

The determinant of the coefficients being $p_n q_{n+1} - q_n p_{n+1} = (-1)^{n+1}$, the system has the unique integral solution

$$\alpha = (-1)^{n+1}(a q_{n+1} - b p_{n+1}),$$

$$\beta = (-1)^{n+1}(b p_n - a q_n).$$

It is well to notice that $\alpha \ne 0$. In fact, $\alpha = 0$ yields $a q_{n+1} = b p_{n+1}$ and, because $\gcd(p_{n+1}, q_{n+1}) = 1$, this means that $q_{n+1} | b$ or $b \ge q_{n+1}$, which is contrary to hypothesis. In the event that $\beta = 0$, the inequality stated in the lemma is clearly true. For $\beta = 0$ leads to $a = p_n \alpha$, $b = q_n \alpha$ and, as a result,

$$|bx - a| = |\alpha| \ |q_n x - p_n| \ge |q_n x - p_n|.$$

Thus, there is no harm in assuming hereafter that $\beta \ne 0$.

When $\beta \ne 0$, we argue that α and β must have opposite signs. If $\beta < 0$, then the equation $q_n \alpha = b - q_{n+1} \beta$ indicates that $q_n \alpha > 0$ and, in turn, $\alpha > 0$. On the other hand if $\beta > 0$, then $b < q_{n+1}$ implies that $b < \beta q_{n+1}$ and therefore $\alpha q_n = b - q_{n+1} \beta < 0$; this makes $\alpha < 0$. We also infer that, because x stands between the consecutive convergents p_n/q_n and p_{n+1}/q_{n+1},

$$q_n x - p_n \quad \text{and} \quad q_{n+1} x - p_{n+1}$$

will have opposite signs. The point of this reasoning is that the numbers

$$\alpha(q_n x - p_n) \quad \text{and} \quad \beta(q_{n+1} x - p_{n+1})$$

must have the same sign; in consequence, the absolute value of their sum equals the sum of their separate absolute values. It is this crucial fact that allows us to complete the proof quickly:

$$|bx - a| = |(q_n \alpha + q_{n+1} \beta)x - (p_n \alpha + p_{n+1} \beta)|$$

$$= |\alpha(q_n x - p_n) + \beta(q_{n+1} x - p_{n+1})|$$

$$= |\alpha| \ |q_n x - p_n| + |\beta| \ |q_{n+1} x - p_{n+1}|$$

$$> |\alpha| \ |q_n x - p_n| \ge |q_n x - p_n|,$$

which is the desired inequality.

The convergents p_n/q_n are best approximations to the irrational number x in that every other rational number with the same or smaller denominator differs from x by a greater amount.

THEOREM 14-8.

If $1 \le b \le q_n$, the rational number a/b satisfies

$$\left| x - \frac{p_n}{q_n} \right| \le \left| x - \frac{a}{b} \right|.$$

Proof: Were it to happen that

$$\left| x - \frac{p_n}{q_n} \right| > \left| x - \frac{a}{b} \right|,$$

then

$$|q_n x - p_n| = q_n \left| x - \frac{p_n}{q_n} \right| > b \left| x - \frac{a}{b} \right| = |bx - a|,$$

violating the conclusion of the lemma.

Historians of mathematics have focused considerable attention on the attempts of early societies to arrive at an approximation to π, perhaps because the increasing accuracy of the results seems to offer a measure of the mathematical skills of different cultures. The first recorded scientific effort to evaluate π appeared in the *Measurement of a Circle* by the great Greek mathematician of ancient Syracuse, Archimedes (287–212 B.C.). Substantially, his method for finding the value of π was to inscribe and circumscribe regular polygons about a circle, determine their perimeters, and use these as lower and upper bounds on the circumference. By this means, and using a polygon of 96 sides, he obtained the two approximations in the inequality

$$223/71 < \pi < 22/7.$$

Theorem 14-8 provides insight into why 22/7, the so-called "Archimedean value of π," was used so frequently in place of π; there is no fraction, given in lowest terms, with smaller denominator which furnishes a better approximation. While

$$\left| \pi - \frac{22}{7} \right| \approx 0.0012645 \quad \text{and} \quad \left| \pi - \frac{223}{71} \right| \approx 0.0007476,$$

Archimedes' value of 223/71, which is not a convergent of π, has a denominator exceeding $q_2 = 7$. Our theorem tells us that 333/106 (a ratio for π employed in Europe in the 16th century) will approximate π more closely than any rational number with denominator less than or equal to 106; indeed,

$$\left| \pi - \frac{333}{106} \right| \approx 0.0000832.$$

Due to the size of $q_4 = 33102$, the convergent $p_3/q_3 = 355/113$ allows one to approximate π with a striking degree of accuracy; from the corollary to Theorem 14-7, we have

$$\left| \pi - \frac{355}{113} \right| < \frac{1}{113 \cdot 33102} < \frac{3}{10^7}.$$

The noteworthy ratio of 355/113 was known to the early Chinese mathematician Tsu Chung-chi (430–501); by some reasoning not stated in his works, he gave 22/7 as an "inaccurate value" of π and 355/113 as the "accurate value." The accuracy of the latter ratio was not equalled in Europe until the end of the 16th century, when Adriaen Anthoniszoon (1527–1617) rediscovered the identical value.

This is a convenient place to record a theorem which says that any "close" (in a suitable sense) rational approximation to x must be a convergent to x. There would be a certain neatness to the theory if

$$\left| x - \frac{a}{b} \right| < \frac{1}{b^2}$$

implied that $a/b = p_n/q_n$ for some n; while this is too much to hope for, a slightly sharper inequality guarantees the same conclusion.

THEOREM 14-9.

Let x be an arbitrary irrational number. If the rational number a/b, where $b \geq 1$ and $\gcd(a, b) = 1$, satisfies

$$\left| x - \frac{a}{b} \right| < \frac{1}{2b^2}$$

then a/b is one of the convergents p_n/q_n in the continued fraction representation of x.

Proof: Assume that a/b is not a convergent of x. Knowing that the q_k form an increasing sequence, there exists unique integer n for which $q_n \leq b, < q_{n+1}$. For this n, the last lemma gives the first inequality in the chain

$$|q_n x - p_n| \leq |bx - a| = b\left| x - \frac{a}{b} \right| < \frac{1}{2b},$$

which may be recast as

$$\left| x - \frac{p_n}{q_n} \right| < \frac{1}{2bq_n}.$$

In view of the supposition that $a/b \neq p_n/q_n$, the difference $bp_n - aq_n$ is a non-zero integer, whence $1 \leq |bp_n - aq_n|$. We are able to conclude at once that

$$\frac{1}{bq_n} \leq \left| \frac{bp_n - aq_n}{bq_n} \right| = \left| \frac{p_n}{q_n} - \frac{a}{b} \right| \leq \left| \frac{p_n}{q_n} - x \right| + \left| x - \frac{a}{b} \right| < \frac{1}{2bq_n} + \frac{1}{2b^2}.$$

This produces the contradiction $b < q_n$, ending the proof.

PROBLEMS 14.3

1. Evaluate each of the following infinite simple continued fractions:
    (a) $[\overline{2; 3}]$
    (b) $[0; \overline{1, 2, 3}]$
    (c) $[2; \overline{1, 2, 1}]$
    (d) $[1; 2, \overline{3, 1}]$
    (e) $[1; 2, 1, 2, \overline{12}]$

2. Prove that if the irrational number $x > 1$ is represented by the infinite continued fraction $[a_0; a_1, a_2, \ldots]$, then $1/x$ has the expansion $[0; a_0, a_1, a_2, \ldots]$. Use this fact to find the value of $[0; 1, 1, 1, \ldots] = [0; \overline{1}]$.

3. Evaluate $[1; 2, \overline{1}]$ and $[1; 2, 3, \overline{1}]$.

4. Determine the infinite continued fraction representation of each irrational number below:
    (a) $\sqrt{5}$
    (b) $\sqrt{7}$
    (c) $\dfrac{1 + \sqrt{13}}{2}$
    (d) $\dfrac{5 + \sqrt{37}}{4}$
    (e) $\dfrac{11 + \sqrt{30}}{13}$

5. (a) For any positive integer n, show that $\sqrt{n^2 + 1} = [n; \overline{2n}]$, $\sqrt{n^2 + 2} = [n; \overline{n, 2n}]$ and $\sqrt{n^2 + 2n} = [n; \overline{1, 2n}]$.
    [*Hint:* Notice that

    $$n + \sqrt{n^2 + 1} = 2n + (\sqrt{n^2 + 1} - n) = 2n + \cfrac{1}{n + \sqrt{n^2 + 1}}.]$$

    (b) Use part (a) to obtain the continued fraction representations of $\sqrt{2}$, $\sqrt{3}$, $\sqrt{15}$, and $\sqrt{37}$.

6. Among the convergents of $\sqrt{15}$, find a rational number that will approximate $\sqrt{15}$ with accuracy to four decimal places.

7. (a) Find a rational approximation to $e = [2; 1, 2, 1, 1, 4, 1, 1, 6, \ldots]$ correct to 4 decimal places.
    (b) If a and b are positive integers, show that the inequality $e < a/b < 87/32$ implies that $b \geq 39$.

8. Prove that of any two consecutive convergents of the irrational number x, at least one, a/b, satisfies the inequality

$$\left| x - \frac{a}{b} \right| < \frac{1}{2b^2}.$$

[*Hint:* Since x lies between any two consecutive convergents,

$$\frac{1}{q_n q_{n+1}} = \left| \frac{p_{n+1}}{q_{n+1}} - \frac{p_n}{q_n} \right| = \left| x - \frac{p_{n+1}}{q_{n+1}} \right| + \left| x - \frac{p_n}{q_n} \right|.$$

Now argue by contradiction.]

9. Given the infinite continued fraction $[1; 3, 1, 5, 1, 7, 1, 9, \ldots]$, find the best rational approximation a/b with
    (a) denominator $b < 25$;   (b) denominator $b < 225$.

10. First show that $|(1 + \sqrt{10})/3 - 18/13| < 1/(2 \cdot 13^2)$, and then verify that $18/13$ is a convergent of $(1 + \sqrt{10})/3$.

11. A famous theorem of A. Hurwitz (1891) says that for any irrational number x, there exist infinitely many rational numbers a/b such that

$$\left| x - \frac{a}{b} \right| < \frac{1}{\sqrt{5}b^2}.$$

Taking $x = \pi$, obtain three rational numbers satisfying this inequality.

12. Assume that the continued fraction representation for the irrational number x ultimately becomes periodic. Mimic the method used in Example 14-4 to prove that x is of the form $r + s\sqrt{d}$, where r and $s \neq 0$ are rational numbers and $d > 0$ is a nonsquare integer.

13. Let x be an irrational number with convergents p_n/q_n. For every $n \geq 0$, verify that
     (a) $1/2q_n q_{n+1} < |x - p_n/q_n| < 1/q_n q_{n+1}$;
     (b) the convergents are successively closer to x in the sense that

$$\left| x - \frac{p_n}{q_n} \right| < \left| x - \frac{p_{n-1}}{q_{n-1}} \right|.$$

[*Hint:* Rewrite the relation

$$x = \frac{x_{n+1}q_n - p_{n-1}}{x_{n+1}q_n + q_{n-1}}$$

as $x_{n+1}(xq_n - p_n) = -q_{n-1}(x - p_{n-1}/q_{n-1})$.]

14.4 PELL'S EQUATION

What little action Fermat took to publicize his discoveries came in the form of challenges to other mathematicians. Perhaps he hoped in this way to convince them that his new style of number theory was worth pursuing. In January of 1657, Fermat proposed as a challenge to the European mathematical community—thinking probably in the first place of John Wallis, England's most renowned practitioner before Newton—a pair of problems:

1. Find a cube which, when increased by the sum of its proper divisors, becomes a square; for example, $7^3 + (1 + 7 + 7^2) = 20^2$.
2. Find a square which, when increased by the sum of its proper divisors, becomes a cube.

On hearing of the contest, Fermat's favorite correspondent, Bernhard Frénicle de Bessy, quickly supplied a number of answers to the first problem; typical of these is $(2 \cdot 3 \cdot 5 \cdot 13 \cdot 41 \cdot 47)^3$, which when increased by the sum of its proper divisors becomes $(2^7 \cdot 3^2 \cdot 5^2 \cdot 7 \cdot 13 \cdot 17 \cdot 29)^2$. While Frénicle advanced to solutions in still larger composite numbers, Wallis dismissed the problems as not worth his effort, writing, "Whatever the details of the matter, it finds me too absorbed by numerous occupations for me to be able to devote my attention to it immediately; but I can make at this moment this response: the number 1 in and of itself satisfies both demands."

Barely concealing his disappointment, Frénicle expressed astonishment that a mathematician as experienced as Wallis would have made only the trivial response when, in view of Fermat's stature, he should have sensed the problem's greater depths.

Fermat's interest, indeed, lay in general methods, not in the wearying computation of isolated cases. Both Frénicle and Wallis overlooked the theoretical aspect that the challenge-problems were meant to reveal on careful analysis. While the phrasing was not entirely precise, it seems clear that Fermat had intended the first of his queries to be solved for cubes of prime numbers. To put it otherwise, the problem called for finding all integral solutions of the equation

$$1 + x + x^2 + x^3 = y^2,$$

or equivalently

$$(1 + x)(1 + x^2) = y^2,$$

where x is an odd integer. Since 2 is the only prime which divides both factors on the left-hand side of this equation, it may be written as

$$ab = \left(\frac{y}{2}\right)^2, \quad \gcd(a, b) = 1.$$

But if the product of two relatively prime integers is a perfect square, then each of them must be a square; hence, $a = u^2$, $b = v^2$ for some u and v, so that

$$1 + x = 2a = 2u^2, \quad 1 + x^2 = 2b = 2v^2.$$

This means that any integer x which satisfies Fermat's first problem must be a solution of the pair of equations

$$x = 2u^2 - 1, \quad x^2 = 2v^2 - 1,$$

the second being a particular case of the equation $x^2 = dy^2 \pm 1$.

In February, 1657, Fermat issued his second challenge, dealing directly with the theoretical point at issue: Find a number y which will make $dy^2 + 1$ a perfect square, where d is a positive integer which is not a square; for example, $3 \cdot 1^2 + 1 = 2^2$ and $5 \cdot 4^2 + 1 = 9^2$. If, said Fermat, a general rule cannot be obtained, find the smallest values of y which will satisfy the equations $61y^2 + 1 = x^2$; or $109y^2 + 1 = x^2$. Frénicle proceeded to calculate the smallest positive solutions of $x^2 - dy^2 = 1$ for all permissible values of d up to 150 and suggested that Wallis extend the table to $d = 200$ or at least solve $x^2 - 151y^2 = 1$ and $x^2 - 313y^2 = 1$, hinting that the second equation might be beyond Wallis' ability. In reply, Wallis' patron Lord William Brouncker of Ireland stated that it had only taken him an hour or so to discover that

$$(126862368)^2 - 313(7170685)^2 = -1$$

and so $y = 2 \cdot 7170685 \cdot 126862368$ gives the desired solution to $x^2 - 313y^2 = 1$; Wallis solved the other concrete case, furnishing

$$(1728148040)^2 - 151(140634693)^2 = 1.$$

The size of these numbers in comparison with those arising from other values of d suggests that Fermat was in possession of a complete solution to the problem, but this was never disclosed (later, he affirmed that his method of infinite descent had been used with success to show the existence of an infinitude of solutions of $x^2 - dy^2 = 1$). Brouncker, under the mistaken impression that rational and not necessarily integral values were allowed, had no difficulty in supplying an answer; he simply divided the relation

$$(r^2 + d)^2 - d(2r)^2 = (r^2 - d)^2$$

by the quantity $(r^2 - d)^2$ to arrive at the solution

$$x = \frac{r^2 + d}{r^2 - d}, \quad y = \frac{2r}{r^2 - d}$$

where $r \neq d$ is an arbitrary rational number. This, needless to say, was rejected by Fermat, who wrote that "solutions in fractions, which can be given at once from the merest elements of arithmetic, do not satisfy me." Now informed of all the conditions of the challenge, Brouncker and Wallis jointly devised a tentative method for solving $x^2 - dy^2 = 1$ in integers, without being able to give a proof that it will always work. Apparently the honors rested with Brouncker, for Wallis congratulated Brouncker with some pride that he had "preserved untarnished the fame that Englishmen have won in former times with Frenchmen."

After having said all this, we should record that Fermat's well-directed effort to institute a new tradition in arithmetic through a mathematical joust was largely a failure. Save for Frénicle, who lacked the talent to vie in intellectual combat with Fermat, number theory had no special appeal to any of his contemporaries. The subject was permitted to fall into disuse, until Euler, after the lapse of nearly a century, picked up where Fermat had left off. Both Euler and Lagrange contributed to the resolution of the celebrated problem of 1657. By converting $\sqrt{d}$ into an infinite continued fraction, Euler (in 1759) invented a procedure for obtaining the smallest integral solution of $x^2 - dy^2 = 1$, but he failed to show that the process leads to a solution other than $x = 1, y = 0$. It was left to Lagrange to clear up this matter. Completing the theory left unfinished by Euler, Lagrange in 1768 published the first rigorous proof that all solutions arise through the continued fraction expansion of $\sqrt{d}$.

As a result of a mistaken reference, the central point of contention, the equation $x^2 - dy^2 = 1$, has gone into the literature with the title "Pell's equation." The erroneous attribution of its solution to the English mathematician John Pell (1611–1685), who had little to do with the problem, was an oversight on Euler's part. On a cursory reading of Wallis' *Opera Mathematica* (1693), in which Brouncker's method of solving the equation is set forth as well as information as to Pell's work on Diophantine analysis, Euler must have confused their contributions. By all rights we should call the equation $x^2 - dy^2 = 1$ "Fermat's equation," for he was the first to deal with it systematically. While the historical error has long been recognized, Pell's name is the one that is indelibly attached to the equation.

Whatever the integral value of d, the equation $x^2 - dy^2 = 1$ is satisfied trivially by $x = \pm 1, y = 0$. If $d < -1$, then $x^2 - dy^2 \geq 1$ (except when $x = y = 0$) so that these exhaust the solutions; when $d = -1$, two more solutions occur, namely $x = 0$, $y = \pm 1$. The case in which d is a perfect square is easily dismissed. For if $d = n^2$ for some n, then $x^2 - dy^2 = 1$ can be written in the form

$$(x + ny)(x - ny) = 1$$

which is possible if and only if $x + ny = x - ny = \pm 1$; it follows that

$$x = \frac{(x + ny) + (x - ny)}{2} = \pm 1$$

and the equation has no solutions apart from the trivial ones $x = \pm 1$, $y = 0$.

From now on, we shall restrict our investigation of the Pell equation $x^2 - dy^2 = 1$ to the only interesting situation, that where d is a positive integer which is not a square. Let us say that a solution x, y of this equation is a *positive solution* provided both x and y are positive. Since solutions beyond those with $y = 0$ can be arranged in sets of four by combinations of signs $\pm x$, $\pm y$, it is clear that all solutions will be known once all positive solutions have been found. For this reason, we seek only positive solutions of $x^2 - dy^2 = 1$.

The result which provides us with a starting point asserts that any pair of positive integers satisfying Pell's equation can be obtained from the continued fraction representing the irrational number $\sqrt{d}$.

THEOREM 14-10.

If p, q is a positive solution of $x^2 - dy^2 = 1$, then p/q is a convergent of the continued fraction expansion of $\sqrt{d}$.

Proof: In light of the hypothesis that $p^2 - dq^2 = 1$, we have

$$(p - q\sqrt{d})(p + q\sqrt{d}) = 1$$

implying that $p > q$ as well as that

$$\frac{p}{q} - \sqrt{d} = \frac{1}{q(p + q\sqrt{d})}.$$

As a result,

$$0 < \frac{p}{q} - \sqrt{d} < \frac{\sqrt{d}}{q(q\sqrt{d} + q\sqrt{d})} = \frac{\sqrt{d}}{2q^2\sqrt{d}} = \frac{1}{2q^2}.$$

A direct appeal to Theorem 14-9 indicates the p/q must be a convergent of $\sqrt{d}$.

In general, the converse of the preceding theorem is false: not all of the convergents p_n/q_n of $\sqrt{d}$ supply solutions to $x^2 - dy^2 = 1$. Nonetheless, we can say something about the size of the values taken on by the sequence $p_n^2 - dq_n^2$.

THEOREM 14-11.

If p/q is a convergent of the continued fraction expansion of $\sqrt{d}$, then $x = p$, $y = q$ is a solution of one of the equations

$$x^2 - dy^2 = k,$$

where $|k| < 1 + 2\sqrt{d}$.

Proof: If p/q is a convergent of $\sqrt{d}$, then the corollary to Theorem 14-7 guarantees that

$$\left| \sqrt{d} - \frac{p}{q} \right| < \frac{1}{q^2}$$

and therefore

$$|p - q\sqrt{d}| < \frac{1}{q}.$$

This being so, we have

$$|p + q\sqrt{d}| = |(p - q\sqrt{d}) + 2q\sqrt{d}| \le |p - q\sqrt{d}| + |2q\sqrt{d}|$$

$$< \frac{1}{q} + 2q\sqrt{d} < (1 + 2\sqrt{d})q.$$

These two inequalities combine to yield

$$|p^2 - dq^2| = |p - q\sqrt{d}| \, |p + q\sqrt{d}| < \frac{1}{q}(1 + 2\sqrt{d})q = 1 + 2\sqrt{d},$$

which is precisely what was to be proved.

In illustration let us take the case of $d = 7$. Using the continued fraction expansion $\sqrt{7} = [2; \overline{1, 1, 1, 4}]$, the first few convergents of $\sqrt{7}$ are determined to be

$$2/1, 3/1, 5/2, 8/3, \, \cdots \cdot$$

Running through the calculations of $p_n{}^2 - 7q_n{}^2$, we find that

$$2^2 - 7 \cdot 1^2 = -3, \, 3^2 - 7 \cdot 1^2 = 2, \, 5^2 - 7 \cdot 2^2 = -3, \, 8^2 - 7 \cdot 3^2 = 1,$$

whence $x = 8$, $y = 3$ provides a positive solution of the equation $x^2 - 7y^2 = 1$.

While a rather elaborate study can be made of periodic continued fractions, it is not our intention to explore this area at any length. The reader may have noticed already that in the examples considered so far, the continued fraction expansions of $\sqrt{d}$ all took the form

$$\sqrt{d} = [a_0; \overline{a_1, a_2, \, \cdots, a_n}];$$

that is, the periodic part starts after one term, this initial term being $[\sqrt{d}\,]$. It is also true that the last term a_n of the period is always equal to $2a_0$ and that the period, with the last term excluded, is symmetrical (the symmetrical part may or may not have a middle term). This is typical of the general situation. Without entering into the details of proof, let us simply record the fact: if d is a positive integer which is not a perfect square, then the continued fraction expansion of $\sqrt{d}$ necessarily has the form

$$\sqrt{d} = [a_0; \overline{a_1, a_2, a_3, \, \cdots, a_3, a_2, a_1, 2a_0}].$$

In the case in which $d = 19$, for instance, the expansion is

$$\sqrt{19} = [4; \overline{2, 1, 3, 1, 2, 8}]$$

while $d = 73$ gives

$$\sqrt{73} = [8; \overline{1, 1, 5, 5, 1, 1, 16}].$$

Among all $d < 100$, the longest period is that of $\sqrt{94}$ which has sixteen terms:

$$\sqrt{94} = [9; \overline{1, 2, 3, 1, 1, 5, 1, 8, 1, 5, 1, 1, 3, 2, 1, 18}].$$

The accompanying table lists the continued fraction expansions of $\sqrt{d}$, where d is a nonsquare integer between 2 and 40.

$\sqrt{2} = [1;\overline{2}]$	$\sqrt{22} = [4; \overline{1, 2, 4, 2, 1, 8}]$
$\sqrt{3} = [1; \overline{1,2}]$	$\sqrt{23} = [4; \overline{1, 3, 1, 8}]$
$\sqrt{5} = [2; \overline{4}]$	$\sqrt{24} = [4; \overline{1, 8}]$
$\sqrt{6} = [2; \overline{2, 4}]$	$\sqrt{26} = [5; \overline{10}]$
$\sqrt{7} = [2; \overline{1, 1, 1, 4}]$	$\sqrt{27} = [5; \overline{5, 10}]$
$\sqrt{8} = [2; \overline{1, 4}]$	$\sqrt{28} = [5; \overline{3, 2, 3, 10}]$
$\sqrt{10} = [3; \overline{6}]$	$\sqrt{29} = [5; \overline{2, 1, 1, 2, 10}]$
$\sqrt{11} = [3; \overline{3, 6}]$	$\sqrt{30} = [5; \overline{2, 10}]$
$\sqrt{12} = [3; \overline{2, 6}]$	$\sqrt{31} = [5; \overline{1, 1, 3, 5, 3, 1, 1, 10}]$
$\sqrt{13} = [3; \overline{1, 1, 1, 1, 6}]$	$\sqrt{32} = [5; \overline{1, 1, 1, 10}]$
$\sqrt{14} = [3; \overline{1, 2, 1, 6}]$	$\sqrt{33} = [5; \overline{1, 2, 1, 10}]$
$\sqrt{15} = [3; \overline{1, 6}]$	$\sqrt{34} = [5; \overline{1, 4, 1, 10}]$
$\sqrt{17} = [4; \overline{8}]$	$\sqrt{35} = [5; \overline{1, 10}]$
$\sqrt{18} = [4; \overline{4, 8}]$	$\sqrt{37} = [6; \overline{12}]$
$\sqrt{19} = [4; \overline{2, 1, 3, 1, 2, 8}]$	$\sqrt{38} = [6; \overline{6, 12}]$
$\sqrt{20} = [4; \overline{2, 8}]$	$\sqrt{39} = [6; \overline{4, 12}]$
$\sqrt{21} = [4; \overline{1, 3, 1, 8}]$	$\sqrt{40} = [6; \overline{3, 12}]$

Theorem 14-10 indicates that if the equation $x^2 - dy^2 = 1$ possesses a solution, then its positive solutions are to be found among $x = p_k$, $y = q_k$, where p_k/q_k are the convergents $\sqrt{d}$. The period of the continued fraction expansion of $\sqrt{d}$ provides the information we need to show that $x^2 - dy^2 = 1$ actually does have a solution in integers; in fact, there are infinitely many solutions, all obtainable from the convergents of $\sqrt{d}$. Our proof relies on a lemma.

LEMMA.

Let the convergents of a continued fraction expansion of $\sqrt{d}$ be p_k/q_k. If n is the length of the period of the expansion of $\sqrt{d}$, then

$$p_{kn-1}^2 - dq_{kn-1}^2 = (-1)^{kn} \quad (k = 1, 2, 3, \ldots).$$

Proof: For $k \geq 1$, the continued fraction expansion of $\sqrt{d}$ can be written in the form

$$\sqrt{d} = [a_0; a_1, a_2, \ldots, a_{kn-1}, x_{kn}]$$

where

$$x_{kn} = [2a_0; \overline{a_1, \ldots, a_{n-1}, 2a_0}] = a_0 + \sqrt{d}.$$

As in the proof of Theorem 14-2, we have

$$\sqrt{d} = \frac{x_{kn}p_{kn-1} + p_{kn-2}}{x_{kn}q_{kn-1} + q_{kn-2}}.$$

Upon substituting $x_{kn} = a_0 + \sqrt{d}$ and simplifying, this reduces to

$$\sqrt{d}(a_0 q_{kn-1} + q_{kn-2} - p_{kn-1}) = a_0 p_{kn-1} + p_{kn-2} - dq_{kn-1}.$$

Because the right-hand side is rational and $\sqrt{d}$ is irrational, the foregoing relation requires that

$$a_0 q_{kn-1} + q_{kn-2} = p_{kn-1}, \quad \text{and} \quad a_0 p_{kn-1} + p_{kn-2} = dq_{kn-1}.$$

The effect of multiplying the first of these equations by p_{kn-1} and the second by $-q_{kn-1}$, and then adding them, is

$$p_{kn-1}^2 - dq_{kn-1}^2 = p_{kn-1}q_{kn-2} - q_{kn-1}p_{kn-2}.$$

But Theorem 14-3 informs us that

$$p_{kn-1}q_{kn-2} - q_{kn-1}p_{kn-2} = (-1)^{kn-2} = (-1)^{kn},$$

and so

$$p_{kn-1}^2 - dq_{kn-1}^2 = (-1)^{kn},$$

which results in our lemma.

We can now describe all positive solutions of $x^2 - dy^2 = 1$, where $d > 0$ is a nonsquare integer. We state our main result as

THEOREM 14-12.

Let p_k/q_k be the convergents of the continued fraction expansion of $\sqrt{d}$ and let n be the length of the period of the expansion.

(1) *If n is even, then all positive solutions of $x^2 - dy^2 = 1$ are given by*

$$x = p_{kn-1}, \quad y = q_{kn-1} \qquad\qquad (k = 1, 2, 3, \ldots).$$

(2) *If n is odd, then all positive solutions of $x^2 - dy^2 = 1$ are given by*

$$x = p_{2kn-1}, \quad y = q_{2kn-1} \qquad\qquad (k = 1, 2, 3, \ldots).$$

Proof: It has already been established in Theorem 14-10 that any positive solution x_0, y_0 of $x^2 - dy^2 = 1$ is of the form $x_0 = p_k, y_0 = q_k$ for some convergent p_k/q_k.

Taking the lemma into account, $x = p_{kn-1}, y = q_{kn-1}$ will furnish a solution if and only if $(-1)^{kn} = 1$. When n is even, this condition is satisfied by all integers k; when n is odd, the condition holds if and only if k is an even integer.

Example 14-7

As a first application of Theorem 14-12, let us again consider the equation $x^2 - 7y^2 = 1$. Because $\sqrt{7} = [2; \overline{1, 1, 1, 4}]$, the initial twelve convergents are

$2/1, 3/1, 5/2, 8/3, 37/14, 45/17, 82/31,$

$$127/48, 590/223, 717/271, 1307/494, 2024/765.$$

Since the continued fraction representation of $\sqrt{7}$ has a period of length 4, the numerator and denominator of any of the convergents p_{4k-1}/q_{4k-1} form a solution of $x^2 - 7y^2 = 1$. Thus, for instance,

$$p_3/q_3 = 8/3, \ p_7/q_7 = 127/48, \ p_{11}/q_{11} = 2024/765$$

give rise to the first three positive solutions; these solutions are $x_1 = 8, y_1 = 3$; $x_2 = 127, y_2 = 48$; $x_3 = 2024, y_3 = 765$.

Example 14-8

To find the solution of $x^2 - 13y^2 = 1$ in the smallest positive integers, we note that $\sqrt{13} = [3; \overline{1, 1, 1, 1, 6}]$ and that there is a period of length 5. The first ten convergents of $\sqrt{13}$ are

$$3/1, 4/1, 7/2, 11/3, 18/5, 119/33, 137/38, 256/71, 393/109, 649/180.$$

With reference to part (2) of Theorem 14-12, the least positive solution of $x^2 - 13y^2 = 1$ is obtained from the convergent $p_9/q_9 = 649/180$, the solution itself being $x_1 = 649, y_1 = 180$.

There is a quick way to generate other solutions from a single solution of Pell's equation. Before discussing this, let us define the *fundamental solution* of the equation $x^2 - dy^2 = 1$ to be its smallest positive solution. That is, it is the positive solution x_0, y_0 with the property that $x_0 < x', y_0 < y'$ for any other positive solution x', y'. Theorem 14-12 furnishes the following fact: if the length of the period of the continued fraction expansion of $\sqrt{d}$ is n, then the fundamental solution of $x^2 - dy^2 = 1$ is given by $x = p_{n-1}, y = q_{n-1}$ when n is even; and by $x = p_{2n-1}, y = q_{2n-1}$ when n is odd. Thus the equation $x^2 - dy^2 = 1$ can be solved in either n or $2n$ steps.

Finding the fundamental solution can be a difficult task, since the numbers in this solution can be unexpectedly large, even for comparatively small values of d. For example, the innocent-looking equation $x^2 - 991y^2 = 1$ has the smallest positive solution

$$x = 379, 516, 400, 906, 811, 930, 638, 014, 896, 080,$$
$$y = 12, 055, 735, 790, 331, 359, 447, 442, 538, 767.$$

The situation is even worse with $x^2 - 1000099y^2 = 1$, where the smallest positive integer x satisfying this equation has 1118 digits. Needless to say, everything depends upon the continued fraction expansion of $\sqrt{d}$ and, in the case of $\sqrt{1000099}$, the period consists of 2174 terms.

It can also happen that the integers needed to solve $x^2 - dy^2 = 1$ are small for a given value of d and very large for the succeeding value. A striking illustration of this variation is provided by the equation $x^2 - 61y^2 = 1$, whose fundamental solution is given by

$$x = 17{,}663{,}319{,}049, \quad y = 226{,}153{,}980.$$

These numbers are enormous when compared with the case $d = 60$, where the solution is $x = 31$, $y = 4$ or with $d = 62$, where the solution is $x = 63$, $y = 8$.

With the help of the fundamental solution—which can be found by means of continued fractions or by successively substituting $y = 1, 2, 3, \ldots$ into the expression $1 + dy^2$ until it becomes a perfect square—we are able to construct all the remaining positive solutions.

THEOREM 14-13.

> Let x_1, y_1 be the fundamental solution of $x^2 - dy^2 = 1$. Then every pair of integers x_n, y_n defined by the condition
>
> $$x_n + y_n\sqrt{d} = (x_1 + y_1\sqrt{d})^n \qquad (n = 1, 2, 3, \ldots)$$
>
> is also a positive solution.

Proof: It is a modest exercise for the reader to check that

$$x_n - y_n\sqrt{d} = (x_1 - y_1\sqrt{d})^n.$$

Further, because x_1 and y_1 are positive, x_n and y_n are both positive integers. Bearing in mind that x_1, y_1 is a solution of $x^2 - dy^2 = 1$, we obtain

$$\begin{aligned}
x_n^2 - dy_n^2 &= (x_n + y_n\sqrt{d})(x_n - y_n\sqrt{d}) \\
&= (x_1 + y_1\sqrt{d})^n(x_1 - y_1\sqrt{d})^n \\
&= (x_1^2 - dy_1^2)^n = 1^n = 1,
\end{aligned}$$

and so x_n, y_n is a solution.

Let us pause for a moment to look at an example. By inspection, it is seen that $x_1 = 6$, $y_1 = 1$ forms the fundamental solution of $x^2 - 35y^2 = 1$. A second positive solution x_2, y_2 can be obtained from the formula

$$x_2 + y_2\sqrt{35} = (6 + \sqrt{35})^2 = 71 + 12\sqrt{35},$$

which implies that $x_2 = 71$, $y_2 = 12$. These integers satisfy the equation $x^2 - 35y^2 = 1$, since

$$71^2 - 35 \cdot 12^2 = 5041 - 5040 = 1.$$

A third positive solution arises from

$$x_3 + y_3\sqrt{35} = (6 + \sqrt{35})^3$$

$$= (71 + 12\sqrt{35})(6 + \sqrt{35}) = 846 + 143\sqrt{35}.$$

This gives $x_3 = 846$, $y_3 = 143$ and in fact

$$846^2 - 35 \cdot 143^2 = 715716 - 715715 = 1,$$

so that these values provide another solution.

Returning to the equation $x^2 - dy^2 = 1$, our final theorem tells us that any positive solution can be calculated from the formula

$$x_n + y_n\sqrt{d} = (x_1 + y_1\sqrt{d})^n,$$

where n takes on integral values; that is, if u, v is a positive solution of $x^2 - dy^2 = 1$, then $u = x_n$, $v = y_n$ for a suitably chosen integer n. We state this as

THEOREM 14-14.

If x_1, y_1 is the fundamental solution of $x^2 - dy^2 = 1$, then every positive solution of the equation is given by x_n, y_n, where x_n and y_n are the integers determined from

$$x_n + y_n\sqrt{d} = (x_1 + y_1\sqrt{d})^n \qquad (n = 1, 2, 3, \ldots)$$

Proof: In anticipation of a contradiction, let us suppose that there exists a positive solution u, v which is not obtainable by the formula $(x_1 + y_1\sqrt{d})^n$. Since $x_1 + y_1\sqrt{d} > 1$, the powers of $x_1 + y_1\sqrt{d}$ become arbitrarily large; this means that $u + v\sqrt{d}$ must lie between two consecutive powers of $x_1 + y_1\sqrt{d}$, say,

$$(x_1 + y_1\sqrt{d})^n < u + v\sqrt{d} < (x_1 + y_1\sqrt{d})^{n+1}$$

or, to phrase it in different terms,

$$x_n + y_n\sqrt{d} < u + v\sqrt{d} < (x_n + y_n\sqrt{d})(x_1 + y_1\sqrt{d}).$$

On multiplying this inequality by the positive number $x_n - y_n\sqrt{d}$ and noting that $x_n^2 - dy_n^2 = 1$, we are led to

$$1 < (x_n - y_n\sqrt{d})(u + v\sqrt{d}) < x_1 + y_1\sqrt{d}.$$

Next define the integers r and s by $r + s\sqrt{d} = (x_n - y_n\sqrt{d})(u + v\sqrt{d})$; that is, let

$$r = x_n u - y_n v d, \quad s = x_n v - y_n u.$$

An easy calculation reveals that

$$r^2 - ds^2 = (x_n^2 - dy_n^2)(u^2 - dv^2) = 1$$

and so r, s is a solution of $x^2 - dy^2 = 1$ satisfying

$$1 < r + s\sqrt{d} < x_1 + y_1\sqrt{d}.$$

Completion of the proof requires us to show that the pair r, s is a positive solution. Because $1 < r + s\sqrt{d}$ and $(r + s\sqrt{d})(r - s\sqrt{d}) = 1$, we find that $0 < r - s\sqrt{d} < 1$. In consequence,

$$2r = (r + s\sqrt{d}) + (r - s\sqrt{d}) > 1 + 0 > 0$$

$$s\sqrt{d} = (r + s\sqrt{d}) - (r - s\sqrt{d}) > 1 - 1 = 0$$

which makes both r and s positive. The upshot is that since x_1, y_1 is the fundamental solution of $x^2 - dy^2 = 1$, we must have $x_1 < r$ and $y_1 < s$; but then $x_1 + y_1\sqrt{d} < r + s\sqrt{d}$, violating an earlier inequality. This contradiction ends our argument.

Pell's equation has attracted mathematicians throughout the ages. There is historical evidence that methods for solving the equation were known to the Greeks some 400 years before the beginning of the Christian era. A famous problem of indeterminate analysis known as the "cattle problem" is contained in an epigram sent by Archimedes to Eratosthenes as a challenge to Alexandrian scholars. In it, one is required to find the number of bulls and cows of each of four colors, the eight unknown quantities being connected by nine conditions. These conditions ultimately involve the solution of the Pell equation

$$x^2 - 4729494y^2 = 1,$$

which leads to enormous numbers; one of the eight unknown quantities is a figure having 206545 digits (assuming that 15 printed digits take up one inch of space, the number would be over $1/5$ of a mile long). While it is generally agreed that the problem originated with the celebrated mathematician of Syracuse, no one contends that Archimedes actually carried through all the necessary computations.

Such equations and dogmatic rules, without any proof, for calculating their solutions spread to India more than a thousand years before they appeared in Europe. In the 7th century, Brahmagupta said that a person who can within a year solve the equation $x^2 - 92y^2 = 1$ is a mathematician; for those days, he would at least have to be a good arithmetician, since $x = 1151$, $y = 120$ is the smallest positive solution. A computationally more difficult task would be to find integers satisfying $x^2 - 94y^2 = 1$, for here the fundamental solution is given by $x = 2143295$, $y = 221064$.

Fermat was not the first therefore to propose solving the equation $x^2 - dy^2 = 1$, or even to devise a general method of solution. He was perhaps the first to assert that the equation has an infinitude of solutions whatever the value of the nonsquare integer d. Moreover, his effort to elicit purely integral solutions to both this and other problems was a watershed in number theory, breaking away as it did from the classical tradition of Diophantus' *Arithmetica*.

PROBLEMS 14.4

1. If x_0, y_0 is a positive solution of the equation $x^2 - dy^2 = 1$, prove that $x_0 > y_0$.

2. By the technique of successively substituting $y = 1, 2, 3, \ldots$ into $dy^2 + 1$, determine the smallest positive solution of $x^2 - dy^2 = 1$ when d is
    (a) 7; (b) 11; (c) 18; (d) 30; (e) 39.

3. Find all positive solutions of the following equations for which $y < 250$:
   (a) $x^2 - 2y^2 = 1$; (b) $x^2 - 3y^2 = 1$; (c) $x^2 - 5y^2 = 1$.

4. Show that there is an infinitude of even integers n with the property that both $n + 1$ and $n/2 + 1$ are perfect squares. Exhibit two such integers.

5. Indicate two positive solutions of each of the equations below:
   (a) $x^2 - 23y^2 = 1$; (b) $x^2 - 26y^2 = 1$; (c) $x^2 - 33y^2 = 1$.

6. Find the fundamental solutions of
   (a) $x^2 - 29y^2 = 1$; (b) $x^2 - 41y^2 = 1$; (c) $x^2 - 74y^2 = 1$.
   [*Hint:* $\sqrt{41} = [6; \overline{2, 2, 12}]$ and $\sqrt{74} = [8; \overline{1, 1, 1, 1, 16}]$.]

7. Exhibit a solution of each of the following equations:
   (a) $x^2 - 13y^2 = -1$; (b) $x^2 - 29y^2 = -1$; (c) $x^2 - 41y^2 = -1$.

8. Establish that if x_0, y_0 is a solution of the equation $x^2 - dy^2 = -1$, then $x = 2x_0^2 + 1$, $y = 2x_0 y_0$ satisfies $x^2 - dy^2 = 1$. Brouncker used this fact in solving $x^2 - 313y^2 = 1$.

9. If d is divisible by a prime $p \equiv 3 \pmod 4$, show that the equation $x^2 - dy^2 = -1$ has no solution.

10. If x_1, y_1 is the fundamental solution of $x^2 - dy^2 = 1$ and
$$x_n + y_n\sqrt{d} = (x_1 + y_1\sqrt{d})^n \quad (n = 1, 2, 3, \ldots),$$
prove that the pair of integers x_n, y_n can be calculated from the formulas
$$x_n = \tfrac{1}{2}\left[(x_1 + y_1\sqrt{d})^n + (x_1 - y_1\sqrt{d})^n\right]$$
$$y_n = \frac{1}{2\sqrt{d}}\left[(x_1 + y_1\sqrt{d})^n - (x_1 - y_1\sqrt{d})^n\right].$$

11. Verify that the integers x_n, y_n in the previous problem can be defined inductively either by
$$x_{n+1} = x_1 x_n + dy_1 y_n$$
$$y_{n+1} = x_1 y_n + x_n y_1,$$
for $n = 1, 2, 3, \ldots$, or by
$$x_{n+1} = 2x_1 x_n - x_{n-1}$$
$$y_{n+1} = 2x_1 y_n - y_{n-1}$$
for $n = 2, 3, \ldots$.

12. Using the information that $x_1 = 15, y_1 = 2$ is the fundamental solution of $x^2 - 56y^2 = 1$, determine two more positive solutions.

13. (a) Prove that whenever the equation $x^2 - dy^2 = c$ is solvable, then it has infinitely many solutions. [*Hint:* If u, v satisfy $x^2 - dy^2 = c$ and r, s satisfy $x^2 - dy^2 = 1$, then $(ur \pm dvs)^2 - d(us \pm vr)^2 = (u^2 - dv^2)(r^2 - ds^2) = c$.]
    (b) Given that $x = 16, y = 6$ is a solution of $x^2 - 7y^2 = 4$, obtain two other positive solutions.
    (c) Given that $x = 18, y = 3$ is a solution of $x^2 - 35y^2 = 9$, obtain two other positive solutions.

14. Apply the theory of this section to confirm that there exist infinitely many primitive Pythagorean triples x, y, z in which x and y are consecutive integers. [*Hint:* Note the identity $(s^2 - t^2) - 2st = (s - t)^2 - 2t^2$.]

15. The *Pell numbers* p_n and q_n are defined by

$$p_0 = 0, \, p_1 = 1, \quad p_n = 2p_{n-1} + p_{n-2} \qquad (n \geq 2),$$

$$q_0 = 1, \, q_1 = 1, \quad q_n = 2q_{n-1} + q_{n-2} \qquad (n \geq 2).$$

This gives us the two sequences

$$0, 1, 2, 5, 12, 29, 70, 169, 408, \ldots$$

$$1, 1, 3, 7, 17, 41, 99, 239, 577, \ldots$$

If $\alpha = 1 + \sqrt{2}$ and $\beta = 1 - \sqrt{2}$, show that the Pell numbers can be expressed as

$$p_n = \frac{\alpha^n - \beta^n}{2\sqrt{2}}, \quad q_n = \frac{\alpha^n + \beta^n}{2}$$

for $n \geq 0$. [*Hint:* Mimic the argument on page 273, noting that α and β are roots of the equation $x^2 - 2x - 1 = 0$.]

16. For the Pell numbers, derive the relations below, where $n \geq 1$:

(a) $p_{2n} = 2p_n q_n$,

(b) $p_n + p_{n-1} = q_n$,

(c) $2q_n^2 - q_{2n} = (-1)^n$,

(d) $p_n + p_{n+1} + p_{n+3} = 3p_{n+2}$,

(e) $q_n^2 - 2p_n^2 = (-1)^n$, hence q_n/p_n are the convergents of $\sqrt{2}$.

15

Some Twentieth-Century Developments

"As with everything else, so with a mathematical theory:
beauty can be perceived, but not explained."
ARTHUR CAYLEY

15.1 HARDY, DICKSON, AND ERDÖS

The vitality of any field of mathematics is maintained only as long as its practitioners continue to ask (and to find answers to) interesting and worthwhile questions. Thus far, our study of number theory has shown how that process has worked from its classical beginnings to the present day. The reader has acquired a working knowledge of how number theory is developed and has seen that the field is still very much alive and growing. This brief closing chapter indicates several of the more promising directions that growth has taken in the twentieth century.

We begin by looking at some contributions of three prominent number theorists from this century, each from a different country: Godfrey H. Hardy, Leonard E. Dickson, and Paul Erdös. In considerably advancing our mathematical knowledge, they are worthy successors to the great masters of the past.

For more than a quarter of a century G. H. Hardy (1877–1947) dominated English mathematics through both the significance of his work and the force of his personality. Hardy entered Cambridge University in 1896 and joined its faculty in 1906 as a lecturer in mathematics, a position he continued to hold until 1919. Perhaps his greatest service to mathematics in this early period was his well-known book *A Course in Pure Mathematics*. England had had a great tradition in applied mathematics, starting with Newton, but in 1900 pure mathematics was at a low ebb there. *A Course in Pure Mathematics* was designed to give the undergraduate student a rigorous exposition of the basic ideas of analysis. Running through numerous editions and translated into several languages, it transformed the trend of university teaching in mathematics.

Godfrey Harold Hardy
(1877–1947)
(Master and Fellows of Trinity College, Cambridge)

Hardy's anti-war stand excited strong negative feelings at Cambridge and, in 1919, he was only too ready to accept the Savilian chair in geometry at Oxford. He was succeeded on the Cambridge staff by John E. Littlewood. Eleven years later Hardy returned to Cambridge, where he remained until his retirement in 1942.

Hardy's name is inevitably linked with that of Littlewood, with whom he carried on the most prolonged (35 years), extensive and fruitful partnership in the history of mathematics. They wrote nearly 100 papers together, the last appearing a year after Hardy's death. It was often joked that there were only three great English mathematicians in those days: Hardy, Littlewood and Hardy-Littlewood. (One mathematician, upon meeting Littlewood for the first time, exclaimed, "I thought that you were merely a name used by Hardy for those papers which he did not think were quite good enough to publish under his own name.")

There are very few areas of number theory to which Hardy did not make a significant contribution. A major interest of his was Waring's Problem; that is, the question of representing an arbitrary positive integer as the sum of at most $g(k)$ k'th powers (see Section 12.3). The general theorem that $g(k)$ is finite for all k was first proved by Hilbert in 1909 using an argument that shed no light on how many k'th powers are needed. In a series of papers published during the 1920s, Hardy and Littlewood obtained upper bounds on $G(k)$, which is defined to be the least number of k'th powers required to represent all sufficiently large integers. They showed (1921) that $g(k) \leq (k-2)2^{k-1} + 5$ for all k; and, more particularly, that $g(4) \leq 19$, $g(5) \leq 41$, $g(6) \leq 87$ and $g(7) \leq 193$. Another of their results (1925) is that for "almost all" positive integers $g(4) \leq 15$, while $g(k) \leq (1/2\, k - 1)2^{k-1} + 3$ when $k = 3$ or $k \geq 5$. Since $79 = 4 \cdot 2^4 + 15 \cdot 1^4$ requires 19 4'th powers, $G(4) \geq 19$; this, together with the bound $g(4) \leq 19$ suggested that $g(4) = 19$ and raised the possibility that its actual value could be settled by computation.

Another topic that drew the attention of the two collaborators was the classical three-primes problem: can every odd integer $n \geq 7$ be written as the sum of three prime numbers? In 1922, Hardy and Littlewood proved that if certain hypotheses are made then there exists a positive number N such that every odd integer $n \geq N$ is a sum of three primes. They also found an approximate formula for the number of such representations of n. I. M. Vinogradov later obtained the Hardy-Littlewood conclusion without invoking their hypotheses. All the Hardy-Littlewood papers stimulated a vast amount of further research by many mathematicians.

L. E. Dickson (1874–1954) was prominent among a small circle of those who greatly influenced the rapid development of American mathematics at the turn of the century. He received the first doctorate in mathematics from the newly founded University of Chicago in 1896, became an assistant professor there in 1900, and remained at Chicago until his retirement in 1939.

Reflecting the abstract interests of his thesis advisor, the distinguished E. H. Moore, Dickson initially pursued the study of finite groups. By 1906, Dickson's prodigious output had already reached 126 papers. He would jokingly remark that, while his honeymoon was a success, he managed to get only two research articles written then. His monumental *History of the Theory of Numbers* (1919), which appeared in three volumes totaling more than 1600 pages, took nine years to complete; by itself this would have been a life's work for an ordinary man. One of the century's most prolific mathematicians, Dickson wrote 267 papers and 18 books covering a broad range of topics in his field. An enduring bit of legend is his barb against applicable mathematics: "Thank God that number theory is unsullied by applications." (Expressing much the same view, Hardy is reported to have made the toast: "Here's to pure mathematics! May it never have any use.") In recognition of his work, Dickson was the first recipient of the Cole Prize in algebra and number theory, awarded in 1928 by the American Mathematical Society.

Dickson stated that he always wished to work in number theory, and that he wrote the *History of the Theory of Numbers* so he could know all that had been done on the subject. He was particularly interested in the existence of perfect numbers, abundant and deficient numbers, and Waring's Problem. A typical result of his investigations was to list (in 1914) all the odd abundant numbers less than 15000.

In a long series of papers beginning in 1927, Dickson gave an almost complete solution of the original form of Waring's problem. His final result (in 1936) was that, for nearly all k, $g(k)$ assumes the ideal value $g(k) = 2^k + [(3/2)^k] - 2$, as was conjectured by Euler in 1772. Dickson obtained a simple arithmetic condition on k for insuring that the foregoing formula for $g(k)$ held, and showed that the condition was satisfied for k between 7 and 400. With the dramatic increase in computer power, it is now known that Euler's conjecture for $g(k)$ holds when k is between 2 and 471,600,000.

Paul Erdös (born 1913–), who is often described as one of the greatest living mathematicians, is unique in mathematical folklore. The son of two high school teachers of mathematics, his genius became apparent at a very early age. Erdös entered the University of Budapest when he was 17 and graduated four years later with a Ph.D. in mathematics. As a college freshman, he published his first paper, which was a simple proof of Bertrand's conjecture that for any $n > 1$ there is always a prime between n and $2n$.

After a four-year fellowship at Manchester University, England, Erdös adopted the lifestyle of a wandering scholar, a "Professor of the Universe." He has traveled the world constantly, often visiting as many as 15 universities and research centers in a month. (Where Gauss's motto was "Few, but ripe," Erdös takes as his the words "Another roof, another proof.") Although Erdös has never held a regular academic appointment, he has standing offers at several institutions where he may pause for short periods. In his total dedication to mathematical research, Erdös has dispensed with the pleasures and possessions of daily life. He has neither property nor fixed address, carries no money and has never cooked anything, not even boiled water for tea; a few close friends handle his financial affairs, including filing his income tax. A generous person, Erdös is apt to give away the small honoraria he picks up from his lectures, or use them to fund two scholarships that he set up for young mathematicians—one in Hungary and one in Israel.

Erdös's work in number theory is always substantial, and frequently monumental. In 1949, he and Alte Selberg independently published "elementary—though not easy—proofs of what is called the Prime Number Theorem. (It asserts that $\pi(x) \approx x \log x$, where $\pi(x)$ is the number of primes $p \leq x$.) This veritable sensation among number theorists helped earn Selberg a Fields Medal (1950) and Erdös a Cole Prize (1952). Erdös received the prestigious Wolf Prize in 1983 for outstanding achievement in mathematics; of the $50,000 award he retained only $750 for himself.

Erdös has published, either alone or jointly, more than 1200 papers. With over 300 co-authors, he has collaborated with more people than any other mathematician. As a spur to his collaborators, Erdös attaches monetary rewards to problems that he has been unable to solve. The rewards generally range from $10 to $10,000, depending on his assessment of the difficulty of the problem. The inducement to obtain a solution is not as much financial as prestigious, for there is a certain notoriety associated with owning a check bearing Erdös's name. The following reflect the range of questions that he would like to see answered:

1. Does there exist an odd integer which is not of the form $2^k + n$, with n square-free?
2. Are there infinitely many primes p (such as $p = 101$) for which $p - k$ is composite whenever $1 \leq k! < p$?
3. Is it true that, for all $k > 8$, 2^k cannot be written as the sum of distinct powers of 3? [Note that $2^8 = 3^5 + 3^2 + 3 + 1$.]
4. If $p(n)$ is the largest prime factor of n, does the inequality $p(n) > p(n + 1) > p(n + 2)$ have an infinite number of solutions?
5. Given an infinite sequence of integers, the sum of whose reciprocals diverges, does the sequence contain arbitrarily long arithmetic progressions? ($3000 offered for an answer)

Through a host of problems and conjectures like these Paul Erdös has stimulated two generations of number theorists.

A word about a current trend: computation has always been an important investigative tool in number theory. Therefore it is not surprising that number theorists were among the first mathematicians to exploit the research potential of modern electronic computers. The general availability of computing machinery has given rise to a new branch of our discipline, called Computational Number Theory. Among its wide spectrum of activities, this subject is concerned with testing the primality of given integers, finding lower bounds for odd perfect numbers, discovering new pairs of twin primes and amicable numbers, and obtaining numerical solutions to certain Diophantine equations (such as $x^2 + 999 = y^3$). Another fruitful line of work is to verify special cases of conjectures, or to produce counterexamples to them; for instance, in regard to the conjecture that there exist pseudoprimes of the form $2^n - 2$, a computer search found the pseudoprime $2^{465794} - 2$. The problem of factoring large composite numbers has been of continuing computational interest. The most dramatic feat of this kind was the recently completed factorization of the eleventh Fermat number F_{11}, an integer of 617 digits. The calculations, which involved over 360 million multiplications, produced a prime factor having 564 digits. No doubt number-theory records will continue to fall with the application of new techniques and new technology.

15.2 THE PRIME NUMBER THEOREM

Although the sequence of prime numbers exhibits great irregularities of detail, a trend is definitely apparent in the large. The celebrated Prime Number Theorem allows one to predict, at least in gross terms, how many primes there are less than a given number. It states that if the number is n, then there are about n divided by log n (here, log n denotes the natural logarithm of n) primes before it. Thus the Prime Number Theorem tells us how the primes are distributed "in the large," or "on the average," or "in a probability sense."

One measure of the distribution of primes is the function $\pi(x)$ which, for any real number x, represents the number of primes that do not exceed x; in symbols, $\pi(x) = \Sigma_{p \leq x} 1$. In Chapter 3, we proved that there are infinitely many primes, which is simply an expression of the fact that $\lim_{x \to \infty} \pi(x) = \infty$. Going in the other direction, it is clear that the prime numbers become on the average more widely spaced in the higher parts of any table of primes; in informal terms, one might say that almost all of the positive integers are composite.

By way of justifying our last assertion, let us show that the limit $\lim_{x \to \infty} \pi(x)/x = 0$. Since $\pi(x)/x \geq 0$ for all $x > 0$, the problem is reduced to proving that $\pi(x)/x$ can be made arbitrarily small by choosing x sufficiently large. In more precise terms, what we shall prove is that if $\epsilon > 0$ is any number, then there must exist some positive integer N such that $\pi(x)/x < \epsilon$ whenever $x \geq N$.

To start, let n be a positive integer and use Bertrand's conjecture to pick a prime p with $2^{n-1} < p \leq 2^n$. Then $p \mid (2^n)!$, but $p \nmid (2^{n-1})!$, so that the binomial coefficient $\binom{2^n}{2^n - 1}$ is divisible by p. This leads to the inequalities

$$2^{2^n} \geq \binom{2^n}{2^n - 1} \geq \prod_{2^{n-1} < p \leq 2^n} p \geq (2^{n-1})^{\pi(2^n) - \pi(2^{n-1})}$$

and, upon taking the exponents of 2 on each side, the subsequent inequality

$$(1) \qquad\qquad \pi(2^n) - \pi(2^{n-1}) \leq \frac{2^n}{n-1}.$$

If we successively set $n = 2k, 2k - 1, 2k - 2, \ldots, 3$ in inequality (1) and add the resulting inequalities, we get

$$\pi(2^{2k}) - \pi(2^2) \leq \sum_{r=3}^{2k} \frac{2^r}{r-1}.$$

But $\pi(2^2) < 2^2$ trivially, so that

$$\pi(2^{2k}) < \sum_{r=2}^{2k} \frac{2^r}{r-1} = \sum_{r=2}^{k} \frac{2^r}{r-1} + \sum_{r=k+1}^{2k} \frac{2^r}{r-1}.$$

In the last two sums, let us replace the denominators $r - 1$ by 1 and k respectively to arrive at

$$\pi(2^{2k}) < \sum_{r=2}^{k} 2^r + \sum_{r=k+1}^{2k} 2^r/k < 2^{k+1} + 2^{2k+1}/k.$$

Since $k < 2^k$, we have $2^{k+1} < 2^{2k+1}/k$ for $k \geq 2$ and so

$$\pi(2^{2k}) < 2(2^{2k+1}/k) = 4(2^{2k}/k),$$

which can be written as

(2) $$\pi(2^{2k})/2^{2k} < 4/k.$$

With this inequality available, our argument proceeds rapidly to its conclusion. Given any real number $x \geq 2$, there exists a unique integer k satisfying $2^{2k-2} < x \leq 2^{2k}$. From (2), it follows that

$$\pi(x)/x \leq \pi(2^{2k})/x < \pi(2^{2k})/2^{2k-2} = 4(\pi(2^{2k})/2^{2k}) < 16/k.$$

If we now take $x \geq N = 2^{2([16/\epsilon]+1)}$, then $k \geq [16/\epsilon] + 1$; hence,

$$\pi(x)/x < 16/([16/\epsilon] + 1) < \epsilon,$$

as desired.

A well-known conjecture of Hardy and Littlewood, dating from 1923, is that

$$\pi(x + y) \leq \pi(x) + \pi(y)$$

for all integers x, y with $2 \leq y \leq x$. Written as $\pi(x + y) - \pi(y) \leq \pi(x)$, the inequality asserts that no interval $y < k \leq x + y$ of length x can contain as many prime numbers as there are in the interval $0 < k \leq x$. While the conjecture has been checked for $x + y \leq 100000$, it appears likely that there will be exceptions which, even though rare, will prove the conjecture false. The computations simply have not gone far enough to produce the first counterexample. Curiously there is no counterexample when $x = y$, since it has been shown (1975) that the inequality $\pi(2x) < 2\pi(x)$ holds for all $x \geq 11$.

It was Euler (probably about 1740) who introduced into analysis the zeta function

$$\zeta(s) = \sum_{n=1}^{\infty} 1/n^s = 1^{-s} + 2^{-s} + 3^{-s} + \cdots,$$

the function on whose properties the proof of the Prime Number Theorem ultimately depended. Euler's fundamental contribution to the subject is the formula representing $\zeta(s)$ as a convergent infinite product; namely,

$$\zeta(s) = \prod_p (1 - 1/p^s)^{-1}, \qquad\qquad (s > 1)$$

where p runs through all primes. Its importance arises from the fact that it asserts equality of two expressions of which one contains the primes explicitly while the other does not. Euler considered $\zeta(s)$ as a function of a real variable only, but his formula nonetheless indicates the existence of a deep-lying connection between the theory of primes and the analytic properties of the zeta function.

Euler's expression for $\zeta(s)$ results from expanding each of the factors in the right-hand member as

$$\frac{1}{1 - 1/p^s} = 1 + 1/p^s + (1/p^s)^2 + (1/p^s)^3 + \cdots$$

and observing that their product is the sum of all terms of the form

$$\frac{1}{(p_1^{k_1} p_2^{k_2} \cdots p_r^{k_r})^s}$$

where $p_1, \cdots, p_r$ are distinct primes. Since every positive integer n can be written uniquely as a product of prime powers, each term $1/n^s$ appears once and only once in this sum; that is, the sum simply is $\sum_{n=1}^{\infty} 1/n^s$.

It turns out that Euler's formula for the zeta function leads to a deceptively short proof of the infinitude of primes: the occurrence of a finite product on the right-hand side would contradict the fact that $\lim_{s \to 1} \zeta(s) = \infty$.

Legendre was the first to make any significant conjecture about functions which give a good approximation to $\pi(x)$ for large values of x. In his book *Essai sur la Théorie des Nombres* (1798), Legendre ventured that $\pi(x)$ is approximately equal to the function

$$\frac{x}{\log x - 1.08366}.$$

By compiling extensive tables on how the primes distribute themselves in blocks of 1000 consecutive integers, Gauss reached the conclusion that $\pi(x)$ increases at roughly the same rate as each of the functions $x/\log x$ and

$$\mathrm{Li}(x) = \int_2^x \frac{du}{\log u}$$

with the logarithmic integral $\mathrm{Li}(x)$ providing a much closer numerical approximation. Gauss' observations were communicated in a letter to the noted astronomer Johann Encke in 1849, and first published in 1863, but appear to have begun as early as 1791 when Gauss was fourteen years old—needless to say, well before Legendre's treatise was written.

It is interesting to compare these remarks with the evidence of the tables:

x	$\pi(x)$	$\dfrac{x}{\log x - 1.08366}$	$x/\log x$	$\mathrm{Li}(x)$	$\pi(x)/(x/\log x)$
1000	168	172	145	178	1.159
10,000	1,229	1,231	1086	1246	1.132
100,000	9,592	9,588	8,686	9,630	1.104
1,000,000	78,498	78,534	72,382	78,628	1.084
10,000,000	664,579	665,138	620,420	664,918	1.071
100,000,000	5,761,455	5,769,341	5,428,681	5,762,209	1.061

The first demonstrable progress towards comparing $\pi(x)$ with $x/\log x$ was made by the Russian mathematician P. L. Tchebychef. In 1850, he proved that there exist positive constants a and b, $a < 1 < b$, such that

$$a(x/\log x) < \pi(x) < b(x/\log x)$$

for sufficiently large x. Tchebychef also showed that if the quotient $\pi(x)/(x/\log x)$ has a limit as x increases, then its value must be 1. Tchebychef's work, fine as it is, is a record of failure: what he could not establish is that the foregoing limit does in fact exist, and, as he failed to do this, he failed to prove the Prime Number Theorem. It was not until some 45 years later that the final gap was filled.

We might observe at this point that Tchebychef's result implies that the series $\Sigma_p \, 1/p$, extended over all primes, diverges. To see this, let p_n be the nth prime, so that $\pi(p_n) = n$. Since we have

$$\pi(x) > a(x/\log x)$$

for sufficiently large x, it follows that the inequality

$$n = \pi(p_n) > a(p_n/\log p_n) > \sqrt{p_n}$$

holds if n is taken sufficiently large. But $n^2 > p_n$ leads to $\log p_n < 2 \log n$ and so we get

$$ap_n < n \log p_n < 2n \log n$$

when n is large. In consequence, the series $\Sigma_{n=1}^{\infty} \, 1/p_n$ will diverge in comparison with the known divergent series $\Sigma_{n=2}^{\infty} \, (1/n \log n)$.

A point of passing interest is that V. Brun, around 1920, showed that the twin primes are so sparse that the sum of their reciprocals converges to $1.9021604 \pm 5\cdot10^{-7}$.

The radically new ideas which were to furnish the key to a proof of the Prime Number Theorem were introduced by Bernhard Riemann in his epoch-making memoir *Über die Anzahl der Primzahlen unter einer gegebenen Grösse* of 1859 (his only paper on the theory of numbers). Where Euler had restricted the zeta function $\zeta(s)$ to real values of s, Riemann recognized the connection between the distribution of primes and the behavior of $\zeta(s)$ as a function of a complex variable $s = a + bi$. He enunciated a number of properties of the zeta function, together with a remarkable identity, known as Riemann's Explicit Formula, relating $\pi(x)$ to the zeroes of $\zeta(s)$ in the s-plane. The result has caught the imagination of most mathematicians because it is so unexpected, connecting two seemingly unrelated areas in mathematics; namely, number theory, which is the study of the discrete, and complex analysis, which deals with continuous processes.

In his memoir, Riemann made a number of conjectures concerning the distribution of the zeroes of the zeta function. The most famous is the so-called Riemann Hypothesis which asserts that all the nonreal zeroes of $\zeta(s)$ are at points $\frac{1}{2} + bi$ of the complex plane; that is, they lie on the "critical line" $Re(s) = 1/2$. In 1914, G. H. Hardy provided the first concrete result by proving that there are infinitely many zeros of $\zeta(s)$ on the critical line. A series of large computations has been made, culminating in the recent verification that the Riemann Hypothesis holds for all of the first $(1.5)10^9$ zeros, an effort that involved over a thousand hours on a modern supercomputer. This famous conjecture has never been proved or disproved, and it is undoubtedly the most important unsolved problem in mathematics today.

Riemann's investigations were exploited by Jacques Hadamard and Charles de la Vallée Poussin who in 1896, independently of each other and almost simultaneously, succeeded in proving that

$$\lim_{x \to \infty} \frac{\pi(x)}{x/\log x} = 1.$$

The result expressed in this formula has since become known as the Prime Number Theorem. De la Vallée Poussin went considerably further in his research. He showed that, for sufficiently large values of x, $\pi(x)$ is more accurately represented by the logarithmic integral $\text{Li}(x)$ than by the function

$$\frac{x}{\log x - A},$$

no matter what value is assigned to the constant A, and that the most favorable choice of A in Legendre's function is 1. This is at variance with Legendre's original contention that $A = 1.08366$, but his estimate (based on tables extending only as far as $x = 400{,}000$) had long been recognized as having little more than historical interest.

Today a good deal more is known about the relationship between $\pi(x)$ and $\text{Li}(x)$. We shall only mention a theorem of Littlewood to the effect that the difference $\pi(x) - \text{Li}(x)$ assumes both positive and negative values infinitely often as x runs over all positive integers. Littlewood's result is a pure "existence theorem" and no numerical value of x for which $\pi(x) - \text{Li}(x)$ is positive has ever been found. It is a curious fact that an upper bound on the size of the first x satisfying $\pi(x) > \text{Li}(x)$ is available; such an x must occur someplace before

$$e^{e^{e^{e^{79}}}} \approx 10^{10^{10^{34}}},$$

a number of incomprehensible magnitude. This upper limit, obtained by S. Skewes in 1933, has gone into the literature under the name of the Skewes number. Somewhat later (1955), Skewes decreased the top exponent in his number from 34 to 3. In 1986, this bound was reduced considerably when it was proved that there are more than 10^{180} successive integers x between $(6.62)10^{370}$ and $(6.69)10^{370}$ for which $\pi(x) > \text{Li}(x)$. However, an explicit numerical value of x is still beyond the reach of any computer. What is perhaps remarkable is that $\pi(x) < \text{Li}(x)$ for all x at which $\pi(x)$ has been calculated exactly; that is, for all x in the range $x < 4 \cdot 10^{16}$. Some values are given in the following table:

x	$\pi(x)$	$\mathrm{Li}(x) - \pi(x)$
10^9	50,847,543	1701
10^{10}	455,052,511	3104
10^{11}	4,118,054,813	11,588
10^{12}	37,607,912,018	38,263
10^{13}	346,065,536,839	108,971
10^{14}	3,204,941,750,802	314,890
10^{15}	29,844,570,422,669	1,052,619
10^{16}	279,238,341,033,925	3,214,632
$4 \cdot 10^{16}$	1,075,292,778,753,150	5,538,861

Although this table gives the impression that $\mathrm{Li}(x) - \pi(x)$ is always positive and gets larger as x increases, negative values will eventually overwhelm the positive ones.

A useful sidelight to the Prime Number Theorem deserves our attention; to wit,

$$\lim_{n \to \infty} \frac{n \log n}{p_n} = 1.$$

For, starting with the relation

$$\lim_{x \to \infty} \frac{\pi(x) \log x}{x} = 1,$$

we may take logarithms and use the fact that the logarithmic function is continuous to obtain

$$\lim_{x \to \infty} \left[\log \pi(x) + \log(\log x) - \log x\right] = 0$$

or equivalently

$$\lim_{x \to \infty} \frac{\log \pi(x)}{\log x} = 1 - \lim_{x \to \infty} \frac{\log(\log x)}{\log x}.$$

But $\lim_{x \to \infty} \log(\log x)/\log x = 0$, which leads to

$$\lim_{x \to \infty} \log \pi(x)/\log x = 1.$$

We then get

$$1 = \lim_{x \to \infty} \frac{\pi(x) \log x}{x} = \lim_{x \to \infty} \frac{\pi(x) \log \pi(x)}{x} \cdot \frac{\log x}{\log \pi(x)}$$

$$= \lim_{x \to \infty} \frac{\pi(x) \log \pi(x)}{x}.$$

Setting $x = p_n$, so that $\pi(p_n) = n$, the result

$$\lim_{n \to \infty} \frac{n \log n}{p_n} = 1$$

follows. This may be interpreted as asserting that if there are n primes in an interval, then the length of the interval is roughly $n \log n$.

Until recent times, the opinion prevailed that the Prime Number Theorem could not be proved without the help of the properties of the zeta function, and without recourse to complex function theory. It came as a great surprise when in 1949 the Norwegian mathematician Atle Selberg discovered a purely arithmetical proof. His paper *An Elementary Proof of the Prime Number Theorem* is "elementary" in the technical sense of avoiding the methods of modern analysis; indeed, its content is exceedingly difficult. Selberg was awarded a Fields medal at the 1950 International Congress of Mathematicians for his work in this area. The Fields Medal is considered to be the equivalent in mathematics of a Nobel Prize. (It is generally believed that Alfred Nobel's bad relations with the Swedish mathematician Gösta Mittag-Leffler was the reason that Nobel did not establish a prize in mathematics.)

Appendixes

General References

Adams, W., and L. Goldstein. 1976. *Introduction to Number Theory*. Englewood Cliffs, N.J.: Prentice-Hall.

Agnew, Jeanne. 1972. *Exploring Number Theory*. Monterey, Calif.: Brooks/Cole.

Andrews, George. 1971. *Number Theory*. Philadelphia: W. B. Saunders.

Archibald, Ralph. 1970. *An Introduction to the Theory of Numbers*. Columbus, Ohio: Charles E. Merrill.

Baker, Alan. 1984. *A Concise Introduction to the Theory of Numbers*. Cambridge, England: Cambridge University Press.

Barnett, I. A. 1972. *Elements of Number Theory*. Rev. ed. Boston: Prindle, Weber & Schmidt.

Beck, A., M. Bleicher, and D. Crowe. 1969. *Excursions into Mathematics*. New York: Worth.

Beiler, A. H. 1966. *Recreations in the Theory of Numbers*. 2d ed. New York: Dover.

Burton, David. 1991. *The History of Mathematics: An Introduction*. 2d ed. Dubuque, Iowa: Wm. C. Brown.

Dantzig, Tobias. 1956. *Number: The Language of Science*. Garden City, N.Y.: Doubleday.

Dickson, Leonard. 1920. *History of the Theory of Numbers*. Vols. 1, 2, 3. Washington, D.C.: Carnegie Institute of Washington. (Reprinted, New York: Chelsea, 1952.)

Dudley, Underwood. 1978. *Elementary Number Theory,* 2d ed. New York: W. H. Freeman.

Edwards, Harold. 1977. *Fermat's Last Theorem*. New York: Springer-Verlag.

Eves, Howard. 1990. *An Introduction to the History of Mathematics*. 6th ed. Philadelphia: Saunders College Publishing.

Guy, Richard. 1981. *Unsolved Problems in Number Theory*. New York: Springer-Verlag.

Hardy, G. H., and E. M. Wright. 1975. *An Introduction to the Theory of Numbers*. 5th ed. London: Oxford University Press.

Heath, Thomas. 1910. *Diophantus of Alexandria*. Cambridge, England: Cambridge University Press. (Reprinted, New York: Dover, 1964.)

Hoggatt, Jr., and E. Verner. 1969. *Fibonacci and Lucas Numbers*. Boston: Houghton Mifflin.

Ireland, K., and M. Rosen. 1990. *A Classical Introduction to Modern Number Theory*. 2d ed. New York: Springer-Verlag.

Landau, E. 1952. *Elementary Number Theory*. Translated by J. Goodman, New York: Chelsea.

Le Veque, William. 1977. *Fundamentals of Number Theory*. Reading, Mass.: Addison-Wesley. (Reprinted, New York: Dover, 1990.)

Long, Calvin. 1972. *Elementary Introduction to Number Theory*. 2d ed. Lexington, Mass.: D. C. Heath.

Maxfield, J., and M. Maxfield. 1972. *Discovering Number Theory*. Philadelphia: W. B. Saunders.

Nagell, Trygve. 1964. *Introduction to Number Theory*. 2d ed. New York: Chelsea.

Niven, I., H. Zuckerman, and H. Montgomery. 1991. *An Introduction to the Theory of Numbers*. 5th ed. New York: John Wiley and Sons.

Ogilvy, C. S., and J. Anderson. 1966. *Excursions in Number Theory*. New York: Oxford University Press.

Olds, Carl D. 1963. *Continued Fractions*. New York: Random House.

Ore, Oystein. 1948. *Number Theory and Its History*. New York: McGraw-Hill. (Reprinted, New York: Dover, 1988.)

———. 1967. *Invitation to Number Theory*. New York: Random House.

Ribenboim, Paulo. 1979. *13 Lectures on Fermat's Last Theorem*. New York: Springer-Verlag.

———. 1988. *The Book of Prime Number Records*. New York: Springer-Verlag.

———. 1991. *The Little Book of Big Primes*. New York: Springer-Verlag.

Riesel, Hans. 1985. *Prime Numbers and Computer Methods for Factorization*. Boston: Birkhauser.

Roberts, Joe. 1977. *Elementary Number Theory*. Cambridge, Mass.: MIT Press.

Rose, H. E. 1988. *A Course in Number Theory*. New York: Oxford University Press.

Rosen, Kenneth. 1992. *Elementary Number Theory and Its Applications*. 3d ed. Reading, Mass.: Addison-Wesley.

Scharlu, W., and H. Opolka. 1984. *From Fermat to Minkowski*. New York: Springer-Verlag.

Schroeder, Manfred. 1987. *Number Theory in Science and Communication*. 2d ed. New York: Springer-Verlag.

Shanks, Daniel. 1985. *Solved and Unsolved Problems in Number Theory*. 3d ed. New York: Chelsea.

Shapiro, Harold. 1983. *Introduction to the Theory of Numbers*. New York: John Wiley and Sons.

Shoemaker, Richard. 1973. *Perfect Numbers*. Washington, D.C.: National Council of Teachers of Mathematics.

Sierpinski, Waclaw. 1988. *Elementary Theory of Numbers*. Translated by A. Hulaniki. 2d ed. Amsterdam: North-Holland.

————. 1962. *Pythagorean Triangles*. Translated by A. Sharma. New York: Academic Press.

Starke, Harold. 1970. *An Introduction to Number Theory*. Chicago: Markham.

Stewart, B. M. 1964. *Theory of Numbers*. 2d ed. New York: Macmillan.

Struik, Dirk. 1969. *A Source Book in Mathematics 1200–1800*. Cambridge: Harvard University Press.

Upensky, J., and M. A. Heaslet. 1939. *Elementary Number Theory*. New York: McGraw-Hill.

Vanen Eyden, Charles. 1987. *Elementary Number Theory*, 2d ed. New York: Random House.

Vaja, S. 1989. *Fibonacci and Lucas Numbers and the Golden Section: Theory and Applications*. Chichester, England: Ellis Horwood.

Vorobyov, N. 1963. *The Fibonacci Numbers*. Boston: D. C. Heath.

Weil, Andre. 1984. *Number Theory: An Approach through History*. Boston: Birkhauser.

Welsh, Dominic. 1988. *Codes and Cryptography*. New York: Oxford University Press.

Suggested Further Reading

Berndt, Bruce. "Ramanujan—100 Years Old (Fashioned) or 100 Years New (Fangled)?" *The Mathematical Intelligencer* 10, No. 3 (1988): 24–29.

Bezuska, Stanley. "Even Perfect Numbers—An Update." *Mathematics Teacher* 74(1981): 460–63.

Brown, Ezra. "The First Proof of the Quadratic Reciprocity Law, Revisited." *American Mathematical Monthly* 88(1981): 257–64.

Collison, Mary Joan. "The Unique Factorization Theorem: From Euclid to Gauss." *Mathematics Magazine* 53(1980): 96–100.

Cox, David. "Quadratic Reciprocity: Its Conjecture and Application." *American Mathematical Monthly* 95(1988): 442–48.

Devlin, Keith. "Factoring Fermat Numbers." *New Scientist* 111, No. 1527(1986): 41–44.

Dixon, John. "Factorization and Primality Tests." *American Mathematical Monthly* 91(1984): 333–51.

Dudley, Underwood. "Formulas for Primes." *Mathematics Magazine* 56(1983): 17–22.

Edwards, Harold. "Euler and Quadratic Recriprocity." *Mathematics Magazine* 56(1983): 285–91.

Erdős, Paul. "Some Unconventional Problems in Number Theory." *Mathematics Magazine* 52(1979): 67–70.

Erdős, Paul. "On Some of My Problems in Number Theory I would Most Like to See Solved." In *Lecture Notes in Mathematics* 1122. New York: Springer-Verlag, 1985: 74–84.

———. "Some Remarks and Problems in Number Theory Related to the Work of Euler." *Mathematics Magazine* 56(1983): 292–98.

Feistel, Horst. "Cryptography and Computer Security." *Scientific American* 228 (May 1973): 15–23.

Francis, Richard. "Mathematical Haystacks: Another Look at Repunit Numbers." *College Mathematics Journal* 19 (1988): 240–46.

Gardner, Martin. "Simple Proofs of the Pythagorean Theorem, and Sundry Other Matters." *Scientific American* 211(Oct. 1964): 118–26.

———. "A Short Treatise on the Useless Elegance of Perfect Numbers and Amicable Pairs." *Scientific American* 218(March 1968): 121–26.

———. "The Fascination of the Fibonacci Sequence." *Scientific American* 220(March 1969): 116–20.

———. "Diophantine Analysis and the Problem of Fermat's Legendary 'Last Theorem.'" *Scientific American* 223(July 1970): 117–19.

———. "On Expressing Integers as the Sums of Cubes and Other Unsolved Number-Theory Problems." *Scientific American* 229(Dec. 1973): 118–21.

———. "A New Kind of Cipher That Would Take Millions of Years to Break." *Scientific American* 237(Aug. 1977): 120–24.

———. "Patterns in Primes Are a Clue to the Strong Law of Small Numbers." *Scientific American* 243(Dec. 1980): 18–28.

Goldstein, Larry. "A History of the Prime Number Theorem." *American Mathematical Monthly* 80(1973): 599–615.

Guy, Richard. "The Strong Law of Small Numbers." *American Mathematical Monthly* 95 (1988): 697–712.

———. "The Second Strong Law of Small Numbers." *Mathematics Magazine* 63 (1990): 1–20.

Hoffman, Paul. "The Man Who Loved Numbers." *The Atlantic* 260 (Nov. 1987): 60–74.

Honsberger, Ross. "An Elementary Gem Concerning $\pi(n)$, the Number of Primes $< n$." *Two-Year College Mathematics Journal* 11(1980): 305–11.

Lee, Elvin, and Joseph Madachy. "The History and Discovery of Amicable Numbers—Part I." *Journal of Recreational Mathematics* 5(1972): 77–93.

Luciano, Dennis, and Gordon Prichett. "Cryptography: From Caesar Ciphers to Public-Key Cryptosystems." *College Mathematics Journal* 18 (1987): 2–17.

Mahoney, Michael. "Fermat's Mathematics: Proofs and Conjectures." *Science* 178(Oct. 1972): 30–36.

Matkovic, David. "The Chinese Remainder Theorem: An Historical Account." *Pi Mu Epsilon Journal* 8 (1988): 493–502.

McCarthy, Paul. "Odd Perfect Numbers." *Scripta Mathematica* 23(1957): 43–47.

Ondrejka, Rudolf. "Ten Extraordinary Primes." *Journal of Recreational Mathematics* 18(1985–86): 87–92.

Pomerance, Carl. "Recent Developments in Primality Testing." *The Mathematical Intelligencer* 3(1981): 97–105.

———. "The Search for Prime Numbers." *Scientific American* 247(Dec. 1982): 122–30.

Reid, Constance. "Perfect Numbers." *Scientific American* 88(March 1953): 84–86.

Ribenboim, Paulo. "Lecture: Recent Results on Fermat's Last Theorem." *Canadian Mathematical Bulletin* 20(1977): 229–42.

Schroeder, Manfred. "Where Is the Next Mersenne Prime Hiding?" *The Mathematical Intelligencer* 5, No. 3 (1983): 31–33.

Sierpinski, Waclaw. "On Some Unsolved Problems of Arithmetic." *Scripta Mathematica* 25(1960): 125–36.

Silverman, Robert. "A Perspective on Computation in Number Theory." *Notices of the American Mathematical Society* 38 (1991): 562–568.

Slowinski, David. "Searching for the 27th Mersenne Prime." *Journal of Recreational Mathematics* 11(1978–79): 258–61.

Small, Charles. "Waring's Problem." *Mathematics Magazine* 50(1977): 12–16.

Stewart, Ian. "The Formula Man." *New Scientist* 1591 (Dec. 17, 1987): 24–28.

Uhler, Horace. "A Brief History of the Investigations on Mersenne Numbers and the Latest Immense Primes." *Scripta Mathematica* 18(1952): 122–31.

Vandiver, H. S. "Fermat's Last Theorem." *American Mathematical Monthly* 53(1946): 555–78.

Wagon, Stan. "Fermat's Last Theorem." *The Mathematical Intelligencer* 8, No. 1 (1986): 59–61.

———. "Carmichael's 'Empirical Theorem.' " *The Mathematical Intelligencer* 8, No. 2 (1986): 61–63.

Yates, Samuel. "Peculiar Properties of Repunits." *Journal of Recreational Mathematics* 2(1969): 139–46.

———. "The Mystique of Repunits." *Mathematics Magazine* 51(1978): 22–28.

Tables

TABLE 1

The least primitive root r of each prime p, where 2 ≤ p < 1000.

p	r	p	r	p	r	p	r	p	r	p	r
2	1	127	3	283	3	467	2	661	2	877	2
3	2	131	2	293	2	479	13	673	5	881	3
5	2	137	3	307	5	487	3	677	2	883	2
7	3	139	2	311	17	491	2	683	5	887	5
11	2	149	2	313	10	499	7	691	3	907	2
13	2	151	6	317	2	503	5	701	2	911	17
17	3	157	5	331	3	509	2	709	2	919	7
19	2	163	2	337	10	521	3	719	11	929	3
23	5	167	5	347	2	523	2	727	5	937	5
29	2	173	2	349	2	541	2	733	6	941	2
31	3	179	2	353	3	547	2	739	3	947	2
37	2	181	2	359	7	557	2	743	5	953	3
41	6	191	19	367	6	563	2	751	3	967	5
43	3	193	5	373	2	569	3	757	2	971	6
47	5	197	2	379	2	571	3	761	6	977	3
53	2	199	3	383	5	577	5	769	11	983	5
59	2	211	2	389	2	587	2	773	2	991	6
61	2	223	3	397	5	593	3	787	2	997	7
67	2	227	2	401	3	599	7	797	2		
71	7	229	6	409	21	601	7	809	3		
73	5	233	3	419	2	607	3	811	3		
79	3	239	7	421	2	613	2	821	2		
83	2	241	7	431	7	617	3	823	3		
89	3	251	6	433	5	619	2	827	2		
97	5	257	3	439	15	631	3	829	2		
101	2	263	5	443	2	641	3	839	11		
103	5	269	2	449	3	643	11	853	2		
107	2	271	6	457	13	647	5	857	3		
109	6	277	5	461	2	653	2	859	2		
113	3	281	3	463	3	659	2	863	5		

TABLE 2

The smallest prime factor of each odd integer n, $3 \leq n \leq 4999$, *not divisible by* 5; *a dash in the table indicates that n is itself prime.*

1		101	—	201	3	301	7	401	—
3	—	103	—	203	7	303	3	403	13
7	—	107	—	207	3	307	—	407	11
9	3	109	—	209	11	309	3	409	—
11	—	111	3	211	—	311	—	411	3
13	—	113	—	213	3	313	—	413	7
17	—	117	3	217	7	317	—	417	3
19	—	119	7	219	3	319	11	419	—
21	3	121	11	221	13	321	3	421	—
23	—	123	3	223	—	323	17	423	3
27	3	127	—	227	—	327	3	427	7
29	—	129	3	229	—	329	7	429	3
31	—	131	—	231	3	331	—	431	—
33	3	133	7	233	—	333	3	433	—
37	—	137	—	237	3	337	—	437	19
39	3	139	—	239	—	339	3	439	—
41	—	141	3	241	—	341	11	441	3
43	—	143	11	243	3	343	7	443	—
47	—	147	3	247	13	347	—	447	3
49	7	149	—	249	3	349	—	449	—
51	3	151	—	251	—	351	3	451	11
53	—	153	3	253	11	353	—	453	3
57	3	157	—	257	—	357	3	457	—
59	—	159	3	259	7	359	—	459	3
61	—	161	7	261	3	361	19	461	—
63	3	163	—	263	—	363	3	463	—
67	—	167	—	267	3	367	—	467	—
69	3	169	13	269	—	369	3	469	7
71	—	171	3	271	—	371	7	471	3
73	—	173	—	273	3	373	—	473	11
77	7	177	3	277	—	377	13	477	3
79	—	179	—	279	3	379	—	479	—
81	3	181	—	281	—	381	3	481	13
83	—	183	3	283	—	383	—	483	3
87	3	187	11	287	7	387	3	487	—
89	—	189	3	289	17	389	—	489	3
91	7	191	—	291	3	391	17	491	—
93	3	193	—	293	—	393	3	493	17
97	—	197	—	297	3	397	—	497	7
99	3	199	—	299	13	399	3	499	—

TABLE 2

501	3	601	—	701	—	801	3	901	17
503	—	603	3	703	19	803	11	903	3
507	3	607	—	707	7	807	3	907	—
509	—	609	3	709	—	809	—	909	3
511	7	611	13	711	3	811	—	911	—
513	3	613	—	713	23	813	3	913	11
517	11	617	—	717	3	817	19	917	7
519	3	619	—	719	—	819	3	919	—
521	—	621	3	721	7	821	—	921	3
523	—	623	7	723	3	823	—	923	13
527	17	627	3	727	—	827	—	927	3
529	23	629	17	729	3	829	—	929	—
531	3	631	—	731	17	831	3	931	7
533	13	633	3	733	—	833	7	933	3
537	3	637	7	737	11	837	3	937	—
539	7	639	3	739	—	839	—	939	3
541	—	641	—	741	3	841	29	941	—
543	3	643	—	743	—	843	3	943	23
547	—	647	—	747	3	847	7	947	—
549	3	649	11	749	7	849	3	949	13
551	19	651	3	751	—	851	23	951	3
553	7	653	—	753	3	853	—	953	—
557	—	657	3	757	—	857	—	957	3
559	13	659	—	759	3	859	—	959	7
561	3	661	—	761	—	861	3	961	31
563	—	663	3	763	7	863	—	963	3
567	3	667	23	767	13	867	3	967	—
569	—	669	3	769	—	869	11	969	3
571	—	671	11	771	3	871	13	971	—
573	3	673	—	773	—	873	3	973	7
577	—	677	—	777	3	877	—	977	—
579	3	679	7	779	19	879	3	979	11
581	7	681	3	781	11	881	—	981	3
583	11	683	—	783	3	883	—	983	—
587	—	687	3	787	—	887	—	987	3
589	19	689	13	789	3	889	7	989	23
591	3	691	—	791	7	891	3	991	—
593	—	693	3	793	13	893	19	993	3
597	3	697	17	797	—	897	3	997	—
599	—	699	3	799	17	899	29	999	3

TABLE 2

1001	7	1101	3	1201	—	1301	—	1401	3
1003	17	1103	—	1203	3	1303	—	1403	23
1007	19	1107	3	1207	17	1307	—	1407	3
1009	—	1109	—	1209	3	1309	7	1409	—
1011	3	1111	11	1211	7	1311	3	1411	17
1013	—	1113	3	1213	—	1313	13	1413	3
1017	3	1117	—	1217	—	1317	3	1417	13
1019	—	1119	3	1219	23	1319	—	1419	3
1021	—	1121	19	1221	3	1321	—	1421	7
1023	3	1123	—	1223	—	1323	3	1423	—
1027	13	1127	7	1227	3	1327	—	1427	—
1029	3	1129	—	1229	—	1329	3	1429	—
1031	—	1131	3	1231	—	1331	11	1431	3
1033	—	1133	11	1233	3	1333	31	1433	—
1037	17	1137	3	1237	—	1337	7	1437	3
1039	—	1139	17	1239	3	1339	13	1439	—
1041	3	1141	7	1241	17	1341	3	1441	11
1043	7	1143	3	1243	11	1343	17	1443	3
1047	3	1147	31	1247	29	1347	3	1447	—
1049	—	1149	3	1249	—	1349	19	1449	3
1051	—	1151	—	1251	3	1351	7	1451	—
1053	3	1153	—	1253	7	1353	3	1453	—
1057	7	1157	13	1257	3	1357	23	1457	31
1059	3	1159	19	1259	—	1359	3	1459	—
1061	—	1161	3	1261	13	1361	—	1461	3
1063	—	1163	—	1263	3	1363	29	1463	7
1067	11	1167	3	1267	7	1367	—	1467	3
1069	—	1169	7	1269	3	1369	37	1469	13
1071	3	1171	—	1271	31	1371	3	1471	—
1073	29	1173	3	1273	19	1373	—	1473	3
1077	3	1177	11	1277	—	1377	3	1477	7
1079	13	1179	3	1279	—	1379	7	1479	3
1081	23	1181	—	1281	3	1381	—	1481	—
1083	3	1183	7	1283	—	1383	3	1483	—
1087	—	1187	—	1287	3	1387	19	1487	—
1089	3	1189	29	1289	—	1389	3	1489	—
1091	—	1191	3	1291	—	1391	13	1491	3
1093	—	1193	—	1293	3	1393	7	1493	—
1097	—	1197	3	1297	—	1397	11	1497	3
1099	7	1199	11	1299	3	1399	—	1499	—

TABLE 2

1501	19	1601	—	1701	3	1801	—	1901	—
1503	3	1603	7	1703	13	1803	3	1903	11
1507	11	1607	—	1707	3	1807	13	1907	—
1509	3	1609	—	1709	—	1809	3	1909	23
1511	—	1611	3	1711	29	1811	—	1911	3
1513	17	1613	—	1713	3	1813	7	1913	—
1517	37	1617	3	1717	17	1817	23	1917	3
1519	7	1619	—	1719	3	1819	17	1919	19
1521	3	1621	—	1721	—	1821	3	1921	17
1523	—	1623	3	1723	—	1823	—	1923	3
1527	3	1627	—	1727	11	1827	3	1927	41
1529	11	1629	3	1729	7	1829	31	1929	3
1531	—	1631	7	1731	3	1831	—	1931	—
1533	3	1633	23	1733	—	1833	3	1933	—
1537	29	1637	—	1737	3	1837	11	1937	13
1539	3	1639	11	1739	37	1839	3	1939	7
1541	23	1641	3	1741	—	1841	7	1941	3
1543	—	1643	31	1743	3	1843	19	1943	29
1547	7	1647	3	1747	—	1847	—	1947	3
1549	—	1649	17	1749	3	1849	43	1949	—
1551	3	1651	13	1751	17	1851	3	1951	—
1553	—	1653	3	1753	—	1853	17	1953	3
1557	3	1657	—	1757	7	1857	3	1957	19
1559	—	1659	3	1759	—	1859	11	1959	3
1561	7	1661	11	1761	3	1861	—	1961	37
1563	3	1663	—	1763	41	1863	3	1963	13
1567	—	1667	—	1767	3	1867	—	1967	7
1569	3	1669	—	1769	29	1869	3	1969	11
1571	—	1671	3	1771	7	1871	—	1971	3
1573	11	1673	7	1773	3	1873	—	1973	—
1577	19	1677	3	1777	—	1877	—	1977	3
1579	—	1679	23	1779	3	1879	—	1979	—
1581	3	1681	41	1781	13	1881	3	1981	7
1583	—	1683	3	1783	—	1883	7	1983	3
1587	3	1687	7	1787	—	1887	3	1987	—
1589	7	1689	3	1789	—	1889	—	1989	3
1591	37	1691	19	1791	3	1891	31	1991	11
1593	3	1693	—	1793	11	1893	3	1993	—
1597	—	1697	—	1797	3	1897	7	1997	—
1599	3	1699	—	1799	7	1899	3	1999	—

TABLE 2

2001	3	2101	11	2201	31	2301	3	2401	7
2003	—	2103	3	2203	—	2303	7	2403	3
2007	3	2107	7	2207	—	2307	3	2407	29
2009	7	2109	3	2209	47	2309	—	2409	3
2011	—	2111	—	2211	3	2311	—	2411	—
2013	3	2113	—	2213	—	2313	3	2413	19
2017	—	2117	29	2217	3	2317	7	2417	—
2019	3	2119	13	2219	7	2319	3	2419	41
2021	43	2121	3	2221	—	2321	11	2421	3
2023	7	2123	11	2223	3	2323	23	2423	—
2027	—	2127	3	2227	17	2327	13	2427	3
2029	—	2129	—	2229	3	2329	17	2429	7
2031	3	2131	—	2231	23	2331	3	2431	11
2033	19	2133	3	2233	7	2333	—	2433	3
2037	3	2137	—	2237	—	2337	3	2437	—
2039	—	2139	3	2239	—	2339	—	2439	3
2041	13	2141	—	2241	3	2341	—	2441	—
2043	3	2143	—	2243	—	2343	3	2443	7
2047	23	2147	19	2247	3	2347	—	2447	—
2049	3	2149	7	2249	13	2349	3	2449	31
2051	7	2151	3	2251	—	2351	—	2451	3
2053	—	2153	—	2253	3	2353	13	2453	11
2057	11	2157	3	2257	37	2357	—	2457	3
2059	29	2159	17	2559	3	2359	7	2459	—
2061	3	2161	—	2261	7	2361	3	2461	23
2063	—	2163	3	2263	31	2363	17	2463	3
2067	3	2167	11	2267	—	2367	3	2467	—
2069	—	2169	3	2269	—	2369	23	2469	3
2071	19	2171	13	2271	3	2371	—	2471	7
2073	3	2173	41	2273	—	2373	3	2473	—
2077	31	2177	7	2277	3	2377	—	2477	—
2079	3	2179	—	2279	43	2379	3	2479	37
2081	—	2181	3	2281	—	2381	—	2481	3
2083	—	2183	37	2283	3	2383	—	2483	13
2087	—	2187	3	2287	—	2387	7	2487	3
2089	—	2189	11	2289	3	2389	—	2489	19
2091	3	2191	7	2291	29	2391	3	2491	47
2093	7	2193	3	2293	—	2393	—	2493	3
2097	3	2197	13	2297	—	2397	3	2497	11
2099	—	2199	3	2299	11	2399	—	2499	3

TABLE 2

2501	41	2601	3	2701	37	2801	—	2901	3
2503	—	2603	19	2703	3	2803	—	2903	—
2507	23	2607	3	2707	—	2807	7	2907	3
2509	13	2609	—	2709	3	2809	53	2909	—
2511	3	2611	7	2711	—	2811	3	2911	41
2513	7	2613	3	2713	—	2813	29	2913	3
2517	3	2617	—	2717	11	2817	3	2917	—
2519	11	2619	3	2719	—	2819	—	2919	3
2521	—	2621	—	2721	3	2821	7	2921	23
2523	3	2623	43	2723	7	2823	3	2923	37
2527	7	2627	37	2727	3	2827	11	2927	—
2529	3	2629	11	2729	—	2829	3	2929	29
2531	—	2631	3	2731	—	2831	19	2931	3
2533	17	2633	—	2733	3	2833	—	2933	7
2537	43	2637	3	2737	7	2837	—	2937	3
2539	—	2639	7	2739	3	2839	17	2939	—
2541	3	2641	19	2741	—	2841	3	2941	17
2543	—	2643	3	2743	13	2843	—	2943	3
2547	3	2647	—	2747	41	2847	3	2947	7
2549	—	2649	3	2749	—	2849	7	2949	3
2551	—	2651	11	2751	3	2851	—	2951	13
2553	3	2653	7	2753	—	2853	3	2953	—
2557	—	2657	—	2757	3	2857	—	2957	—
2559	3	2659	—	2759	31	2859	3	2959	11
2561	13	2661	3	2761	11	2861	—	2961	3
2563	11	2663	—	2763	3	2863	7	2963	—
2567	17	2667	3	2767	—	2867	47	2967	3
2569	7	2669	17	2769	3	2869	19	2969	—
2571	3	2671	—	2771	17	2871	3	2971	—
2573	31	2673	3	2773	47	2873	13	2973	3
2577	3	2677	—	2777	—	2877	3	2977	13
2579	—	2679	3	2779	7	2879	—	2979	3
2581	29	2681	7	2781	3	2881	43	2981	11
2583	3	2683	—	2783	11	2883	3	2983	19
2587	13	2687	—	2787	3	2887	—	2987	29
2589	3	2689	—	2789	—	2889	3	2989	7
2591	—	2691	3	2791	—	2891	7	2991	3
2593	—	2693	—	2793	3	2893	11	2993	41
2597	7	2697	3	2797	—	2897	—	2997	3
2599	23	2699	—	2799	3	2899	13	2999	—

TABLE 2

3001	—	3101	7	3201	3	3301	—	3401	19
3003	3	3103	29	3203	—	3303	3	3403	41
3007	31	3107	13	3207	3	3307	—	3407	—
3009	3	3109	—	3209	—	3309	3	3409	7
3011	—	3111	3	3211	13	3311	7	3411	3
3013	23	3113	11	3213	3	3313	—	3413	—
3017	7	3117	3	3217	—	3317	31	3417	3
3019	—	3119	—	3219	3	3319	—	3419	13
3021	3	3121	—	3221	—	3321	3	3421	11
3023	—	3123	3	3223	11	3323	—	3423	3
3027	3	3127	53	3227	7	3327	3	3427	23
3029	13	3129	3	3229	—	3329	—	3429	3
3031	7	3131	31	3231	3	3331	—	3431	47
3033	3	3133	13	3233	53	3333	3	3433	—
3037	—	3137	—	3237	3	3337	47	3437	7
3039	3	3139	43	3239	41	3339	3	3439	19
3041	—	3141	3	3241	7	3341	13	3441	3
3043	17	3143	7	3243	3	3343	—	3443	11
3047	11	3147	3	3247	17	3347	—	3447	3
3049	—	3149	47	3249	3	3349	17	3449	—
3051	3	3151	23	3251	—	3351	3	3451	7
3053	43	3153	3	3253	—	3353	7	3453	3
3057	3	3157	7	3257	—	3357	3	3457	—
3059	7	3159	3	3259	—	3359	—	3459	3
3061	—	3161	29	3261	3	3361	—	3461	—
3063	3	3163	—	3263	13	3363	3	3463	—
3067	—	3167	—	3267	3	3367	7	3467	—
3069	3	3169	—	3269	7	3369	3	3469	—
3071	37	3171	3	3271	—	3371	—	3471	3
3073	7	3173	19	3273	3	3373	—	3473	23
3077	17	3177	3	3277	29	3377	11	3477	3
3079	—	3179	11	3279	3	3379	31	3479	7
3081	3	3181	—	3281	17	3381	3	3481	59
3083	—	3183	3	3283	7	3383	17	3483	3
3087	3	3187	—	3287	19	3387	3	3487	11
3089	—	3189	3	3289	11	3389	—	3489	3
3091	11	3191	—	3291	3	3391	—	3491	—
3093	3	3193	31	3293	37	3393	3	3493	7
3097	19	3197	23	3297	3	3397	43	3497	13
3099	3	3199	7	3299	—	3399	3	3499	—

TABLE 2

3501	3	3601	13	3701	—	3801	3	3901	47
3503	31	3603	3	3703	7	3803	—	3903	3
3507	3	3607	—	3707	11	3807	3	3907	—
3509	11	3609	3	3709	—	3809	13	3909	3
3511	—	3611	23	3711	3	3811	37	3911	—
3513	3	3613	—	3713	47	3813	3	3913	7
3517	—	3617	—	3717	3	3817	11	3917	—
3519	3	3619	7	3719	—	3819	3	3919	—
3521	7	3621	3	3721	61	3821	—	3921	3
3523	13	3623	—	3723	3	3823	—	3923	—
3527	—	3627	3	3727	—	3827	43	3927	3
3529	—	3629	19	3729	3	3829	7	3929	—
3531	3	3631	—	3731	7	3831	3	3931	—
3533	—	3633	3	3733	—	3833	—	3933	3
3537	3	3637	—	3737	37	3837	3	3937	31
3539	—	3639	3	3739	—	3839	11	3939	3
3541	—	3641	11	3741	3	3841	23	3941	7
3543	3	3643	—	3743	19	3843	3	3943	—
3547	—	3647	7	3747	3	3847	—	3947	—
3549	3	3649	41	3749	23	3849	3	3949	11
3551	53	3651	3	3751	11	3851	—	3951	3
3553	11	3653	13	3753	3	3853	—	3953	59
3557	—	3657	3	3757	13	3857	7	3957	3
3559	—	3659	—	3759	3	3859	17	3959	37
3561	3	3661	7	3761	—	3861	3	3961	17
3563	7	3663	3	3763	53	3863	—	3963	3
3567	3	3667	19	3767	—	3867	3	3967	—
3569	43	3669	3	3769	—	3869	53	3969	3
3571	—	3671	—	3771	3	3871	7	3971	11
3573	3	3673	—	3773	7	3873	3	3973	29
3577	7	3677	—	3777	3	3877	—	3977	41
3579	3	3679	13	3779	—	3879	3	3979	23
3581	—	3681	3	3781	19	3881	—	3981	3
3583	—	3683	29	3783	3	3883	11	3983	7
3587	17	3687	3	3787	7	3887	13	3987	3
3589	37	3689	7	3789	3	3889	—	3989	—
3591	3	3691	—	3791	17	3891	3	3991	13
3593	—	3693	3	3793	—	3893	17	3993	3
3597	3	3697	—	3797	—	3897	3	3997	7
3599	59	3699	3	3799	29	3899	7	3999	3

TABLE 2

4001	—	4101	3	4201	—	4301	11	4401	3
4003	—	4103	11	4203	3	4303	13	4403	7
4007	—	4107	3	4207	7	4307	59	4407	3
4009	19	4109	7	4209	3	4309	31	4409	—
4011	3	4111	—	4211	—	4311	3	4411	11
4013	—	4113	3	4213	11	4313	19	4413	3
4017	3	4117	23	4217	—	4317	3	4417	7
4019	—	4119	3	4219	—	4319	7	4419	3
4021	—	4121	13	4221	3	4321	29	4421	—
4023	3	4123	7	4223	41	4323	3	4423	—
4027	—	4127	—	4227	3	4327	—	4427	19
4029	3	4129	—	4229	—	4329	3	4429	43
4031	29	4131	3	4231	—	4331	61	4431	3
4033	37	4133	—	4233	3	4333	7	4433	11
4037	11	4137	3	4237	19	4337	—	4437	3
4039	7	4139	—	4239	3	4339	—	4439	23
4041	3	4141	41	4241	—	4341	3	4441	—
4043	13	4143	3	4243	—	4343	43	4443	3
4047	3	4147	11	4247	31	4347	3	4447	—
4049	—	4149	3	4249	7	4349	—	4449	3
4051	—	4151	7	4251	3	4351	19	4451	—
4053	3	4153	—	4253	—	4353	3	4453	61
4057	—	4157	—	4257	3	4357	—	4457	—
4059	3	4159	—	4259	—	4359	3	4459	7
4061	31	4161	3	4261	—	4361	7	4461	3
4063	17	4163	23	4263	3	4363	—	4463	—
4067	7	4167	3	4267	17	4367	11	4467	3
4069	13	4169	11	4269	3	4369	17	4469	41
4071	3	4171	43	4271	—	4371	3	4471	17
4073	—	4173	3	4273	—	4373	—	4473	3
4077	3	4177	—	4277	7	4377	3	4477	11
4079	—	4179	3	4279	11	4379	29	4479	3
4081	7	4181	37	4281	3	4381	13	4481	—
4083	3	4183	47	4283	—	4383	3	4483	—
4087	61	4187	53	4287	3	4387	41	4487	7
4089	3	4189	59	4289	—	4389	3	4489	67
4091	—	4191	3	4291	7	4391	—	4491	3
4093	—	4193	7	4293	3	4393	23	4493	—
4097	17	4197	3	4297	—	4397	—	4497	3
4099	—	4199	13	4299	3	4399	53	4499	11

TABLE 2

4501	7	4601	43	4701	3	4801	—	4901	13
4503	3	4603	—	4703	—	4803	3	4903	—
4507	—	4607	17	4707	3	4807	11	4907	7
4509	3	4609	11	4709	17	4809	3	4909	—
4511	13	4611	3	4711	7	4811	17	4911	3
4513	—	4613	7	4713	3	4813	—	4913	17
4517	—	4617	3	4717	53	4817	—	4917	3
4519	—	4619	31	4719	3	4819	61	4919	—
4521	3	4621	—	4721	—	4821	3	4921	7
4523	—	4623	3	4723	—	4823	7	4923	3
4527	3	4627	7	4727	29	4827	3	4927	13
4529	7	4629	3	4729	—	4829	11	4929	3
4531	23	4631	11	4731	3	4831	—	4931	—
4533	3	4633	41	4733	—	4833	3	4933	—
4537	13	4637	—	4737	3	4837	7	4937	—
4539	3	4639	—	4739	7	4839	3	4939	11
4541	19	4641	3	4741	11	4841	47	4941	3
4543	7	4643	—	4743	3	4843	29	4943	—
4547	—	4647	3	4747	47	4847	37	4947	3
4549	—	4649	—	4749	3	4849	13	4949	7
4551	3	4651	—	4751	—	4851	3	4951	—
4553	29	4653	3	4753	7	4853	23	4953	3
4557	3	4657	—	4757	67	4857	3	4957	—
4559	47	4659	3	4759	—	4859	43	4959	3
4561	—	4661	59	4761	3	4861	—	4961	11
4563	3	4663	—	4763	11	4863	3	4963	7
4567	—	4667	13	4767	3	4867	31	4967	—
4569	3	4669	7	4769	19	4869	3	4969	—
4571	7	4671	3	4771	13	4871	—	4971	3
4573	17	4673	—	4773	3	4873	11	4973	—
4577	23	4677	3	4777	17	4877	—	4977	3
4579	19	4679	—	4779	3	4879	7	4979	13
4581	3	4681	31	4781	7	4881	3	4981	17
4583	—	4683	3	4783	—	4883	19	4983	3
4587	3	4687	43	4787	—	4887	3	4987	—
4589	13	4689	3	4789	—	4889	—	4989	3
4591	—	4691	—	4791	3	4891	67	4991	7
4593	3	4693	13	4793	—	4893	3	4993	—
4597	—	4697	7	4797	3	4897	59	4997	19
4599	3	4699	37	4799	—	4899	3	4999	—

TABLE 3

The prime numbers between 5000 *and* 10,000.

5003	5387	5693	6053	6367	6761	7103
5009	5393	5701	6067	6373	6763	7109
5011	5399	5711	6073	6379	6779	7121
5021	5407	5717	6079	6389	6781	7127
5023	5413	5737	6089	6397	6791	7129
5039	5417	5741	6091	6421	6793	7151
5051	5419	5743	6101	6427	6803	7159
5059	5431	5749	6113	6449	6823	7177
5077	5437	5779	6121	6451	6827	7187
5081	5441	5783	6131	6469	6829	7193
5087	5443	5791	6133	6473	6833	7207
5099	5449	5801	6143	6481	6841	7211
5101	5471	5807	6151	6491	6857	7213
5107	5477	5813	6163	6521	6863	7219
5113	5479	5821	6173	6529	6869	7229
5119	5483	5827	6197	6547	6871	7237
5147	5501	5839	6199	6551	6883	7243
5153	5503	5843	6203	6553	6899	7247
5167	5507	5849	6211	6563	6907	7253
5171	5519	5851	6217	6569	6911	7283
5179	5521	5857	6221	6571	6917	7297
5189	5527	5861	6229	6577	6947	7307
5197	5531	5867	6247	6581	6949	7309
5209	5557	5869	6257	6599	6959	7321
5227	5563	5879	6263	6607	6961	7331
5231	5569	5881	6269	6619	6967	7333
5233	5573	5891	6271	6637	6971	7349
5237	5581	5903	6277	6653	6977	7351
5261	5591	5923	6287	6659	6983	7369
5273	5623	5927	6299	6661	6991	7393
5279	5639	5939	6301	6673	6997	7411
5281	5641	5953	6311	6679	7001	7417
5297	5647	5981	6317	6689	7013	7433
5303	5651	5987	6323	6691	7019	7451
5309	5653	6007	6329	6701	7027	7457
5323	5657	6011	6337	6703	7039	7459
5333	5659	6029	6343	6709	7043	7477
5347	5669	6037	6353	6719	7057	7481
5351	5683	6043	6359	6733	7069	7487
5381	5689	6047	6361	6737	7079	7489

TABLE 3

7499	7759	8111	8431	8741	9049	9377	9679
7507	7789	8117	8443	8747	9059	9391	9689
7517	7793	8123	8447	8753	9067	9397	9697
5723	7817	8147	8461	8761	9091	9403	9719
7529	7823	8161	8467	8779	9103	9413	9721
7537	7829	8167	8501	8783	9109	9419	9733
7541	7841	8171	8513	8803	9127	9421	9739
7547	7853	8179	8521	8807	9133	9431	9743
7549	7867	8191	8527	8819	9137	9433	9749
7559	7873	8209	8537	8821	9151	9437	9767
7561	7877	8219	8539	8831	9157	9439	9769
7573	7879	8221	8543	8837	9161	9461	9781
7577	7883	8231	8563	8839	9173	9463	9787
7583	7901	8233	8573	8849	9181	9467	9791
7589	7907	8237	8581	8861	9187	9473	9803
7591	7919	8243	8597	8863	9199	9479	9811
7603	7927	8263	8599	8867	9203	9491	9817
7607	7933	8269	8609	8887	9209	9497	9829
7621	7937	8273	8623	8893	9221	9511	9833
7639	7949	8287	8627	8923	9227	9521	9839
7643	7951	8291	8629	8929	9239	9533	9851
7649	7963	8293	8641	8933	9241	9539	9857
7669	7993	8297	8647	8941	9257	9547	9859
7673	8009	8311	8663	8951	9277	9551	9871
7681	8011	8317	8669	8963	9281	9587	9883
7687	8017	8329	8677	8969	9283	9601	9887
7691	8039	8353	8681	8971	9293	9613	9901
7699	8053	8363	8689	8999	9311	9619	9907
7703	8059	8369	8693	9001	9319	9623	9923
7717	8069	8377	8699	9007	9323	9629	9929
7723	8081	8387	8707	9011	9337	9631	9931
7727	8087	8389	8713	9013	9341	9643	9941
7741	8089	8419	8719	9029	9343	9649	9949
7753	8093	8423	8731	9041	9349	9661	9967
7757	8101	8429	8737	9043	9371	9677	9973

TABLE 4

The number of primes and the number of pairs of twin primes in the indicated intervals.

Interval	Number of primes	Number of pairs of twin primes
1–100	25	8
101–200	21	7
201–300	16	4
301–400	16	2
401–500	17	3
501–600	14	2
601–700	16	4
701–800	14	0
801–900	15	5
901–1000	14	0
2501–2600	11	2
2601–2700	15	2
2701–2800	14	3
2801–2900	12	1
2901–3000	11	1
10001–10100	11	4
10101–10200	12	1
10201–10300	10	1
10301–10400	12	2
10401–10500	10	2
29501–29600	10	1
29601–29700	8	1
29701–29800	7	1
29801–29900	10	1
29901–30000	7	0
100001–100100	6	0
100101–100200	9	1
100201–100300	8	0
100301–100400	9	2
100401–100500	8	0
299501–299600	7	1
299601–299700	8	1
299701–299800	8	1
299801–299900	6	0
299901–300000	9	0

TABLE 5

The squares and cubes of integers, n, where $1 \le n \le 499$.

n	n^2	n^3	n	n^2	n^3
			35	1 225	42 875
1	1	1	36	1 296	46 656
2	4	8	37	1 369	50 653
3	9	27	38	1 444	54 872
4	16	64	39	1 521	59 319
5	25	125	40	1 600	64 000
6	36	216	41	1 681	68 921
7	49	343	42	1 764	74 088
8	64	512	43	1 849	79 507
9	81	729	44	1 936	85 184
10	100	1 000	45	2 025	91 125
11	121	1 331	46	2 116	97 336
12	144	1 728	47	2 209	103 823
13	169	2 197	48	2 304	110 592
14	196	2 744	49	2 401	117 649
15	225	3 375	50	2 500	125 000
16	256	4 096	51	2 601	132 651
17	289	4 913	52	2 704	140 608
18	324	5 832	53	2 809	148 877
19	361	6 859	54	2 916	157 464
20	400	8 000	55	3 025	166 375
21	441	9 261	56	3 136	175 616
22	484	10 648	57	3 249	185 193
23	529	12 167	58	3 364	195 112
24	576	13 824	59	3 481	205 379
25	625	15 625	60	3 600	216 000
26	676	17 576	61	3 721	226 981
27	729	19 683	62	3 844	238 328
28	784	21 952	63	3 969	250 047
29	841	24 389	64	4 096	262 144
30	900	27 000	65	4 225	274 625
31	961	29 791	66	4 356	287 496
32	1 024	32 768	67	4 489	300 763
33	1 089	35 937	68	4 624	314 432
34	1 156	39 304	69	4 761	328 509

TABLE 5

n	n^2	n^3	n	n^2	n^3
70	4 900	343 000	110	12 100	1 331 000
71	5 041	357 911	111	12 321	1 367 631
72	5 184	373 248	112	12 544	1 404 928
73	5 329	389 017	113	12 769	1 442 897
74	5 476	405 224	114	12 996	1 481 544
75	5 625	421 875	115	13 225	1 520 875
76	5 776	438 976	116	13 456	1 560 896
77	5 929	456 533	117	13 689	1 601 613
78	6 084	474 552	118	13 924	1 643 032
79	6 241	493 039	119	14 161	1 685 159
80	6 400	512 000	120	14 400	1 728 000
81	6 561	531 441	121	14 641	1 771 561
82	6 724	551 368	122	14 884	1 815 848
83	6 889	571 787	123	15 129	1 860 867
84	7 056	592 704	124	15 376	1 906 624
85	7 225	614 125	125	15 625	1 953 125
86	7 396	636 056	126	15 876	2 000 376
87	7 569	658 503	127	16 129	2 048 383
88	7 744	681 472	128	16 384	2 097 152
89	7 921	704 969	129	16 641	2 146 689
90	8 100	729 000	130	16 900	2 197 000
91	8 281	753 571	131	17 161	2 248 091
92	8 464	778 688	132	17 424	2 299 968
93	8 649	804 357	133	17 689	2 352 637
94	8 836	830 584	134	17 956	2 406 104
95	9 025	857 375	135	18 225	2 460 375
96	9 216	884 736	136	18 496	2 515 456
97	9 409	912 673	137	18 769	2 571 353
98	9 604	941 192	138	19 044	2 628 072
99	9 801	970 299	139	19 321	2 685 619
100	10 000	1 000 000	140	19 600	2 744 000
101	10 201	1 030 301	141	19 881	2 803 221
102	10 404	1 061 208	142	20 164	2 863 288
103	10 609	1 092 727	143	20 449	2 924 207
104	10 816	1 124 864	144	20 736	2 985 984
105	11 025	1 157 625	145	21 025	3 048 625
106	11 236	1 191 016	146	21 316	3 112 136
107	11 449	1 225 043	147	21 609	3 176 523
108	11 664	1 259 712	148	21 904	3 241 792
109	11 881	1 295 029	149	22 201	3 307 949

TABLE 5

n	n^2	n^3	n	n^2	n^3
150	22 500	3 375 000	190	36 100	6 859 000
151	22 801	3 442 951	191	36 481	6 967 871
152	23 104	3 511 808	192	36 864	7 077 888
153	23 409	3 581 577	193	37 249	7 189 057
154	23 716	3 652 264	194	37 636	7 301 384
155	24 025	3 723 875	195	38 025	7 414 875
156	24 336	3 796 416	196	38 416	7 529 536
157	24 649	3 869 893	197	38 809	7 645 373
158	24 964	3 944 312	198	39 204	7 762 392
159	25 281	4 019 679	199	39 601	7 880 599
160	25 600	4 096 000	200	40 000	8 000 000
161	25 921	4 173 281	201	40 401	8 120 601
162	26 244	4 251 528	202	40 804	8 242 408
163	26 569	4 330 747	203	41 209	8 365 427
164	26 896	4 410 944	204	41 616	8 489 664
165	27 225	4 492 125	205	42 025	8 615 125
166	27 556	4 574 296	206	42 436	8 741 816
167	27 889	4 657 463	207	42 849	8 869 743
168	28 224	4 741 632	208	43 264	8 998 912
169	28 561	4 826 809	209	43 681	9 129 329
170	28 900	4 913 000	210	44 100	9 261 000
171	29 241	5 000 211	211	44 521	9 393 931
172	29 584	5 088 448	212	44 944	9 528 128
173	29 929	5 117 717	213	45 369	9 663 597
174	30 276	5 268 024	214	45 796	9 800 344
175	30 625	5 359 375	215	46 225	9 938 375
176	30 976	5 451 776	216	46 656	10 077 696
177	31 329	5 545 233	217	47 089	10 218 313
178	31 684	5 639 752	218	47 524	10 360 232
179	32 041	5 735 339	219	47 961	10 503 459
180	32 400	5 832 000	220	48 400	10 648 000
181	32 761	5 929 741	221	48 841	10 793 861
182	33 124	6 028 568	222	49 284	10 941 048
183	33 489	6 128 487	223	49 729	11 089 567
184	33 856	6 229 504	224	50 176	11 239 424
185	34 225	6 331 625	225	50 625	11 390 625
186	34 596	6 434 856	226	51 076	11 543 176
187	34 969	6 539 203	227	51 529	11 697 083
188	35 344	6 644 672	228	51 984	11 852 352
189	35 721	6 751 269	229	52 441	12 008 989

TABLE 5

n	n^2	n^3	n	n^2	n^3
230	52 900	12 167 000	270	72 900	19 683 000
231	53 361	12 326 391	271	73 441	19 902 511
232	53 824	12 487 168	272	73 984	20 123 648
233	54 289	12 649 337	273	74 529	20 346 417
234	54 756	12 812 904	274	75 076	20 570 824
235	55 225	12 977 875	275	75 625	20 796 875
236	55 696	13 144 256	276	76 176	21 024 576
237	56 169	13 312 053	277	76 729	21 253 933
238	56 644	13 481 272	278	77 284	21 484 952
239	57 121	13 651 919	279	77 841	21 717 639
240	57 600	13 824 000	280	78 400	21 952 000
241	58 081	13 997 521	281	78 961	22 188 041
242	58 564	14 172 488	282	79 524	22 425 768
243	59 049	14 348 907	283	80 089	22 665 187
244	59 536	14 526 784	284	80 656	22 906 304
245	60 025	14 706 125	285	81 225	23 149 125
246	60 516	14 886 936	286	81 796	23 393 656
247	61 009	15 069 223	287	82 369	23 639 903
248	61 504	15 252 992	288	82 944	23 887 872
249	62 001	15 438 249	289	83 521	24 137 569
250	62 500	15 625 000	290	84 100	24 389 000
251	63 001	15 813 251	291	84 681	24 642 171
252	63 504	16 003 008	292	85 264	24 897 088
253	64 009	16 194 277	293	85 849	25 153 757
254	64 516	16 387 064	294	86 436	25 412 184
255	65 025	16 581 375	295	87 025	25 672 375
256	65 536	16 777 216	296	87 616	25 934 336
257	66 049	16 974 593	297	88 209	26 198 073
258	66 564	17 173 512	298	88 804	26 463 592
259	67 081	17 373 979	299	89 401	26 730 899
260	67 600	17 576 000	300	90 000	27 000 000
261	68 121	17 779 581	301	90 601	27 270 901
262	68 644	17 984 728	302	91 204	27 543 608
263	69 169	18 191 447	303	91 809	27 818 127
264	69 696	18 399 744	304	92 416	28 094 464
265	70 225	18 609 625	305	93 025	28 372 625
266	70 756	18 821 096	306	93 636	28 652 616
267	71 289	19 034 163	307	94 249	28 934 443
268	71 824	19 248 832	308	94 864	29 218 112
269	72 361	19 465 109	309	95 481	29 503 629

TABLE 5

n	n^2	n^3	n	n^2	n^3
310	96 100	29 791 000	350	122 500	42 875 000
311	96 721	30 080 231	351	123 201	43 243 551
312	97 344	30 371 328	352	123 904	43 614 208
313	97 969	30 664 297	353	124 609	43 986 977
314	98 596	30 959 144	354	125 316	44 361 864
315	99 225	31 255 875	355	126 025	44 738 875
316	99 856	31 554 496	356	126 736	45 118 016
317	100 489	31 855 013	357	127 449	45 499 293
318	101 124	32 157 432	358	128 164	45 882 712
319	101 761	32 461 759	359	128 881	46 268 279
320	102 400	32 768 000	360	129 600	46 656 000
321	103 041	33 076 161	361	130 321	47 045 881
322	103 684	33 386 248	362	131 044	47 437 928
323	104 329	33 698 267	363	131 769	47 832 147
324	104 976	34 012 224	364	132 496	48 228 544
325	105 625	34 328 125	365	133 225	48 627 125
326	106 276	34 645 976	366	133 956	49 027 896
327	106 929	34 965 783	367	134 689	49 430 863
328	107 584	35 287 552	368	135 424	49 836 032
329	108 241	35 611 289	369	136 161	50 243 409
330	108 900	35 937 000	370	136 900	50 653 000
331	109 561	36 264 691	371	137 641	51 064 811
332	110 224	36 594 368	372	138 384	51 478 848
333	110 889	36 926 037	373	139 129	51 895 117
334	111 556	37 259 704	374	139 876	52 313 624
335	112 225	37 595 375	375	140 625	52 734 375
336	112 896	37 933 056	376	141 376	53 157 376
337	113 569	38 272 753	377	142 129	53 582 633
338	114 244	38 614 472	378	142 884	54 010 152
339	114 921	38 958 219	379	143 641	54 439 939
340	115 600	39 304 000	380	144 400	54 872 000
341	116 281	39 651 821	381	145 161	55 306 341
342	116 964	40 001 688	382	145 924	55 742 968
343	117 649	40 353 607	383	146 689	56 181 887
344	118 336	40 707 584	384	147 456	56 623 104
345	119 025	41 063 625	385	148 225	57 066 625
346	119 716	41 421 736	386	148 996	57 512 456
347	120 409	41 781 923	387	149 769	57 960 603
348	121 104	42 144 192	388	150 544	58 411 072
349	121 801	42 508 549	389	151 321	58 863 869

TABLE 5

n	n²	n³	n	n²	n³
390	152 100	59 319 000	430	184 900	79 507 000
391	152 881	59 776 471	431	185 761	80 062 991
392	153 664	60 236 288	432	186 624	80 621 568
393	154 449	60 698 457	433	187 489	81 182 737
394	155 236	61 162 984	434	188 356	81 746 504
395	156 025	61 629 875	435	189 225	82 312 875
396	156 816	62 099 136	436	190 096	82 881 856
397	157 609	62 570 773	437	190 969	83 453 453
398	158 404	63 044 792	438	191 844	84 027 672
399	159 201	63 521 199	439	192 721	84 604 519
400	160 000	64 000 000	440	193 600	85 184 000
401	160 801	64 481 201	441	194 481	85 766 121
402	161 604	64 964 808	442	195 364	86 350 888
403	162 409	65 450 827	443	196 249	86 938 307
404	163 216	65 939 264	444	197 136	87 528 384
405	164 025	66 430 125	445	198 025	88 121 125
406	164 836	66 923 416	446	198 916	88 716 536
407	165 649	67 419 143	447	199 809	89 314 623
408	166 464	67 917 312	448	200 704	89 915 392
409	167 281	68 417 929	449	201 601	90 518 849
410	168 100	68 921 000	450	202 500	91 125 000
411	168 921	69 426 531	451	203 401	91 733 851
412	169 744	69 934 528	452	204 304	92 345 408
413	170 569	70 444 997	453	205 209	92 959 677
414	171 396	70 957 944	454	206 116	93 576 664
415	172 225	71 473 375	455	207 025	94 196 375
416	173 056	71 991 296	456	207 936	94 818 816
417	173 889	72 511 713	457	208 849	95 443 993
418	174 724	73 034 632	458	209 764	96 071 912
419	175 561	73 560 059	459	210 681	96 702 579
420	176 400	74 088 000	460	211 600	97 336 000
421	177 241	74 618 461	461	212 521	97 972 181
422	178 084	75 151 448	462	213 444	98 611 128
423	178 929	75 686 967	463	214 369	99 252 847
424	179 776	76 225 024	464	215 296	99 897 344
425	180 625	76 765 625	465	216 225	100 544 625
426	181 476	77 308 776	466	217 156	101 194 696
427	182 329	77 854 483	467	218 089	101 847 563
428	183 184	78 402 752	468	219 024	102 503 232
429	184 041	78 953 589	469	219 961	103 161 709

TABLE 5

n	n^2	n^3	n	n^2	n^3
470	220 900	103 823 000	485	235 225	114 084 125
471	221 841	104 487 111	486	236 196	114 791 256
472	222 784	105 154 048	487	237 169	115 501 303
473	223 729	105 823 817	488	238 144	116 214 272
474	224 676	106 496 424	489	239 121	116 930 169
475	225 625	107 171 875	490	240 100	117 649 000
476	226 576	107 850 176	491	241 081	118 370 771
477	227 529	108 531 333	492	242 064	119 095 488
478	228 484	109 215 352	493	243 049	119 823 157
479	229 441	109 902 239	494	244 036	120 553 784
480	230 400	110 592 000	495	245 025	121 287 375
481	231 361	111 284 641	496	246 016	122 023 936
482	232 324	111 980 168	497	247 009	122 763 473
483	233 289	112 678 587	498	248 004	123 505 992
484	234 256	113 379 904	499	249 001	124 251 499

TABLE 6

The values of $\tau(n)$, $\sigma(n)$, $\phi(n)$ and $\mu(n)$, where $1 \leq n \leq 100$

n	$\tau(n)$	$\sigma(n)$	$\phi(n)$	$\mu(n)$	n	$\tau(n)$	$\sigma(n)$	$\phi(n)$	$\mu(n)$
1	1	1	1	1	41	2	42	40	-1
2	2	3	1	-1	42	8	96	12	-1
3	2	4	2	-1	43	2	44	42	-1
4	3	7	2	0	44	6	84	20	0
5	2	6	4	-1	45	6	78	24	0
6	4	12	2	1	46	4	72	22	1
7	2	8	6	-1	47	2	48	46	-1
8	4	15	4	0	48	10	124	16	0
9	3	13	6	0	49	3	57	42	0
10	4	18	4	1	50	6	93	20	0
11	2	12	10	-1	51	4	72	32	1
12	6	28	4	0	52	6	98	24	0
13	2	14	12	-1	53	2	54	52	-1
14	4	24	6	1	54	8	120	18	0
15	4	24	8	1	55	4	72	40	1
16	5	31	8	0	56	8	120	24	0
17	2	18	16	-1	57	4	80	36	1
18	6	39	6	0	58	4	90	28	1
19	2	20	18	-1	59	2	60	58	-1
20	6	42	8	0	60	12	168	16	0
21	4	32	12	1	61	2	62	60	-1
22	4	36	10	1	62	4	96	30	1
23	2	24	22	-1	63	6	104	36	0
24	8	60	8	0	64	7	127	32	0
25	3	31	20	0	65	4	84	48	1
26	4	42	12	1	66	8	144	20	-1
27	4	40	18	0	67	2	68	66	-1
28	6	56	12	0	68	6	126	32	0
29	2	30	28	-1	69	4	96	44	1
30	8	72	8	-1	70	8	144	24	-1
31	2	32	30	-1	71	2	72	70	-1
32	6	63	16	0	72	12	195	24	0
33	4	48	20	1	73	2	74	72	-1
34	4	54	16	1	74	4	114	36	1
35	4	48	24	1	75	6	124	40	0
36	9	91	12	0	76	6	140	36	0
37	2	38	36	-1	77	4	96	60	1
38	4	60	18	1	78	8	168	24	-1
39	4	56	24	1	79	2	80	78	-1
40	8	90	16	0	80	10	186	32	0

TABLE 6

n	$\tau(n)$	$\sigma(n)$	$\phi(n)$	$\mu(n)$	n	$\tau(n)$	$\sigma(n)$	$\phi(n)$	$\mu(n)$
81	5	121	54	0	91	4	112	72	1
82	4	126	40	1	92	6	168	44	0
83	2	84	82	−1	93	4	128	60	1
84	12	224	24	0	94	4	144	46	1
85	4	108	64	1	95	4	120	72	1
86	4	132	42	1	96	12	252	32	0
87	4	120	56	1	97	2	98	96	−1
88	8	180	40	0	98	6	171	42	0
89	2	90	88	−1	99	6	156	60	0
90	12	234	24	0	100	9	217	40	0

Answers to Selected Problems

Section 1.1

5. (a) 4, 5, and 7
(b) $(3 \cdot 2)! \neq 3!2!$, $(3 + 2)! \neq 3! + 2!$

Section 1.3

5. (a) $t_6 = 21$ and $t_5 = 15$
6. (b) $1^2 = t_1$, $6^2 = t_8$, $204^2 = t_{288}$
7. (b) $t_6 = t_3 + t_5$, $t_{10} = t_4 + t_9$

Section 2.3

1. 1, 9, and 17

2. (a) $x = 4$, $y = -3$.   (b) $x = 6$, $y = -1$.
(c) $x = 7$, $y = -3$.   (d) $x = 39$, $y = -29$.
8. 32,461,   22,338,   and 23,664

12. $x = 171$, $y = 114$, $z = -2$

Section 2.4

2. (a) $x = 20 + 9t$, $y = -15 - 7t$
(b) $x = 18 + 23t$, $y = -3 - 4t$
(c) $x = 176 + 35t$, $y = -1111 - 221t$
3. (a) $x = 1$, $y = 6$
(b) $x = 2$, $y = 38$; $x = 9$, $y = 20$; $x = 16$, $y = 2$
(c) No solutions
(d) $x = 17 - 57t$, $y = 47 - 158t$, where $t \leq 0$
5. (b) $x = 8 + 2k$, $y = -48 - 15k - 5t$, $z = 16 + 5k + 2t$
6. (a) The fewest coins are 3 dimes and 17 quarters, while 43 dimes and 1 quarter give the largest number. It is possible to have 13 dimes and 13 quarters.
(b) There may be 40 adults and 24 children, or 45 adults and 12 children, or 50 adults.
(c) six 6s and ten 9s.
7. There may be 5 calves, 41 lambs, and 54 pigs; or 10 calves, 22 lambs, and 68 pigs; or 15 calves, 3 lambs, and 82 pigs.

8. $10.21

9. (b) 28 pieces is one answer.
(d) One answer is 1 man, 5 women, and 14 children.
(e) 56 and 44

Section 3.1

2. 25 is a counterexample.

7. All primes ≤ 47.

11. (a) $2^{13} - 1$ is prime.

Section 3.2

11. $59 - 53 = 53 - 47, \quad 157 - 151 = 163 - 157$

13. (b) $R_{10} = 11 \cdot 41 \cdot 271 \cdot 9091$

Section 3.3

3. 2 and 5

11. $h(22) = 23 \cdot 67$

14. 71, 13859

16. $37 = -1 + 2 + 3 + 5 + 7 + 11 - 13 + 17 - 19 + 23 - 29 + 31,$
$31 = -1 + 2 - 3 + 5 - 7 - 11 + 13 + 17 - 19 - 23 + 2(29).$

19. $81 = 3 + 5 + 73$
$125 = 5 + 13 + 107$

28. (b) $n = 1$

Section 4.2

4. (a) 4 and 6   (b) 0

Section 4.3

2. 89

5. (a) 9, (b) 4, (c) 5, (d) 9
8. 7

10. $x = 7, \quad y = 8$

11. 143

14. $n = 1, 3$

20. $R_6 = 3 \cdot 7 \cdot 11 \cdot 13 \cdot 37$

22. $x = 3, \quad y = 2$

23. $x = 8, \quad y = 0, \quad z = 6$

Section 4.4

1. (a) $x \equiv 18 \pmod{29}$
   (c) $x \equiv 6, 13,$ and $20 \pmod{21}$
   (e) $x \equiv 45, 94 \pmod{98}$

   (b) $x \equiv 16 \pmod{26}$
   (d) No solutions
   (f) $x \equiv 16, 59, 102, 145, 188,$
   $231,$ and $274 \pmod{301}$

2. (a) $x = 15 + 51t, \quad y = -1 - 4t$
   (b) $x = 13 + 25t, \quad y = 7 - 12t$
   (c) $x = 14 + 53t, \quad y = 1 + 5t$

3. $x \equiv 11 + t \pmod{13}, \qquad y \equiv 5 + 6t \pmod{13}$

4. (a) $x \equiv 52 \pmod{105}$
   (c) $x \equiv 785 \pmod{1122}$

   (b) $x \equiv 4944 \pmod{9889}$
   (d) $x \equiv 653 \pmod{770}$

5. $x \equiv 99 \pmod{210}$

6. 62

7. (a) 548, 549, 550 (b) $5^2 \mid 350, \quad 3^3 \mid 351, \quad 2^4 \mid 352$

8. 119

9. 301

10. 3930

14. 838

15. (a) 17, (b) 59, (c) 1103

16. $n \equiv 1, 7, 13 \pmod{15}$

17. $x \equiv 7, \quad y \equiv 9 \pmod{13}$

18. $x \equiv 59, 164 \pmod{210}$

Section 5.2

1. (b) $127 \cdot 83,$ (c) $691 \cdot 493$

3. $89 \cdot 23$

4. $29 \cdot 17, \quad 2925 \cdot 13$

5. (a) $2911 = 71 \cdot 41$
   (b) $4573 = 17 \cdot 269$
   (c) $6923 = 23 \cdot 301$

6. $13561 = 71 \cdot 191$

7. (a) $4537 = 13 \cdot 349$
   (b) $14429 = 47 \cdot 307$

Section 5.3

6. 1

8. (b) $x \equiv 16 \pmod{31}, \quad x \equiv 10 \pmod{11}, \quad x \equiv 25 \pmod{29}$.

Section 5.4

8. 5, 13

11. 12,   17;   6,   31

Section 6.1

2. 6;   6,300,402

12. p^9 and p^4q; $48 = 2^4 \cdot 3$

Section 6.3

3. 249, 330

5. (b) 150!,   151!,   152!,   153!,   154!

8. (b) 36,   396

9. 405

Section 7.2

1. 720,   1152,   9600

18. $\phi(n) = 16$ when $n = 17, 32, 34, 40, 48,$ and 60. $\phi(n) = 24$ when $n = 35, 39, 45, 52, 56, 70, 72, 78, 84,$ and 90.

Section 7.3

7. 1

8. (b) $x \equiv 19 \pmod{26}$, $x \equiv 34 \pmod{40}$, $x \equiv 7 \pmod{49}$

Section 7.4

10. 29348,   29349,   29350,   29351

Section 7.5

4. 1747,   157

5. 253

6. 2318   1932   1106   2197   1631   0337   1728

7. REPLY NOW

8. SELL SHORT

9. $x_2 = x_4 = x_6 = 1$, $x_1 = x_3 = x_5 = 0$.
$x_3 = x_4 = x_5 = 1$, $x_1 = x_2 = x_6 = 0$.
$x_1 = x_2 = x_4 = x_5 = 1$, $x_3 = x_6 = 0$.
$x_1 = x_2 = x_3 = x_6 = 1$, $x_4 = x_5 = 0$.

10. (a) and (c) are superincreasing

11. (a) $x_1 = x_2 = x_3 = x_6 = 1$, $x_4 = x_5 = 0$.
(b) $x_2 = x_3 = x_5 = 1$, $x_1 = x_4 = 0$.
(c) $x_3 = x_4 = x_6 = 1$, $x_1 = x_2 = x_5 = 0$.

13. 3, 4, 10, 21

14. CIPHER

15. (a) 14, 21, 9, 11, 29
(b) 25 9 59 40 30 41 59 9 35

Section 8.1

1. (a) 8, 16, 16
(b) 18, 18, 9
(c) 11, 11, 22

8. (c) $2^{17} - 1$ is prime; $233 \mid 2^{29} - 1$.

12. (a) 3, 7. (b) 3, 5, 6, 7, 10, 11, 12, 14.

13. (b) 41, 239.

Section 8.2

2. 1, 4, 11, 14; 8, 18, 47, 57; 8, 14, 19, 25.

3. $2, 6 \equiv 2^9$, $7 \equiv 2^7$, 8;
$2, 3 \equiv 2^{13}$, $10 \equiv 2^{17}$, $13 \equiv 2^5$, $14 \equiv 2^7$, $15 \equiv 2^{11}$;
$5, 7 \equiv 5^{19}$, $10 \equiv 5^3$, $11 \equiv 5^9$, $14 \equiv 5^{21}$, $15 \equiv 5^{17}$, $17 \equiv 5^7$, $19 \equiv 5^{15}$, $20 \equiv 5^5$, $21 \equiv 5^{13}$.

4. (a) 7, 37. (b) 9, 10, 13, 14, 15, 17, 23, 24, 25, 31, 38, 40.

5. 11, 50.

Section 8.3

1. (a) 7, 11, 15, 19; 2, 3, 8, 12, 13, 17, 22, 23.
(b) 2, 5;
2, 5, 11, 14, 20, 23;
2, 5, 11, 14, 20, 23, 29, 32, 38, 41, 47, 50, 56, 59, 65, 68, 74, 77.

4. (b) 3.

5. 6, 7, 11, 12, 13, 15, 17, 19, 22, 24, 26, 28, 29, 30, 34, 35;
7, 11, 13, 15, 17, 19, 29, 35, 47, 53, 63, 65, 67, 69, 71, 75.

11. (b) $x \equiv 34 \pmod{40}$, $x \equiv 30 \pmod{77}$.

Section 8.4

1. $\text{ind}_2 \, 5 = 9$, $\text{ind}_6 \, 5 = 9$, $\text{ind}_7 \, 5 = 3$, $\text{ind}_{11} \, 5 = 3$.

2. (a) $x \equiv 7 \pmod{11}$. (b) $x \equiv 5, 6 \pmod{11}$.
   (c) No solutions.

3. (a) $x \equiv 6, 7, 10, 11 \pmod{17}$. (b) $x \equiv 5 \pmod{17}$.
   (c) $x \equiv 3, 5, 6, 7, 10, 11, 12, 14 \pmod{17}$. (d) $x \equiv 1 \pmod{16}$.

4. 14.

8. (a) In each case, $a = 2, 5, 6$.
   (b) 1, 2, 4; 1, 3, 4, 5, 9; 1, 3, 9.

12. Only the first equation has a solution.

16. (b) $x \equiv 3, 7, 11, 15 \pmod{16}$; $x \equiv 8, 17 \pmod{18}$.

17. $b \equiv 1, 3, 9 \pmod{13}$.

Section 9.1

1. (a) $x \equiv 6, 9 \pmod{11}$. (b) $x \equiv 4, 6 \pmod{13}$.
   (c) $x \equiv 9, 22 \pmod{23}$.

8. (b) 6, 11; 17, 24

11. (a) 1, 4, 5, 6, 7, 9, 11, 16, 17
    (b) 1, 4, 5, 6, 7, 9, 13, 16, 20, 22, 23, 24, 25, 28;
        1, 2, 4, 5, 7, 8, 9, 10, 14, 16, 18, 19, 20, 25, 28.

Section 9.2

1. (a) -1, (b) 1, (c) 1, (d) -1, (e) 1

2. (a) $(-1)^3$, (b) $(-1)^3$, (c) $(-1)^4$,
   (d) $(-1)^5$, (e) $(-1)^9$

Section 9.3

1. (a) 1, (b) -1, (c) -1, (d) 1, (e) 1

3. (a) Solvable, (b) Not solvable, (c) Solvable

6. $p = 2$ or $p \equiv 1 \pmod{4}$; $p = 2$ or $p \equiv 1$ or $3 \pmod{8}$;
   $p = 2$ or $p \equiv 1 \pmod{6}$.

8. 73

14. $x \equiv 9, 16, 19, 26 \pmod{35}$

16. -1, -1, 1

20. Not solvable

Section 9.4

1. (b) $x \equiv 57,\quad 68 \pmod{5^3}$.
2. (a) $x \equiv 13, 14 \pmod{3^3}$. (b) $x \equiv 42, 83 \pmod{5^3}$.
   (c) $x \equiv 108, 135 \pmod{7^3}$.
3. $x \equiv 5008, 9633 \pmod{11^4}$.
4. $x \equiv 122, 123 \pmod{5^3};\quad x \equiv 11, 15 \pmod{3^3}$.
6. $x \equiv 41, 87, 105 \pmod{2^7}$.
7. (a) When $a = 1,\quad x \equiv 1, 7, 9, 15 \pmod{2^4}$.
     When $a = 9,\quad x \equiv 3, 5, 11, 13 \pmod{2^4}$.
   (b) When $a = 1,\quad x \equiv 1, 15, 17, 31 \pmod{2^5}$.
     When $a = 9,\quad x \equiv 3, 13, 19, 29 \pmod{2^5}$.
     When $a = 17,\quad x \equiv 7, 9, 23, 25 \pmod{2^5}$.
   (c) When $a = 1,\quad x \equiv 1, 31, 33, 63 \pmod{2^6}$.
     When $a = 9,\quad x \equiv 3, 29, 35, 61 \pmod{2^6}$.
     When $a = 17,\quad x \equiv 7, 25, 39, 57 \pmod{2^6}$.
     When $a = 25,\quad x \equiv 5, 27, 37, 59 \pmod{2^6}$.
     When $a = 33,\quad x \equiv 15, 17, 47, 49 \pmod{2^6}$.
     When $a = 41,\quad x \equiv 13, 19, 45, 51 \pmod{2^6}$.
     When $a = 49,\quad x \equiv 7, 25, 39, 57 \pmod{2^6}$.
     When $a = 57,\quad x \equiv 11, 21, 43, 53 \pmod{2^6}$.
9. (a) 4, 8. (b) $x \equiv 3, 147, 153, 297, 303, 447, 453, 597 \pmod{2^3 \cdot 3 \cdot 5^2}$.
10. (b) $x \equiv 51, 70 \pmod{11^2}$.

Section 10.1

1. $\sigma(n) = 2160(2^{11} - 1) \neq 2048(2^{11} - 1)$
8. 56
11. p^3, pq
14. (b) There are none
16. No

Section 10.2

3. $233 \mid M_{29}$.

Section 10.3

3. (b) $3 \mid 2^{2^n} + 5$.
7. $2^{58} + 1 = (2^{29} - 2^{15} + 1)(2^{29} + 2^{15} + 1) = 5 \cdot 107367629 \cdot 536903681$.
9. (c) $83 \mid 2^{41} + 1$ and $59 \mid 2^{29} + 1$.
10. $n = 315,\quad p = 71,\quad$ and $q = 73$.
11. $3 \mid 2^3 + 1$.

Section 11.1

1. (a) (16, 12, 20), (16, 63, 65), (16, 30, 34).
    (b) (40, 9, 41), (40, 399, 401); (60, 11, 61), (60, 91, 109), (60, 221, 229), (60, 899, 901).
8. (12, 5, 13), (8, 6, 10).
12. (a) (3, 4, 5), (20, 21, 29), (119, 120, 169), (696, 697, 985), (4059, 4060, 5741).
    (b) $(t_6, t_7, 35)$, $(t_{40}, t_{41}, 1189)$, $(t_{238}, t_{239}, 30391)$.
13. $t_1 = 1^2$, $t_8 = 6^2$, $t_{49} = 35^2$, $t_{288} = 204^2$, $t_{1681} = 1189^2$.

Section 12.2

1. $113 = 7^2 + 8^2$, $229 = 2^2 + 15^2$, $373 = 7^2 + 18^2$.
2. (a) $17^2 + 18^2 = 613$, $4^2 + 5^2 = 41$, $5^2 + 6^2 = 61$, $9^2 + 10^2 = 181$.
5. (b) $3185 = 56^2 + 7^2$, $39690 = 189^2 + 63^2$, $62920 = 242^2 + 66^2$.
6. $1105 = 5 \cdot 13 \cdot 17 = 9^2 + 32^2 = 12^2 + 31^2 = 23^2 + 24^2$; note that $325 = 5^2 \cdot 13 = 1^2 + 18^2 = 6^2 + 17^2 = 10^2 + 15^2$.
14. $45 = 7^2 - 2^2 = 9^2 - 6^2 = 23^2 - 22^2$.
18. $1729 = 1^3 + 12^3 = 9^3 + 10^3$.

Section 12.3

3. $(2870)^2 = (1^2 + 2^2 + 3^2 + \cdots + 20^2)^2$ leads to $574^2 = 414^2 + 8^2 + 16^2 + 24^2 + 32^2 + \cdots + 152^2$.
6. $509 = 12^2 + 13^2 + 14^2$.
7. $459 = 15^2 + 15^2 + 3^2$.
10. $61 = 5^3 - 4^3$, $127 = 7^3 - 6^3$.   :
13. $231 = 15^2 + 2^2 + 1^2 + 1^2$, $391 = 15^2 + 9^2 + 9^2 + 2^2$, $2109 = 44^2 + 12^2 + 5^2 + 2^2$.
17. $t_{13} = 3^3 + 4^3 = 6^3 - 5^3$.
18. $290 = 13^2 + 11^2 = 16^2 + 5^2 + 3^2 = 14^2 + 9^2 + 3^2 + 2^2 = 15^2 + 6^2 + 4^2 + 3^2 + 2^2$.

Section 13.1

7. 2, 5, 144.
8. $u_1, u_2, u_3, u_4, u_6, u_{12}$.
11. $u_{11} = 2u_9 + u_8, u_{12} = 6u_8 + (u_8 - u_4)$.
12. $u_1, u_2, u_4, u_8, u_{10}$.

Section 13.2

7. $50 = u_4 + u_7 + u_9$, $\quad 75 = u_3 + u_5 + u_7 + u_{10}$, $\quad 100 = u_1 + u_3 + u_6 + u_{11}$,
$120 = u_3 + u_9 + u_{11}$.

9. $(3, 4, 5)$, $\quad (5, 12, 13)$, $\quad (8, 15, 17)$, $\quad (39, 80, 89)$, $\quad (105, 208, 233)$.

Section 14.2

1. (a) $[-1; 1, 1, 1, 2, 6]$. (b) $[3; 3, 1, 1, 3, 2]$. (c) $[1; 3, 2, 3, 2]$.
(d) $[0; 2, 1, 1, 3, 5, 3]$.
2. (a) $-710/457$. (b) $741/170$. (c) $321/460$.
4. (a) $[0; 3, 1, 2, 2, 1]$. (b) $[-1; 2, 1, 7]$. (c) $[2; 3, 1, 2, 1, 2]$.
5. (a) 1, $3/2$, $10/7$, $33/23$, $76/53$, $109/76$.
(b) -3, -2, $-5/2$, $-7/3$, $-12/5$, $-43/18$.
(c) 0, $1/2$, $4/9$, $5/11$, $44/97$, $93/205$.
6. (b) $225 = 4 \cdot 43 + 4 \cdot 10 + 3 \cdot 3 + 2 \cdot 1 + 2$.
7. (a) 1, $3/2$, $7/5$, $17/12$, $41/29$, $99/70$, $239/169$, $577/408$, $1393/985$.
(b) 1, 2, $5/3$, $7/4$, $19/11$, $26/15$, $71/41$, $97/56$, $265/153$.
(c) 2, $9/4$, $38/17$, $161/72$, $682/305$, $2889/1292$, $12238/5473$, $51841/23184$, $219602/98209$.
(d) 2, $5/2$, $22/9$, $49/20$, $218/89$, $485/198$, $2158/881$, $4801/1960$, $21362/8721$.
(e) 2, 3, $5/2$, $8/3$, $37/14$, $45/17$, $82/31$, $127/48$, $590/223$.
9. $[3; 7, 16, 11]$, $\quad [3; 7, 15, 1, 25, 1, 7, 4]$.
11. (a) $x = -8 + 51t$, $\quad y = 3 - 19t$
(b) $x = 58 + 227\,t$, $\quad y = -93 - 364t$
(c) $x = 48 + 5t$, $\quad y = -168 - 18t$
(d) $x = -22 - 57t$, $\quad y = -61 - 158t$

Section 14.3

1. (a) $\dfrac{3 + \sqrt{15}}{3}$. (b) $\dfrac{-4 + \sqrt{37}}{3}$. (c) $\dfrac{5 + \sqrt{10}}{3}$.
(d) $\dfrac{19 - \sqrt{21}}{10}$. (e) $\dfrac{314 + \sqrt{37}}{233}$.
2. $\dfrac{\sqrt{5} - 1}{2}$.
3. $\dfrac{5 - \sqrt{5}}{2}$, $\dfrac{87 + \sqrt{5}}{62}$.
4. (a) $[2; \overline{4}]$. (b) $[2; \overline{1, 1, 1, 4}]$. (c) $[2; \overline{3}]$. (d) $[\overline{2; 1, 3}]$.
(e) $[1; 3, \overline{1, 2, 1, 4}]$.
5. (b) $[1; \overline{2}]$, $\quad [1; \overline{1, 2}]$, $\quad [3; \overline{1, 6}]$, $\quad [6; \overline{12}]$.
6. $1677/433$
7. (a) $1264/465$.
9. (a) $29/23$. (b) $267/212$.
11. 3, $355/113$.

Section 14.4

2. (a) $x = 8$, $y = 3$.   (b) $x = 10$, $y = 3$.   (c) $x = 17$, $y = 4$.
    (d) $x = 11$, $y = 2$.   (e) $x = 25$, $y = 4$.
3. (a) $x = 3$, $y = 2$; $x = 17$, $y = 12$; $x = 99$, $y = 70$.
    (b) $x = 2$, $y = 1$; $x = 7$, $y = 4$, $x = 26$, $y = 15$; $x = 97$, $y = 56$;
         $x = 362$, $y = 209$.
    (c) $x = 9$, $y = 4$; $x = 161$, $y = 72$.
4. 48,   1680.
5. (a) $x = 24$, $y = 5$; $x = 1151$, $y = 240$.
    (b) $x = 51$, $y = 10$; $x = 5201$, $y = 1020$.
    (c) $x = 23$, $y = 4$; $x = 1057$, $y = 184$.
6. (a) $x = 9801$, $y = 1820$.   (b) $x = 2049$, $y = 320$.
    (c) $x = 3699$, $y = 430$.
7. (a) $x = 18$, $y = 5$.   (b) $x = 70$, $y = 13$.   (c) $x = 32$, $y = 5$.
12. $x = 449$, $y = 60$; $x = 13455$, $y = 1798$.

13. (b) $x = 254$, $y = 96$; $x = 4048$, $y = 1530$.
    (c) $x = 213$, $y = 36$; $x = 2538$, $y = 429$.

Index

MODERN PERIOD

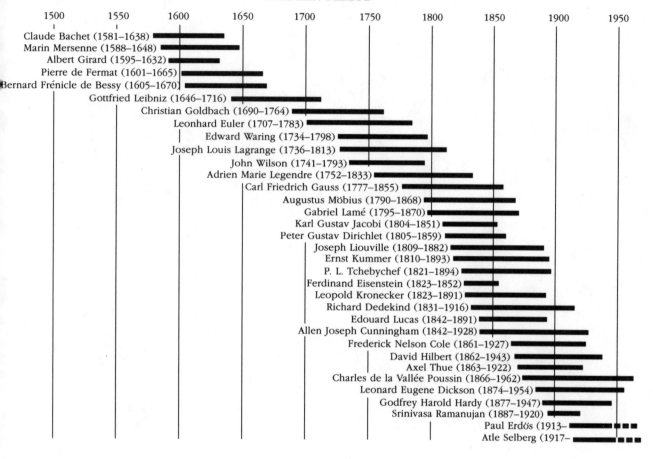

	1500	1550	1600	1650	1700	1750	1800	1850	1900	1950

Claude Bachet (1581–1638)
Marin Mersenne (1588–1648)
Albert Girard (1595–1632)
Pierre de Fermat (1601–1665)
Bernard Frénicle de Bessy (1605–1670)
Gottfried Leibniz (1646–1716)
Christian Goldbach (1690–1764)
Leonhard Euler (1707–1783)
Edward Waring (1734–1798)
Joseph Louis Lagrange (1736–1813)
John Wilson (1741–1793)
Adrien Marie Legendre (1752–1833)
Carl Friedrich Gauss (1777–1855)
Augustus Möbius (1790–1868)
Gabriel Lamé (1795–1870)
Karl Gustav Jacobi (1804–1851)
Peter Gustav Dirichlet (1805–1859)
Joseph Liouville (1809–1882)
Ernst Kummer (1810–1893)
P. L. Tchebychef (1821–1894)
Ferdinand Eisenstein (1823–1852)
Leopold Kronecker (1823–1891)
Richard Dedekind (1831–1916)
Edouard Lucas (1842–1891)
Allen Joseph Cunningham (1842–1928)
Frederick Nelson Cole (1861–1927)
David Hilbert (1862–1943)
Axel Thue (1863–1922)
Charles de la Vallée Poussin (1866–1962)
Leonard Eugene Dickson (1874–1954)
Godfrey Harold Hardy (1877–1947)
Srinivasa Ramanujan (1887–1920)
Paul Erdös (1913–
Atle Selberg (1917–

Elementary Number Theory